Fundamentals of Ocean Renewable Energy

Fundamentals of Ocean Renewable Energy

Generating Electricity from the Sea

Simon P. Neill

M. Reza Hashemi

Dear President Dooley

Many thanks for supporting offshore energy through various initiatives in URI,

Sincerely

Reza Hashemi

June 18, 2019

ACADEMIC PRESS

An imprint of Elsevier

Academic Press is an imprint of Elsevier
125 London Wall, London EC2Y 5AS, United Kingdom
525 B Street, Suite 1650, San Diego, CA 92101, United States
50 Hampshire Street, 5th Floor, Cambridge, MA 02139, United States
The Boulevard, Langford Lane, Kidlington, Oxford OX5 1GB, United Kingdom

Cover photos provided by Innogy Renewables UK Limited, Deepwater Wind, OpenHydro, and Ocean Power Technologies.

Library of Congress Cataloging-in-Publication Data
A catalog record for this book is available from the Library of Congress

British Library Cataloguing-in-Publication Data
A catalogue record for this book is available from the British Library

ISBN 978-0-12-810448-4

For information on all Academic Press publications visit our website at https://www.elsevier.com/books-and-journals

Publisher: Joe Hayton
Acquisition Editor: Raquel Zanol
Editorial Project Manager: Mariana Kuhl
Production Project Manager: Sruthi Satheesh
Cover Designer: Mark Rogers

Typeset by SPi Global, India

The real act of discovery consists not in finding new lands, but in seeing with new eyes.

Marcel Proust

Contents

Preface

Around the time of the industrial revolution, there were less than one billion humans living on Earth. Today the population is approaching eight billion. Such an explosion in the global population has put a large strain on the Earth's natural resources. Coincident with this expansion, we have witnessed a continuous improvement in human development: people are generally healthier and living longer, are more knowledgeable, and have a higher standard of living than they did at the beginning of the industrial revolution. Such an improvement in the quality of life, and especially technological advancement, requires large amounts of energy, particularly for electricity generation. Large-scale electrical grid systems originated towards the end of the 20th century, and we have relied on thermal power stations—power stations that generate electricity mostly from the combustion of fossil fuels—since that time. However, with the depletion of the Earth's natural resources, and a recognition that this form of energy conversion has led to increased concentrations of CO_2 in the atmosphere, and hence the phenomenon of global warming, the world is turning away from fossil fuels and towards more sustainable renewable energy sources for our insatiable electricity demands.

The largest contributor to renewable energy conversion around the world is presently hydropower, with both wind energy and solar energy rapidly gaining popularity. However, one of the largest potential sources of energy conversion surrounds us—the ocean. Covering 70% of the surface of the Earth, the ocean, with tides driven by astronomical forces (Moon and Sun) in conjunction with the Earth's rotation, winds, and wind-generated waves (ultimately originating in solar radiation), is a vast natural resource that could potentially meet all of the world's demand for electricity several times over. However, due to a lack of proven and cost-effective marine renewable energy technologies, only a tiny fraction of this potential has been realized, with the delivery of some offshore wind farms, a few tidal energy and wave projects, and a limited number of tidal range power plants. There is a strong appetite for ocean energy, with many government-funded R&D and industrial commercial projects being developed around the world, and it is very likely that we are on the cusp of very rapid growth in the marine renewable energy sector. We therefore feel that the time is right for this textbook, that integrates and conveys knowledge across a wide range of ocean renewable energy topics, covering tidal energy, offshore wind, and wave energy. The motivation for this book partly stems from both authors, many collaborations with marine renewable energy developers, and students and researchers of marine renewable energy, over the last decade. We felt that a unified text would be suited to those transferring into the marine renewable energy sector from related disciplines, for example, other engineering or energy sectors, where a detailed explanation of the marine renewable energy resource, and ways that the resource can be measured and modelled, would be useful. In addition, students enrolled at both undergraduate and postgraduate levels would find a single text invaluable in helping with those aspects of their studies that relate to marine renewable energy; for example, students of civil engineering, energy engineering, mechanical engineering, ocean engineering, or oceanography.

Both authors are from civil engineering backgrounds, with particular expertise in fluid dynamics and modelling, but have evolved over the last 15–20 years into practicing shelf sea oceanographers. Simon Neill is a Reader in Physical Oceanography in the School of Ocean Sciences, Bangor University (UK), and is founder and course director of an MSc in Marine Renewable Energy. He has published around 70 peer-reviewed journal articles, half of which are on the topic of marine renewable energy. He is involved in many national and international projects, and is a committee member of the International Electrotechnical Commission (IEC), working on revising IEC Technical Specification 62600-201: Tidal Energy Resource Assessment and Characterization. M. Reza Hashemi is an Assistant Professor in the Department of Ocean Engineering and Graduate School of Oceanography at the University of Rhode Island, USA. The University of Rhode Island is the birth place of the first offshore wind farm in the United States: the Block Island Offshore Wind Farm, and has several centres for teaching/research regarding renewable energy. Hashemi has published over 40 peer-reviewed journal articles, mainly focused on renewable energy and coastal engineering. He has developed and taught several undergraduate and graduate courses on Ocean Renewable Energy, and one of the motivations for this book was to provide a textbook for these courses. The material in this book is a culmination of both authors teaching and research, in addition to their experience in working in collaboration with industry.

We thank our colleagues at Bangor University and the University of Rhode Island for discussions on various topics presented in this book, and for providing some of the photographs that appear in the chapters. In particular, we wish to acknowledge the expertise of Matt Lewis (Bangor University) for advice and discussions on several topics during development of the book, and Annette Grilli (University of Rhode Island) and Grover Fugate (Rhode Island Coastal Resources Management Council) for providing insights into offshore wind energy. We also thank the individuals external to our organizations who have provided photographs and original high-resolution figures—these are acknowledged individually within the chapters.

This book is arranged into ten chapters that could logically map onto taught programs in the sequence presented. References are provided at the end of each chapter for further reading. The introduction (Chapter 1) provides the context for renewable energy, and the role of ocean renewable energy within the energy mix, and covers fundamental topics and principles such as energy and power, electrical grid systems, and the cost of energy. Although much of the content of this book could be considered entirely from a descriptive perspective, the reader will gain much more insight by considering ocean renewable energy from a more mathematical perspective, and Chapter 2 introduces the physics and mathematics of fluid dynamics. Chapters 3–6 cover the fundamentals of each ocean renewable energy resource in turn: tidal energy, offshore wind, wave energy, and then (briefly) other forms of ocean energy—ocean currents, OTEC, and salinity gradients. An essential step in the development of any ocean renewable energy project is field measurements of the resource, and this is covered in detail in Chapter 7. At early stages of project development, or to understand how the resource varies over timescales that extend beyond observations, validated models are required, and this is comprehensively covered in Chapter 8. Some applications of ocean renewable energy optimization are covered in Chapter 9 and, finally, more contemporary research topics relating to ocean renewable energy, such as multiple resource interactions, introduced in Chapter 10.

Chapter 1

Introduction

There has been a considerable increase in global electricity consumption over the last few decades, yet with vast differences between countries and regions. In addition, the global energy mix has changed significantly over time—the world still relies on coal for over 40% of its electricity generation, but the amount of electricity that is generated from renewable sources has risen rapidly over the last decade.

In this chapter, we introduce the global energy mix and demonstrate how electricity consumption per capita is linked to quality of life. We discuss the pressures of climate change, and dwindling fossil fuel reserves, and how these two issues are driving the transition towards low carbon renewable sources of energy. However, renewable energy generation presents a challenge to electrical grid systems, and we discuss the challenges of accommodating a high penetration of renewable energy into existing grid infrastructure. Finally, we introduce the topic of marine energy and the fundamental concepts of energy and power.

1.1 THE GLOBAL ENERGY MIX

Global electricity production was 23,950 TWh in 2015 (Fig. 1.1), which, with a world population of 7.35 billion, translates as an annual power output of 3.3 MWh per capita (Fig. 1.2). Electricity generation has doubled since 1990, when global production was 11,854 TWh, representing an annual power output of 2.3 MWh per capita.[1] Energy consumption is one of the most accurate indicators of wealth, and so more affluent countries will generally have a higher electric power consumption per capita (Table 1.1). For example, it is fairly striking to note from this table that the United States, with an electricity usage of 12,988 kWh per capita in 2013, has 17 times the electricity consumption (per capita) as India (765 kWh per capita). The Human Development Index (HDI) is a summary measure of average achievement in key dimensions of human development: a long and healthy life, being knowledgeable, and having a good standard of living [1]. The HDI is the geometric mean of normalized indices for each of the three dimensions. The trend of electricity consumption per capita against HDI is very clear (Fig. 1.3). Noting that the x-axis in this figure is logarithmic, Norway tops Iceland (in contrast to Table 1.1) for HDI, despite

1. Global population was 5.28 billion in 1990 (http://data.worldbank.org).

Fundamentals of Ocean Renewable Energy. https://doi.org/10.1016/B978-0-12-810448-4.00001-X

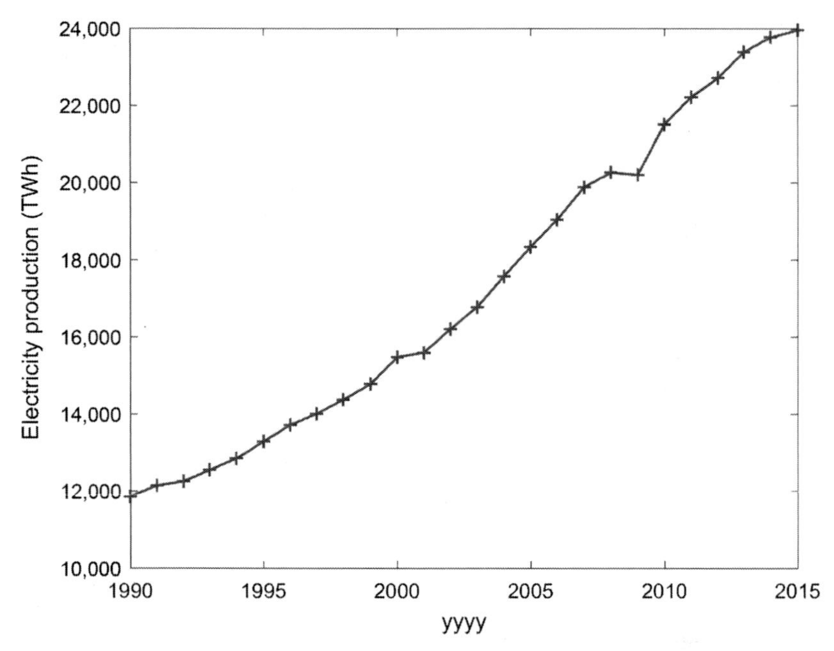

FIG. 1.1 Global electricity production, 1990–2015. *(Data from the International Energy Agency, Key World Energy Statistics, 2016.)*

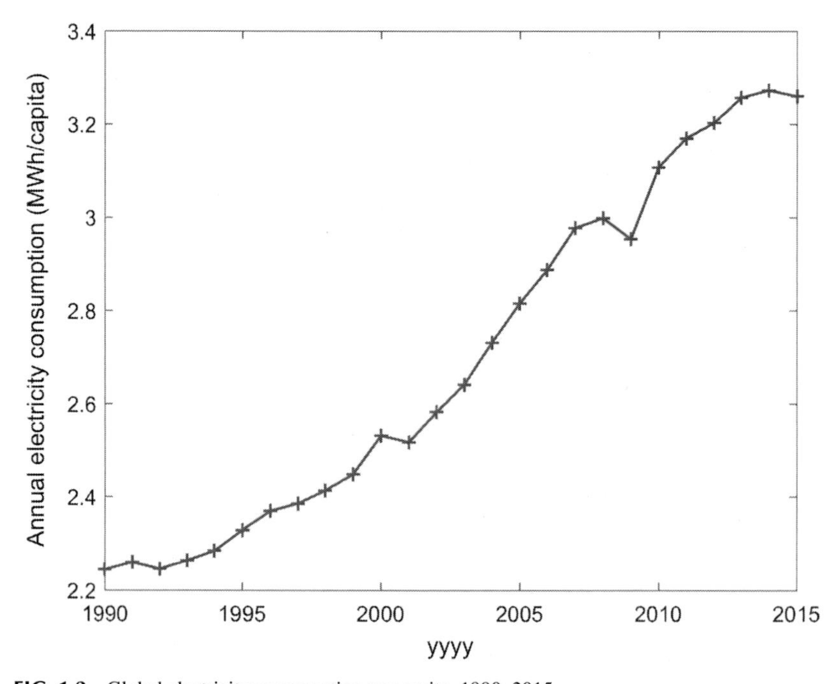

FIG. 1.2 Global electricity consumption per capita, 1990–2015.

TABLE 1.1 Electric Power Consumption per Capita in 2013 for Selected Countries

Country	kWh per Capita
Iceland	54,799
Norway	23,326
Canada	15,519
Qatar	15,471
Sweden	13,870
United States	**12,988**
United Arab Emirates	10,904
Australia	10,134
Saudi Arabia	8741
Japan	7836
France	7374
Russia	6539
Ireland	5702
United Kingdom	5407
Spain	5401
Chile	3879
China	3762
Iran	2899
Brazil	2529
Mexico	2057
Costa Rica	1955
Peru	1270
Indonesia	788
India	**765**
Pakistan	450
Bangladesh	293
Sudan	159
Nigeria	142
Ethiopia	65
Niger	49
South Sudan	39

Source: Data from the World Bank (www.worldbank.org)

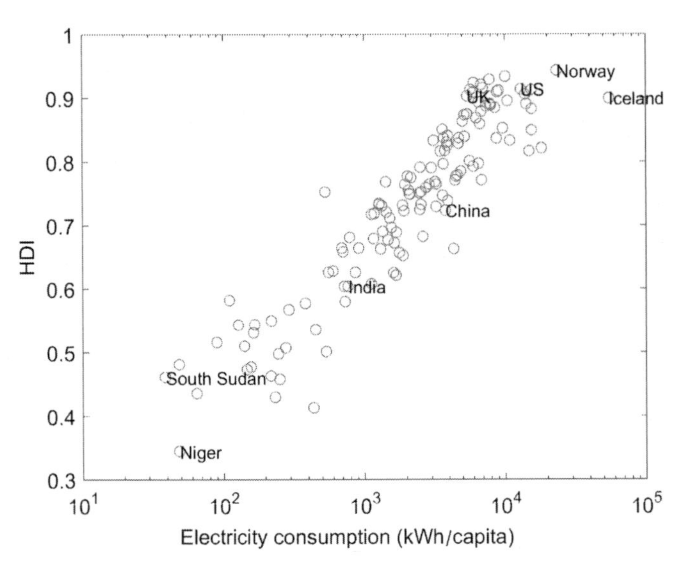

FIG. 1.3 Electricity consumption per capita per country plotted against Human Development Index (HDI) in 2013. *(Electricity data from the World Bank, and HDI data from the United Nations Development Programme.)*

having almost half of the electricity consumption per capita, but observe that the relationship of electricity consumption per capita between the United States and India is reflected in the HDI.

Take a look at the image of Earth's city lights in Fig. 1.4, based on satellite data processed by NASA. The countries and continents of the world can generally be recognized in the image, because populations tend to be concentrated close to coastlines. The US highway network and the connected cities are very prominent in the image. Similarly, Europe, Japan, and many other countries are brightly lit, but so too is India. The brightest areas of the Earth are the most urbanized, but not necessarily the most populated.

Recently, around 1.2 billion people (around 16% of the world's population) do not have access to electricity [2]. This figure has reduced in the last two decades, mainly as a result of increased urbanization. Modern energy services are crucial to human well-being, and to a country's economic development. Access to electricity is essential for the provision of reliable and efficient lighting, heating, cooking, mechanical power, telecommunications, and, in part, transport services.

Snapshots of the global electricity mix for 1973 and 2015 are shown in Fig. 1.5. The main change over these four decades has been a reduction on the reliance of oil for generating electricity (−21%), and this generation has been displaced in the energy mix by an increased share of natural gas (+10%) and nuclear (+8%). Interestingly, the global share of renewable energy generation remained the same (22%) between these two time slices, but the percentage

FIG. 1.4 Image of Earth's city lights created with data from the Defense Meteorological Satellite Program (DMSP) Operational Linescan System (OLS). Originally designed to view clouds by moonlight, the OLS is also used to map the locations of permanent lights on the Earth's surface. *(Data are courtesy of Marc Imhoff (NASA GSFC) and Christopher Elvidge (NOAA NGDC). Image by Craig Mayhew and Robert Simmon, NASA GSFC.)*

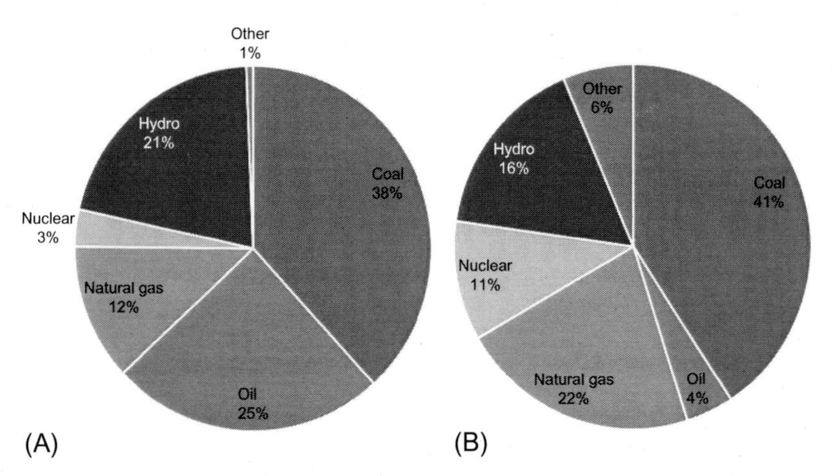

FIG. 1.5 World electricity generation by fuel in (A) 1973 and (B) 2014. Note that coal includes peat and oil shale, and 'other' includes geothermal, solar, and wind. *(Data from the International Energy Agency. Enerdata, Global Energy Statistical Yearbook, 2016.)*

contribution of hydroelectricity reduced significantly (−5%). Note that the largest sector is coal, and that its share in the global electricity mix actually increased from 38% in 1973 to 41% in 2014. Although many countries (e.g. within Europe) have reduced their reliance on coal since 1990 (Fig. 1.6), coal consumption steadily increased from 1965 to 2014, with a slight reduction in

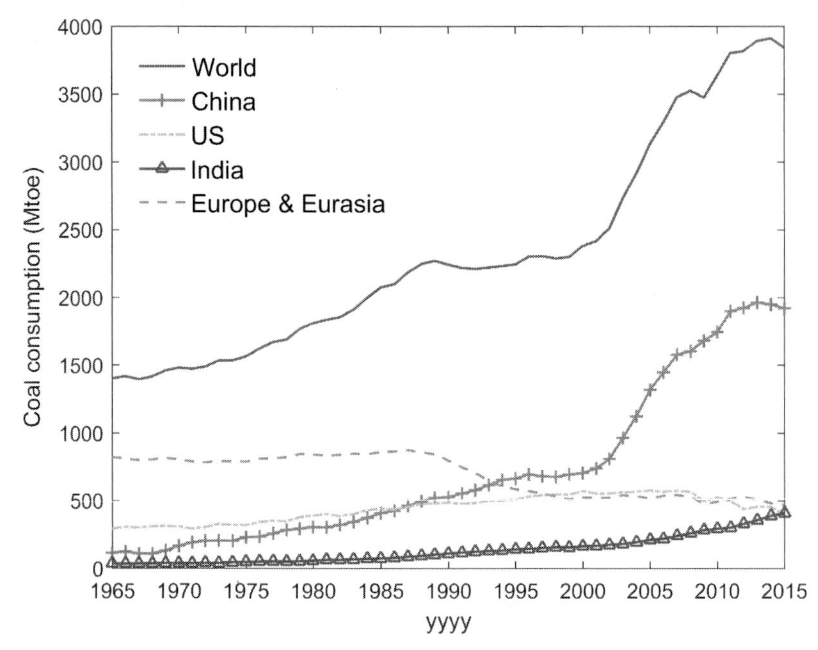

FIG. 1.6 Global and regional coal consumption 1965–2015. Mtoe is million tonnes oil equivalent. *(Data from British Petroleum, BP Statistical Review of World Energy, British Petroleum, London, 2016.)*

2015 [3]. The three largest consumers of coal at present are China (50% of world consumption in 2015), India (10.6%), and the United States (10.3%). Most notable has been the rapid growth in China, since 2002, towards their current position as a major coal consumer.[2] This corresponds with the rapid economic expansion of China (and India) over the last two decades—such economic expansion requires the rapid development of electricity generators. A coal-fired power station takes around 4 years to build, compared with a nuclear power plant, which takes around 5–7 years. Therefore, coal-fired power stations represent a relatively quick, albeit unsustainable, means to increase electricity generating capacity, particularly when a local source of fuel is abundant.

Examining temporal trends in the renewable energy sector in more detail, specifically hydroelectricity, solar, and wind, we see that there has been a relatively steady rise in hydroelectricity consumption over the last 50 years, characterized by more recent (2000 onward) accelerated consumption (Fig. 1.7). In 1965, global hydroelectric power consumption was around 1000 TWh, and

2. On a more sustainable note, China is currently the World's second largest consumer of wind energy, with a global share of 20% in 2015 (the largest is the United States, with a share of 20.3%).

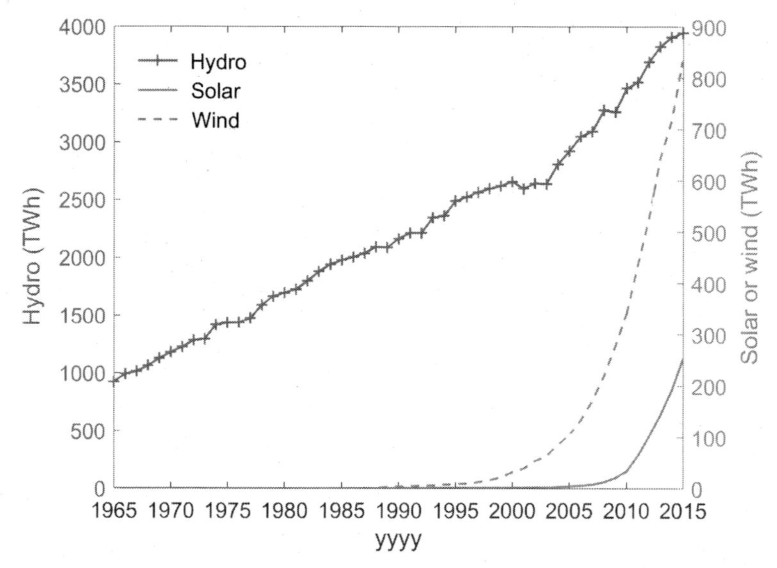

FIG. 1.7 Global electricity consumption of hydroelectricity, solar, and wind from 1965 to 2015. *(Data from British Petroleum, BP Statistical Review of World Energy, British Petroleum, London, 2016.)*

this increased by a factor of 4 to around 4000 TWh in 2015. Although solar and wind energy technologies have existed for a long time,[3] these power sources did not really have an impact on the world stage until the 21st century. However, there has been high growth of these technologies over the last decade, with wind (and solar) contributing around 850 TWh (and 250 TWh), respectively, to the global energy mix in 2015.

1.2 CLIMATE CHANGE AND SUSTAINABILITY

The Earth's climate varies over many timescales, as a response to both natural processes and human influences. The global climate is governed by the planet's radiation balance, and there are three main ways in which this balance can be altered:

1. changes in the incoming radiation (e.g. changes in the Earth's orbit);
2. changes in the reflected radiation (albedo); and
3. changes in long-wave radiation emitted from the Earth (changing greenhouse gas concentrations).

The ice ages that have occurred periodically over the geological past have been linked to regular variations in the Earth's orbit around the Sun, known

3. For example, the world's first wind turbine originates from 1887.

as Milankovitch cycles, which have a period of around 21,000 years. Although these Milankovitch cycles have minimal influence on the global annual mean solar radiation received by the Earth, they alter the solar radiation that is received at each latitude. It has been suggested that when the summer sunshine on the northern continents drops below a critical threshold, snow from the previous winter does not melt, and this triggers an ice age [e.g. 4]. In addition to Milankovitch cycles, incoming radiation varies because the energy that is output from the Sun is not constant. In particular, changes in sunspot activity have been linked to prolonged changes in winter temperatures [5], leading to periods such as the Maunder Minimum (1645–1715), when sunspots were rare, and temperatures across Europe were below average.

Of the Sun's energy that reaches the top of the atmosphere, around one quarter of this energy is reflected by clouds and 'aerosols',[4] and a smaller amount is reflected by the Earth's surface, mainly by light coloured surfaces such as snow and ice. Together, this reflected solar radiation is known as *albedo*. The energy that has not been reflected is absorbed by the atmosphere and, particularly, the surface of the Earth. To balance this incoming solar energy, the Earth emits *long-wave radiation*. Greenhouse gases (GHGs) in the atmosphere act as a partial 'blanket' for this emitted long-wave radiation, and the most important GHGs are water vapour and carbon dioxide (CO_2). The amount of CO_2 in the atmosphere has increased dramatically since the industrial revolution. Since 1960, for example, observations show that the concentration of CO_2 in the atmosphere has increased from around 320 ppm to over 400 ppm (2016)—an increase in concentration of almost 30% in 56 years (Fig. 1.8). This increased concentration of CO_2 in the atmosphere has led to the phenomenon of *global warming*. Feedbacks in the climate system exacerbate global warming. For example, increased concentrations of GHGs in the atmosphere warm the Earth's climate, melting snow and ice. This melting reduces the Earth's albedo, and so these darker surfaces that are revealed absorb more of the Sun's heat in a feedback cycle known as the 'ice-albedo feedback'.

The evidence that GHGs in the atmosphere have drastically increased is incontrovertible (e.g. Fig. 1.8). Much scientific research has been invested in trying to determine the causes and consequences of these increases [e.g. 6]. The cause has primarily been identified as the combustion of fossil fuels, but deforestation also has a role. Looking at Fig. 1.9A, global CO_2 emissions have increased dramatically from close to zero (in 1880) to 9449 million metric tonnes in 2011. This has contributed to an increase in global temperature of over 1°C between 1880 and 2016 (Fig. 1.9B), with consequences, amongst others, of a global rise in sea level of around 200 mm over the same time period (Fig. 1.9C) due to thermal expansion caused by warming of the ocean, and increased melting of land-based ice.

4. Small particles in the atmosphere.

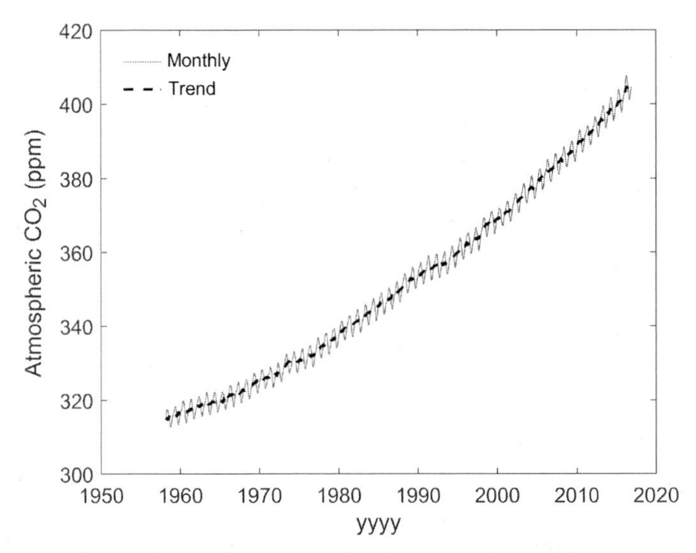

FIG. 1.8 Monthly mean atmospheric carbon dioxide concentrations at Mauna Loa Observatory, Hawaii, 1958–2016—the longest record of atmospheric CO_2 measurements in the world. *(Data from NOAA Earth System Research Laboratory.)*

As mentioned, the Earth's climate has always varied. Climate change sceptics often use this fact to argue that the recent change in the Earth's climate may be due to natural processes (that we may or may not know), and it is not necessarily due to the combustion of fossil fuels. It is true that the Earth has experienced warmer periods, and higher sea levels in the past, when humans did not exist. Scientists have rejected this argument by studying the concentration of carbon dioxide in the atmosphere over many thousands of years. This has been achieved by the analysis of air bubbles trapped in Antarctic ice cores that extend back to 800,000 years. The Carbon Dioxide Information Analysis Center (CDIAC), which serves as the primary climate-change data and information analysis centre of the United States, keeps the record of carbon dioxide data collected at several locations in Antarctica. Based on these data, we can plot the time series of atmospheric carbon dioxide concentration over thousands of years (Fig. 1.10). As we can see, before the industrial revolution, the concentration was always below 300 ppm. The current level of CO_2 in the atmosphere (406 ppm), which is well above 300 ppm, is directly the result of burning fossil fuels. Because we are sure that carbon dioxide is a GHG, we can conclude with certainty that human activities have resulted in global warming.

It has now been accepted that it is too late to *stop* global warming, and much policy is now focussed on trying to *limit* future global warming. The 2016 Paris Agreement, signed by 194 countries, aims to keep the increase in global average temperature to well below 2°C above preindustrial levels, and to pursue efforts to limit the temperature increase to 1.5°C above preindustrial levels, recognizing

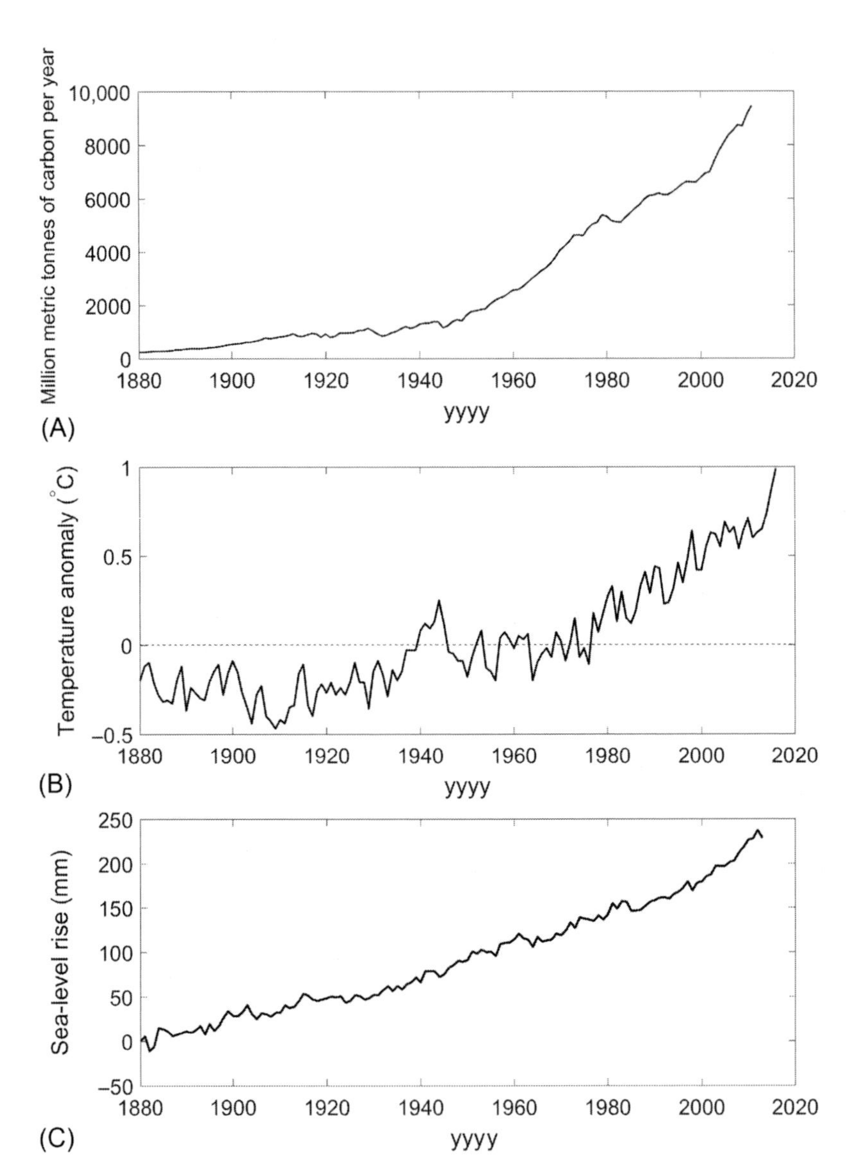

FIG. 1.9 Changes in global (A) carbon dioxide emissions [7], (B) combined land-surface air and sea-surface temperature anomaly—land-ocean temperature index (LOTI) (relative to the 1951–80 mean), and (C) sea-level rise, since 1880. *(LOTI data from NASA Goddard Institute for Space Studies, and sea-level data from the US Environmental Protection Agency.)*

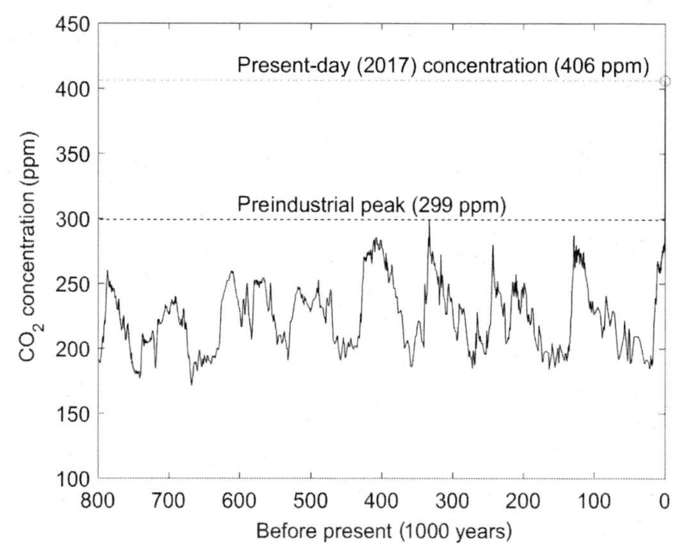

FIG. 1.10 Time series of the carbon dioxide concentration in the atmosphere for a period of 800,000 years based on the ice cores at Dome C in Antarctica. The current concentration of carbon dioxide is above 400 ppm. *(The data have been extracted from the NOAA database.)*

that this would significantly reduce the risks and impacts of climate change. Clearly, one important step, perhaps the most important step, is to reduce the CO_2 that is emitted by electricity generation, that is, by thermal power plants that rely on the combustion of fossil fuels. However, sustainable power plants that are based on renewable 'fuels' such as wind, solar, hydro, and marine will require considerable investment and changes in lifestyle (e.g. increased cost to the consumer, or possibly changes in patterns of consumption), and poses significant challenges, such as variability in the electricity that is generated from renewables, grid integration, and storage. Some of these challenges are introduced in Section 1.3.

Fossil Fuel Reserves

As if global warming was not enough of an incentive to seek low carbon (renewable) sources of electricity generation, the other major reason is the finite nature of fossil fuel reserves. The geographical distribution of estimated oil, coal, and natural gas reserves is plotted in Fig. 1.11, with the actual values cited in the figure caption. Around half of the world's oil reserves are in the Middle East (Fig. 1.11A). However, the fossil fuels that are currently used for significant levels of electricity generation around the world are coal and natural gas (Fig. 1.5). Although the Middle East contains large reserves of natural gas, Europe and Eurasia contains almost one-third of the world's reserves, with a much lower proportion (7%) in North America. However, coal reserves are fairly evenly distributed between Asia Pacific (where the largest reserves are in China—12.8% of the world's reserves), North America (where the largest reserves are in the United States—26.6% of the world's reserves), and Europe and Eurasia. Although the numbers shown in the caption of Fig. 1.11 are staggering (e.g. 187 trillion cubic metres of natural gas and 891,531 million tonnes of coal), demand for these fossil fuel reserves is similarly staggering (e.g. Fig. 1.6). At current rates of consumption, it is estimated that these fossil fuel reserves will run out in [3]

- 114 years (coal)
- 53 years (natural gas)
- 51 years (oil)

The final figure is particularly worrying, as much of our modern transport needs are governed by oil. However, there is clearly a strong case for replacing fossil fuel power generating plants with low carbon (renewable) power stations long before coal and natural gas reserves are depleted.

It can be clearly demonstrated that our reliance on fossil fuels is unsustainable by comparing the timescale of fossil fuel formation, and the estimated time that they will be completely consumed by the human race. The age of the Earth is around 4.5 billion years. The age of fossil fuels is in the range 150–650 million years, which is a significant timescale compared with the age of the Earth. However, it is only in the last 300 years that human beings have started exploiting fossil fuels using modern technologies, and at current rates they will be completely consumed in about a century. Obviously, the age of the human race, and particularly the industrial revolution, is almost zero compared with the time for fossil fuels to form. By analogy, if you imagine the Earth as a person aged 80 years, we can say that this person has made fossil fuels in 3 to 10 years. The modern exploitation of fossil fuels would have begun in the last 3 min of this person's life, and the fossil fuels will be all used up in the next minute.

In conclusion, the Earth is not able to provide fossil fuel resources to maintain the energy requirement of societies for future generations. Therefore, sustainable development cannot be based solely on the fossil fuel industry, and moving towards renewable energy technologies is the only way forward.

1.3 ELECTRICAL GRID SYSTEMS

The world's first public power station, the 'Edison Electric Light Station', was built in London in 1881. Power was generated by a steam engine, and a localized grid system supplied electricity to neighbouring customers. Interestingly, this early grid system supplied direct current (DC), in contrast to the alternating current (AC) that is used today. This early power station produced 160 kW of electricity, for lighting and running electric motors [8]. In the 1880s and 1890s, a so-called 'war of currents' arose, with Thomas Edison backing low-voltage DC

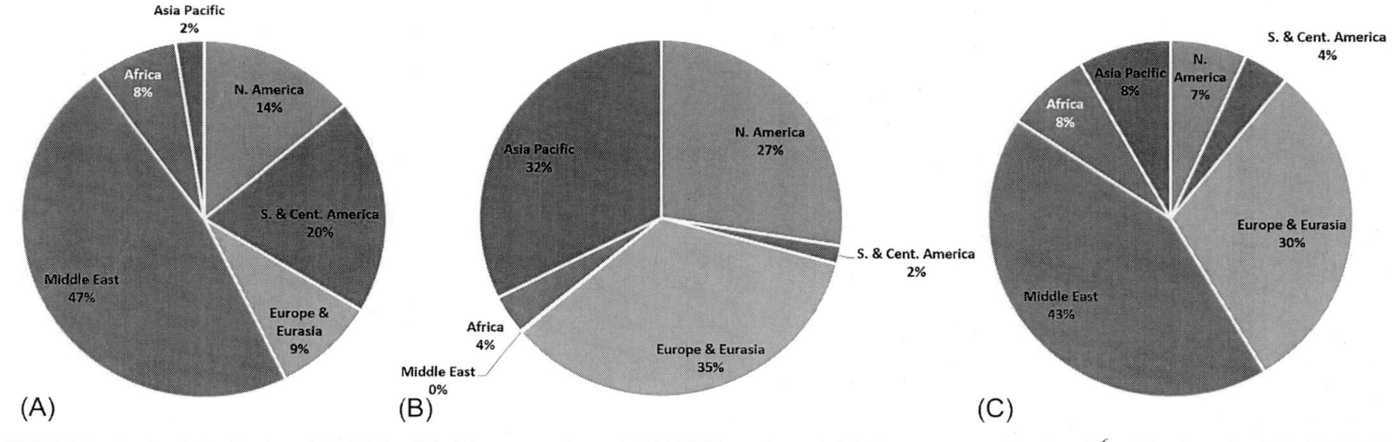

FIG. 1.11 Regional distribution of global fossil fuel reserves at the end of 2014. The estimated global reserves are (A) 1.7×10^6 million barrels of oil, (B) 891,531 million tonnes of coal, and (C) 187 trillion cubic metres of natural gas. *(Data from British Petroleum, BP Statistical Review of World Energy, British Petroleum, London, 2016.)*

systems, and George Westinghouse backing the rival AC current. DC current had the advantage that it transmitted electrical power at the same voltage as that used by the lamps and motors of the customers. However, this required the use of large costly distribution wires, and DC current is only suitable when transmission distances are small (e.g. a service radius of up to a kilometre). Although the press at the time, with the help of Thomas Edison, instilled fear in the public relating to the dangers of high-voltage AC current, it had several advantages over DC systems. AC could be transmitted over long distances using relatively small wires at high voltage, then the voltage reduced using a transformer to make the electricity suitable for public consumption. In addition, AC power stations could be larger and more efficient than their DC counterparts, and the distribution network was less costly. AC won the war of currents.

Early grid systems were inefficient and fragmented. However, in 1926, the UK National Grid was formed as a result of the Electricity Supply Act, which standardized electricity supply, transmitting AC current at 132 kV and 50 Hz. Today, most countries transmit electricity at a frequency of 50 Hz,[5] but the transmission voltage varies within and between grids (although 132 kV is still commonly used). However, the main thing that has changed is the scale. Modern grid systems are many thousands of kilometres in length: the UK grid has a length of around 25,000 km, and in China the grid is over 500,000 km long (although this is formed of six separate grids). These complex systems connect the electricity that is produced by a large number of diverse electricity generators to consumers.

Most large-scale power generators are thermal power stations, where heat energy is converted into electrical power. In general, this is achieved by producing steam, which spins a steam turbine and drives an electrical generator. The greatest variation between thermal power stations is the fuel source. As mentioned in Section 1.1, coal is the most common source of fuel for thermal power stations around the world, followed by natural gas, then nuclear (Fig. 1.5). Note that in the case of natural gas, power stations generally use a combined cycle, where a gas turbine is combined with a steam turbine. Other power stations that are connected to electrical grids include hydroelectric, solar, wind, and of course, marine.

It is desirable for the electricity that is fed into grid systems to have certain characteristics. In particular, the electricity that a power station produces should ideally be:

1. predictable
2. reliable
3. dispatchable

5. Example exceptions are the United States, Canada, Brazil, Mexico, and Saudi Arabia, where the grid frequency is 60 Hz.

Predictable

The electricity that is generated by thermal power stations is *predictable*, because the electricity output can be controlled by altering the fuel supply. This makes the electricity *controllable*, but for coal and nuclear (generally known as 'base load' power plants), this controllability cannot be managed over short-time periods—it will likely take a minimum of several hours to adjust the power output of a large thermal power station. Therefore, coal and nuclear power stations are not *dispatchable*.[6] By contrast, many renewable power stations are not predictable, notably wind and solar. These sources of energy depend on environmental conditions, for example, wind speed or the amount of sunlight striking a solar panel, and so are *intermittent*. An exception is tidal power (both tidal stream and tidal range), which, because tides are governed by astronomical forces, is extremely predictable (Chapter 3). Note that at relatively low levels of renewable energy penetration, errors in forecasts are likely to remain within the parameters that the system is designed to cope with [9]. However, for higher levels of wind and solar penetration, these forecast errors could become significant.

Reliable

It is important for an electrical grid system to have a secure level of base load power—insufficient electricity to meet peak demand, but sufficient to 'keep the lights on'. Currently, such continuous base load tends to be supplied by coal and nuclear power plants. Wind and solar power are weather-dependant, and hydroelectric can be influenced by droughts; therefore, these forms of renewable energy are not reliable. Some form of storage would improve the reliability of renewable energy sources, and tidal lagoons (Chapter 3) present one opportunity for energy storage.

Dispatchable

Electricity generating plants that can be turned on or off, or which can vary their power output in response to changes in demand, are known as *dispatchable*. Base load power plants such as coal and nuclear are not dispatchable, because it will take many hours to adjust their electrical output, and so they cannot closely match demand. However, open cycle gas turbine (OCGT) power plants are a good example of dispatchable electricity, because the power output of an OCGT plant can be adjusted within a short-time frame. For example, from cold start, an OCGT power plant can be fully operational in around 30 min [10]. It is difficult for renewable sources to be dispatchable, but conventional hydroelectric schemes are one exception, because it is possible to control the volume of water that flows through the penstock. In addition, the 1.7 GW

6. By contrast, gas power stations are dispatchable, because their electrical output can be varied at short timescales to respond to changes in demand.

Dinorwig pumped power station in North Wales (UK) can attain full power output, from standby, in under 20 s. Although this power station can only operate for 6 h before exhausting its 'fuel' supply, this is an excellent renewable example of dispatchable power. Further, although wind energy is nondispatchable, it does provide a useful degree of flexibility when the 'fuel' is abundant—for example, under windy conditions, wind turbines can easily be controlled to provide less than maximum output [9].

1.3.1 Supply vs. Demand

At any point in time, the electricity that is supplied to a grid must equal demand. Demand for electricity follows a reasonably predictable pattern. For example, demand is generally greater in winter months when increased electricity is required for heating, and during office hours when workplaces are lit and heated, and workers use electrical equipment such as computers. Those tasked with controlling electricity grids, for example, the National Grid in the UK, forecast demand using a number of factors such as:

- weather patterns
- time of year
- day of week
- time of day

Typical electricity demand time series for the United Kingdom are shown in Fig. 1.12 for a year, a month, and a week. There is a clear seasonal trend, with more demand for electricity during winter months, but considerable variability within each month, week, and day. For example, demand over the week October 19–25 shown in Fig. 1.12C varies from around 25 GW at around 04:00, and peaks at over 45 GW at around 18:00 on most week days.

At the scale of a national grid system, demand throughout the day will fluctuate gradually, because increased demand in one part of the grid (e.g. an electrical heater switched on in one household) is offset by reduced demand in another part of the grid (e.g. an electrical oven switched off in another household). This gradual variation in demand for electricity is known as the 'diversity of demand'.

1.3.2 Grid Inertia

System inertia is considered to be one of the most important system parameters upon which the synchronized operation of present day power systems is based [11]. Synchronization is the process of matching the speed and frequency of a power source to a running network. An AC generator cannot deliver power to an electrical grid unless it is running at the same frequency as the network (i.e. 50 Hz for most grid systems). Presently, the majority of electricity is

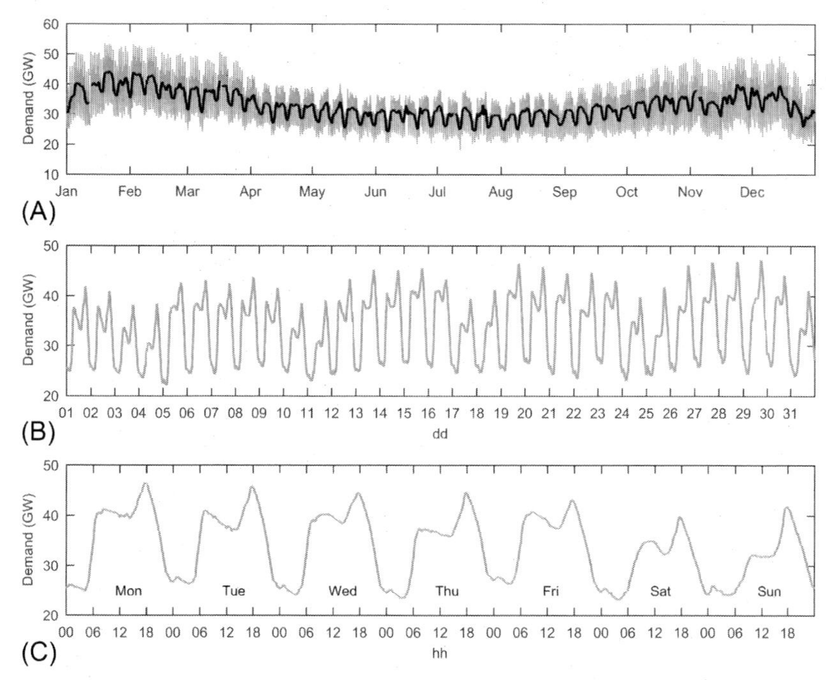

FIG. 1.12 Time series of UK electricity demand in 2015. The raw data (*light gray lines*) are at 5-min time intervals, and the *thick black line* in (A) is the data smoothed over a 24-h averaging period. (A) 2015. (B) October 2015. (C) October 19–25, 2015. *(Data from http://www.gridwatch.templar.co.uk/.)*

produced using thermal power stations. Such power stations work with the help of heavy rotating masses that have *inertia*. This inertia in the rotating masses of synchronous generators determines the immediate frequency response with respect to inequalities in the overall power balance [11]. When a frequency event occurs, for example, a large surge or reduction in demand, the synchronous machines will either inject or absorb kinetic energy into or from the grid to offset the frequency deviation. A grid that has very low system inertia will find it difficult to react to large changes in supply or demand—the grid is said to be 'nervous'.

Renewable energy power plants tend to have very low or zero inertia. Although there is a significant amount of kinetic energy stored in the rotating blades of wind turbines, this inertia is electrically decoupled from the grid, and so does not contribute to system inertia [9]. Solar power plants have no moving parts, and so have zero inertia. As levels of renewable energy penetration increase significantly, it will therefore be necessary to think about the problems that renewables will encounter with respect to grid inertia.

1.3.3 Interconnectors and Grid Storage

In many regions, electricity grids do not function in isolation. For example, there is an integrated North American grid, which links the United States and Canada, and the UK grid has interconnectors that can import/export electricity from/to France, Ireland, and the Netherlands. The French interconnector can import up to 2 GW from France (e.g. during summer months when France has a surplus of nuclear power), and export electricity during winter months. This interconnectedness can help alleviate some issues associated with the intermittency of renewable generation. For example, Germany currently has around 45 GW of wind and 40 GW of solar capacity,[7] and is on track to increase its installed capacity to 59 GW (wind) and 62 GW (solar) by 2030 [12]. Germany's electricity grid is very much interconnected with the neighbouring grids of France, the Czech Republic, Norway, Austria, Switzerland, and the Netherlands. Wind and solar power installations are, by their nature, intermittent. Therefore, during hours of darkness or during times of low wind speed, it will be necessary for Germany to import electricity under a scenario of a high penetration of renewables. Increased interconnectedness could be one way of coping with a high penetration of renewables. For example, over a continent such as Europe, there will be significant geographical and temporal (phase) differences in the wind resource. A large interconnected grid could help smooth out such differences in generation, capitalizing on spatial differences in the resource. Note that tidal energy, which has the major advantage of being predictable, is a strong candidate for such geographical phasing of renewable energy [13,14].

Another way to reduce intermittency in electricity generation is through storage. At present, there is 193 GW of grid storage installed throughout the world, with around half of this capacity fairly equally shared by China, the United States, and Japan (Fig. 1.13). Within each of these countries, pumped hydro accounts for the greatest contribution to grid storage (China: 99.6%, United States: 90.4%, Japan: 99.1%), but the United States has a more diverse blend of storage than either China or Japan, including substantial levels of electrochemical (4.2%) and thermal (2.7%) storage. Clearly, increased grid storage capacity would be consistent with a high penetration of renewables, and would also provide a means of backup power, even in the context of conventional (thermal) power generation.

1.3.4 Levelized Cost of Energy

A key factor that influences the development of (or the investment in) a renewable energy technology is its cost compared with other, conventional, power plants. Many policy discussions over the costs of different technologies for electricity generation are based on the concept of *levelized costs*. Levelized

7. See https://www.carbonbrief.org/how-germany-generates-its-electricity.

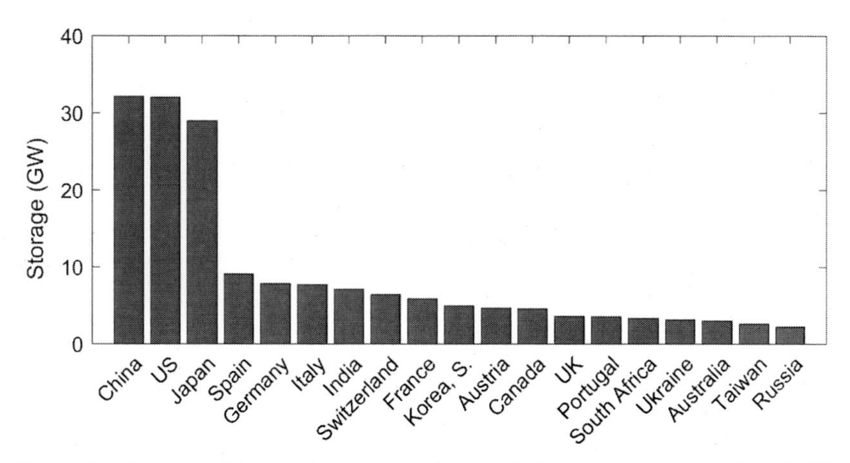

FIG. 1.13 Countries with the highest capacity of installed grid storage at the beginning of 2017. Only countries where storage capacity exceeds 2 GW are listed. *(Data from the US Department of Energy Global Energy Storage Database; http://www.energystorageexchange.org/.)*

costs are defined by the International Energy Agency as the "ratio of total lifetime expenses versus total expected outputs, expressed in terms of the present value equivalent" [15]. Levelized costs therefore provide a mechanism to compare the competitiveness of different technologies, and a transparent method to show how key factors (e.g. capital cost or the cost of fuel) affect the cost of different technologies. Costs generally include

- investment costs (i.e. construction costs)
- operation and maintenance (O&M) costs
- fuel expenditure[8]

Additional costs that can be factored into levelized cost calculations include

- costs of standby generation
- costs of carbon emissions

However, generally the costs of capital, fuel, and O&M dominate the levelized cost.

The calculation of the levelized cost of energy (LCOE) is based on the equivalence of the present value of the sum of discounted revenues and the present value of the sum of discounted costs [16]. Future costs and outputs must be discounted when compared with costs and outputs today. This is for two main reasons:

- time preference for money
- interest

8. One major advantage of renewable technologies is that the fuel is free.

With respect to the time preference for money, there is generally a preference for current consumption built into every individual and organization. In addition, the future is uncertain, and there may be other opportunities to invest, which an organization or individual may have to forego due nonavailability of cash. Discount rate varies per technology [17], but 10% is a value commonly used in LCOE calculations for renewable energy projects [e.g. 16].

To calculate LCOE, first, cost and generation data/estimates should be gathered. Next, the net present value (NPV) of total expected costs for each year should be summed:

$$\text{NPV of total costs} = \sum_{n} \frac{\text{Total capex and opex costs}_n}{(1 + \text{Discount rate})^n} \tag{1.1}$$

where n is the time period. The NPV of expected generation for each year should next be summed:

$$\text{NPV of electricity generation} = \sum_{n} \frac{\text{Net electricity generation}_n}{(1 + \text{Discount rate})^n} \tag{1.2}$$

Finally, the LCOE can be calculated by dividing Eq. (1.1) by Eq. (1.2)

$$\text{LCOE} = \frac{\text{NPV of total costs}}{\text{NPV of electricity generation}} \tag{1.3}$$

Discounting physical values such as power output may not seem intuitive (Eq. 1.2), because physical units neither change in magnitude over time nor do they pay interest [16]. However, what is discounted is the *value* of the output, rather than the output itself.

Table 1.2 shows the estimated values of LCOE for various technologies. Note, for instance, that in general, solar energy is more expensive than onshore wind. Natural gas is very competitive compared with other forms of electricity generation. Well-developed renewable energy technologies like wind can compete with fossil fuels, although they are not as reliable. Because marine renewable technologies (tidal and wave) are immature, their LCOE is much higher. For instance, an NREL (National Renewable Energy Laboratory of the United States) report estimates their LCOE to be of order $1000/MWh [18]. Offshore wind is currently much more financially attractive ($197/MWh). Note that the LCOE for wind energy has reduced significantly over time in response to advances in wind energy technology. It is expected that wave and tidal energy technologies will follow the same trend.

1.4 OCEAN RENEWABLE ENERGY

In the following chapters, we will discuss the ocean renewable energy resource in detail, but here we wish to place ocean energy in context, and to chart commercial progress against progress in an analogous renewable energy sector, wind.

TABLE 1.2 Estimated Levelized Cost of Electricity (LCOE) in the United States

Plant Type	CF[a]	Capital Cost	Fixed O&M	Variable O&M[b]	Transmission	LCOE$/MWh
Conventional coal	85	60.4	4.2	29.4	1.2	95.1
Natural gas	87	14.4	1.7	57.8	1.2	75.2
Advanced nuclear	90	70.1	11.8	12.2	1.1	95.2
Geothermal	92	34.1	12.3	0.0	1.4	47.8
Biomass	83	47.1	14.5	37.6	1.2	100.5
Wind	36	57.7	12.8	0.0	3.1	73.6
Wind—offshore	38	168.6	22.5	0.0	5.8	196.9
Solar PV	25	109.8	11.4	0.0	4.1	125.3
Solar thermal	20	191.6	42.1	0.0	6.0	239.7
Hydroelectric	54	70.7	3.9	7.0	2.0	83.5

[a] *Capacity factor—see Section 1.5.*
[b] *This include the fuel cost which is zero for some technologies such as wind.*
Source: U.S. Energy Information Administration, Annual Energy Outlook 2015, April 2015, DOE/EIA-0383 (2015).

1.4.1 The Nature of Ocean Energy

Many oceanography textbooks include a diagram similar to Fig. 1.14, based on a figure first sketched by Kinsman in 1965 [19]. This figure gives an indication of how energy is distributed across various scales of waves that occur in the ocean, including the wind waves that everyone will be familiar with, through long-period waves such as seiches and tsunamis, to tidal waves. Tidal waves in this sense are *long-period waves* that are governed by astronomical tide generating forces (i.e. due to the Sun-Earth-Moon system). It may surprise some people to hear of tides discussed as waves, but in fact wind waves and tidal waves have many characteristics in common, particularly 'shallow water' wind waves and tidal waves. More details on tidal waves and wind waves are provided in Chapters 3 and 5, respectively. However, what is most evident from Fig. 1.14 is that wind waves and tidal waves contain the most energy across the ocean wave spectrum; indeed, globally there is about 2 TW of each (see Chapters 3 and 5) and so, theoretically, there is an equal and globally significant potential for generating electricity from both waves and tides. Tidal waves and wind waves each contain potential and kinetic energy, and both energy forms are exploited in ocean energy electricity generation. For example, surface point

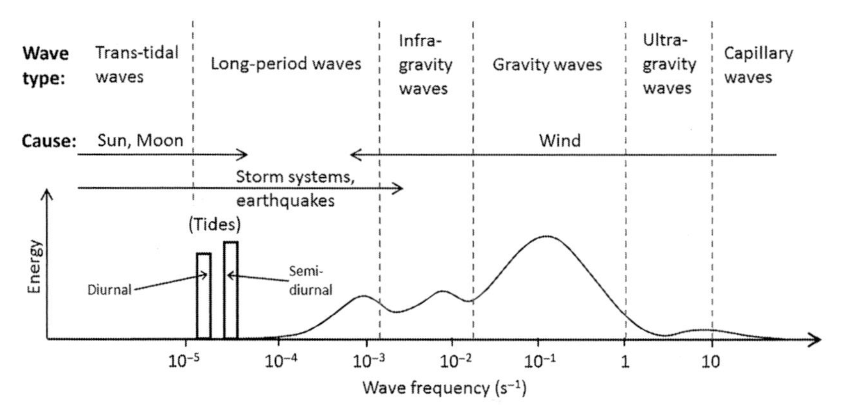

FIG. 1.14 Schematic representation of the energy contained in ocean surface waves. *(Based on a figure first presented by B. Kinsman, Wind Waves: Their Generation and Propagation on the Ocean Surface, Courier Corporation, 1965.)*

absorbers exploit the potential energy of waves, and tidal lagoons have the possibility for significant scale (e.g. >1 GW) tidal range power plants (based on the potential energy of tides). By contrast, oscillating wave converters and tidal stream converters exploit the kinetic energy of waves and tides, respectively.

1.4.2 Lessons From the Wind Energy Industry

The ocean renewable energy industry has many lessons to learn from the firmly established wind energy sector. The history of wind power extends back over a thousand years: the earliest wind-powered grain mills were used by the Persians in AD 500–900, and the Chinese in AD 1200. The first wind turbine that generated electricity was built in Scotland in 1887 by Prof. James Blyth. However, it was not until the 1970s that the deployment of modern wind turbines commenced, as a response to the 1973 oil crisis. Looking at subsequent milestones in the development of the wind energy industry, it took over 40 years to move from prototype to industrial roll-out (Fig. 1.15). In 1991, the average

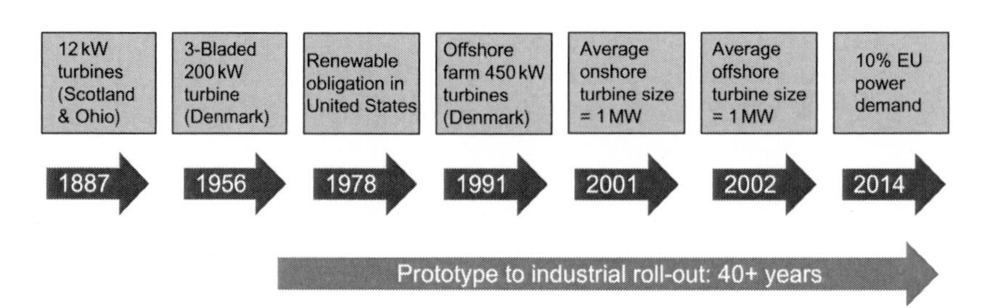

FIG. 1.15 Development of wind turbines, from early prototypes to industrial roll-out. *(Based on a figure presented in the Ocean Energy Forum, Ocean Energy Strategic Roadmap—Building Ocean Energy for Europe, 2016.)*

size of a wind turbine was 224 kW, but it was not until 2001 that turbines exceeded the 1 MW milestone [20]. This transition was based on the concept of establishing small-scale working generators, scaling up over time, and learning from both successes and failures en route. In general, this path is being followed for marine renewable energy devices, but there are important differences. First, high wind speeds can occur in almost any region. These events could be rare in some regions, and more frequent in others, but it is highly likely that at some time in the lifetime of a wind turbine, an extreme 'storm' event will occur, that the wind turbine must be designed to withstand. By contrast, fairly accurate upper bounds can be placed on the extreme conditions that will be experienced by either wave, and particularly tidal, energy convertors during their lifetime. For example, because the tides are governed by astronomical tide generating forces, we can predict with a high level of certainty the highest tidal currents that will be experienced in a tidal channel over a 25-year period.[9] Therefore, robust (yet not overly engineered) tidal energy devices that are suitable for economically exploiting this resource can be designed. Second, it is noticeable that many device developers are working, at prototype and demonstration stages, towards building fairly large precommercial turbines (e.g. the 1.2-MW SeaGen device and the 2-MW OpenHydro device). This is understandable, because tidal power is proportional to velocity cubed, and so there are high rewards from successfully exploiting very high tidal streams. However, such scales of ocean energy conversion could be considered to be fairly far along the analogous wind turbine development timeline (Fig. 1.15), and so relatively high risk. Therefore, it could be prudent to develop, in parallel, a range of devices suitable for deployment in less challenging conditions, an example of which is the 100-kW Nova Innovation M100 turbine. Such smaller devices, in addition to being less challenging to deploy and operate, will have lower cut-in speeds, and so are likely to achieve higher capacity factors when matched with an appropriate resource (Section 1.5).

1.4.3 Roadmaps and Progress

Technology readiness level (TRL) is a system used to estimate technology maturity, and is popular with NASA and the US Department of Defense, etc. TRL is based on a scale from 1 to 9, with 9 being the most mature technology. The use of TRLs enables consistent, uniform discussions of technical maturity across different types of technology. Ocean Energy Europe[10] is a network of ocean energy professionals with the common objective of creating a strong environment for the development of ocean energy. Within Ocean Energy Europe,

9. The expected lifetime of a marine energy device.
10. See https://www.oceanenergy-europe.eu/.

there is an Ocean Energy Forum, who have published a strategic roadmap for the development of ocean energy in Europe [20]. The roadmap outlines a TRL system that can be used to quantify the development of ocean energy devices from the R&D stage through to industrial roll-out, noting that there is some degree of overlap between the phases of development (Table 1.3). The Ocean Energy Forum strategic roadmap presents a timeline for the industrial roll-out of various ocean energy technologies (Fig. 1.16). Noticeable from the timeline is that tidal range is already considered to be at the precommercial stage, whereas tidal stream is at the demonstration stage, and wave energy at R&D stage. In December 2016, tidal range power plants (i.e. tidal lagoons) received a considerable boost in the form of the 'Hendry Review' [21]. This report presented a very positive case for a 320-MW tidal range power plant in Swansea Bay, UK, in the form of a tidal lagoon, and the scheme seems likely to proceed

TABLE 1.3 Phases of Technology Readiness Levels (TRL) Within the Context of Ocean Renewable Energy [20]

Development Phase	TRL	Indicators
R&D	1–4	Small-scale device validated in the lab
		Component testing and validation
		Small/medium-scale pilots
Prototype	3–6	Representative single-scale devices with full-scale components
		Deployed in relevant sea conditions
		Ability to evidence energy generation
Demonstration	5–7	Series or small array of full-scale devices
		Deployed in relevant sea conditions
		Ability to evidence power generation to Grid
Precommercial	6–8	Medium-scale array of full-scale devices experiencing interactions
		Grid connected to a hub or substation (array)
		Deployed in relevant/operational sea conditions
Industrial roll-out	7–9	Full-scale commercial ocean energy power plant or farms
		Deployed in operational real sea conditions
		Mass production of off-the-shelf components and devices

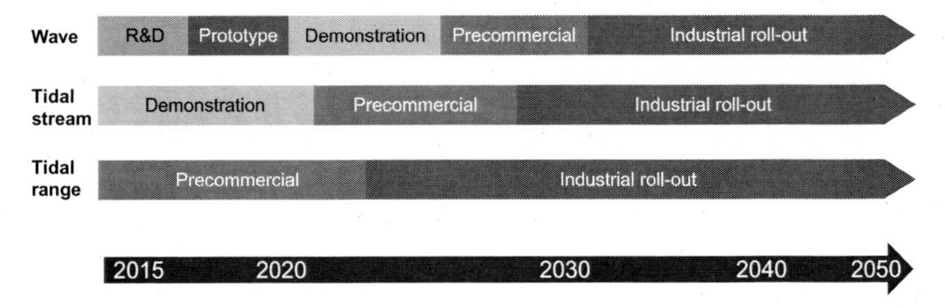

FIG. 1.16 Timeframe for the development of ocean energy technologies. *(Based on a timeline presented in the Ocean Energy Forum, Ocean Energy Strategic Roadmap—Building Ocean Energy for Europe, 2016.)*

to construction. The wave energy industry has suffered several setbacks over the last few years, not least of which was the demise, in 2014, of Pelamis Wave Power—previously considered to be one of the leading developers with their Pelamis P2 device. The wave energy sector is challenging, because sites that are economically viable from a resource perspective are, by their very nature, rather energetic. However, following on from the discontinuation of previously planned large-scale developments, the focus now appears to have shifted more towards the implementation of smaller wave energy projects [22]. The outlook at present is optimistic for the tidal stream industry, with single demonstration devices grid connected, for example, the 1.2-MW SeaGen device that was installed in the Strangford Narrows, Northern Ireland,[11] and the development of the 6 MW phase 1A of the MeyGen project in the Pentland Firth (Scotland), amongst other projects.

1.5 ENERGY AND POWER

Energy is difficult to describe, but a popular and practical definition is:

Energy is the ability of a system to perform work.

Work is energy transferred to or from a body—it requires an applied force moving a certain distance. Therefore, work requires an expenditure of energy, and energy spent performs work.

According to the first law of thermodynamics, energy can neither be created nor destroyed—it can only be transformed from one form to another. It is therefore important when discussing energy, within the context of electricity generation, that the term *energy* is used in the correct way. We do not 'generate energy', we *convert* energy, and generate *power*. Therefore, phrases such as

11. The SeaGen device generated over 8 GWh of electricity; www.seageneration.co.uk.

'wind energy' and 'wind power' that are often used synonymously by the press are inaccurate. In contrast to energy, *power* is the rate at which energy is transformed from one form to another, or the rate that energy is transferred from one place to another.

Units of Energy and Power

To pick up a 5-kg block and raise it 2 m is work, and it requires energy. A force must be exerted, which must be sufficient to overcome gravity, g. The force must be applied directly upwards, against the pull of g, for 2 m [23]. Because

$$\text{Work} = \text{Force} \times \text{Distance} \tag{1.4}$$

and

$$\text{Force} = 5\,\text{kg} \times 9.81\,\text{m/s}^2 = 49.05\,\text{N} \tag{1.5}$$

then

$$\text{Energy} = 49.05\,\text{N} \times 2\,\text{m} = 98.1\,\text{J} \tag{1.6}$$

where J is a Joule, the unit of energy.[12] A Joule can therefore be defined as the work done when a force of 1 N is applied over a distance of 1 m. Referring to the circular definition of work and energy described earlier, 1 J is also the energy expended in performing this task. Energy is stored and converted over a vast range of scales, and some examples are provided in Table 1.4, describing the energy required to melt 1 g of ice (330 J) and the daily energy output of the Sun (3.3×10^{31} J). By way of illustration, global electricity production was 23,950 TWh during 2015 (Fig. 1.1). Therefore, energy is

$$23,950 \times 10^{12} \times 3600 = 8.6 \times 10^{19}\,\text{J} \tag{1.7}$$

Power is the rate at which energy is converted from one form to another. Power is measured in Watts[13] (W), where 1 W = 1 J/s. If we consider what is becoming a fairly obsolete technology, the incandescent light bulb, a 100 W light bulb converts 100 J of electrical energy into light (and 'waste' heat) every second. Consider again the case described earlier, where a 5 kg block was raised 2 m. We would need to raise this block 2 m every second (and somehow convert this energy into electrical energy) if we were to power a 100 W light bulb! As you will have noticed, electricity generation (or consumption) can either by reported as instantaneous power (e.g. in Watts) or as the electricity generated over a period of time (e.g. MWh) (i.e. energy). A simple way of looking at this is, if we have a 1000 W microwave and run it on full power for 1 h (although I dread to think what state the food would be in after an hour!), we would have consumed

---

12. The unit of energy, the Joule, is named after the English physicist James Joule.

13. The unit of power, the Watt, is named after Scottish inventor James Watt.

TABLE 1.4 Energy Orders of Magnitude

Energy (J)	Description
1.0×10^{1}	Energy required to heat 1 g of dry, cool air by 1°C
3.3×10^{2}	Energy to melt 1 g of ice
9.0×10^{3}	Energy in an alkaline AA battery
8.4×10^{6}	Recommended food energy intake per day for a moderately active woman (2000 food calories)
4.0×10^{7}	Energy from the combustion of 1 m^3 of natural gas
1.1×10^{9}	Energy in an average lightning bolt
4.5×10^{9}	Average annual energy usage of a standard refrigerator
7.3×10^{10}	Energy consumed by the average US automobile in the year 2000
6.0×10^{14}	Energy released by an average hurricane in 1 s
1.7×10^{17}	Total energy from the Sun that strikes the face of the Earth each second
2.0×10^{18}	Electricity production of South Korea in 2015
1.6×10^{19}	Electricity production of the United States in 2015
8.6×10^{19}	World electricity production in 2015
5.0×10^{20}	Total world annual energy consumption in 2010
1.5×10^{22}	Total energy from the Sun that strikes the face of the Earth each day
3.9×10^{22}	Estimated energy contained in the world's fossil fuel reserves as of 2010
5.5×10^{24}	Total energy from the Sun that strikes the face of the Earth each year
2.1×10^{29}	Rotational energy of the Earth
3.3×10^{31}	Total energy output of the Sun each day
1.2×10^{34}	Total energy output of the Sun each year
6.6×10^{39}	Theoretical total mass-energy of the Moon
5.4×10^{41}	Theoretical total mass-energy of the Earth
1.2×10^{44}	Approximate lifetime energy output of the Sun
4.0×10^{58}	Visible mass-energy in our galaxy, the Milky Way

Source: Modified from Wikipedia [24].

$$1000\,\mathrm{W} \times 1\,\mathrm{h} = 1\,\mathrm{kWh} \tag{1.8}$$

of electricity.

Capacity Factor

Capacity factor is defined as the actual electricity production divided by the maximum possible electricity output of a power plant, over a period of time. In the example earlier, the microwave consumed a constant 1 kW of electricity over 1 h. However, if we consider a renewable energy power plant, its power output is unlikely to be constant—it will be *intermittent*. We will take the total UK wind generation in 2016 as an example (Fig. 1.17A). As you can see, there is a lot of variability in wind generation over a year, but the mean for 2016 (shown as a black dashed line) was 2.4 GW. Given the information that total operational wind capacity at the end of 2016 was 14.5 GW,[14] this represents a *capacity factor* of

$$100 \times \frac{\text{Actual output}}{\text{Potential output}} = 100 \times \frac{2.4}{14.5} = 17\% \tag{1.9}$$

By contrast, nuclear generation in the United Kingdom was less variable (Fig. 1.17B), with a mean of 7.6 GW in 2016. With an installed nuclear capacity

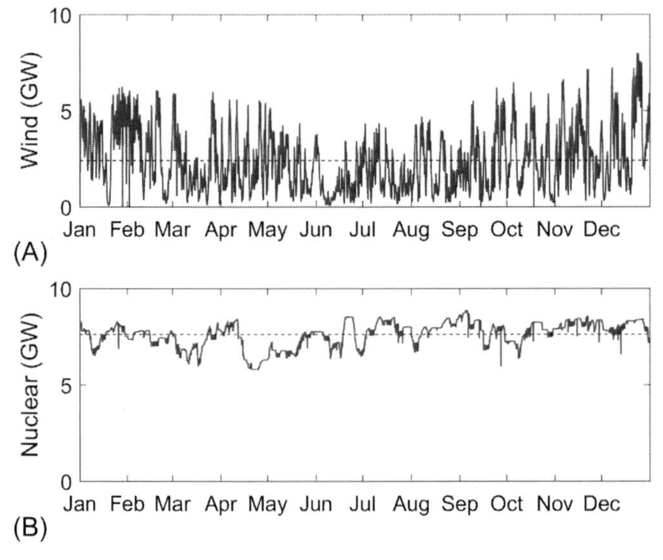

FIG. 1.17 Time series of total UK (A) wind and (B) nuclear generation in 2016. *Dashed lines* are the annual means for each power source. *(Data from http://www.gridwatch.templar.co.uk/.)*

14. See http://www.renewableuk.com/page/UKWEDhome.

TABLE 1.5 Capacity Factors in 2013 for Various UK Power Plants

Plant Type	Capacity Factor (%)
Nuclear power plants	73.8
Combined cycle gas turbine stations	27.9
Coal-fired power plants	58.4
Hydroelectric power stations	31.7
Wind power plants	32.3
Photovoltaic power stations	10.2
Marine (wave and tidal power stations)	9.7
Bioenergy power stations	58.0

Source: Data from the Department of Energy and Climate Change (DECC).

of 8.9 GW, this represents a capacity factor of 85%. Clearly, intermittency is an undesirable and challenging aspect of renewable energy conversion.

Capacity factor varies depending on the type of fuel that is used, the design of the plant, its reliability, maintenance schedule, and can also be subject to market forces. For example, if the wind is blowing strongly, there may be less demand for electricity generated by natural gas, and so the capacity factor of both wind and natural gas will be affected. Capacity factors for a variety of UK power plants are listed in Table 1.5.

REFERENCES

[1] United Nations Development Programme, Human Development Report, 2015.

[2] International Energy Agency, World Energy Outlook, 2016.

[3] British Petroleum, BP Statistical Review of World Energy, British Petroleum, London, 2016.

[4] J. Imbrie, A. Berger, E.A. Boyle, S.C. Clemens, A. Duffy, W.R. Howard, G. Kukla, J. Kutzbach, D.G. Martinson, A. McIntyre, A.C. Mix, B. Molfino, J.J. Morley, L.C. Peterson, N.G. Pisias, W.L. Prell, M.E. Raymo, N.J. Shackleton, J.R. Toggweiler, On the structure and origin of major glaciation cycles 2. The 100,000-year cycle, Paleoceanography 8 (6) (1993) 699–735.

[5] C. De Jager, Solar forcing of climate. 1: Solar variability, Space Sci. Rev. 120 (3–4) (2005) 197–241.

[6] T. Stocker, Climate Change 2013: The Physical Science Basis: Working Group I Contribution to the Fifth Assessment Report of the Intergovernmental Panel on Climate Change, Cambridge University Press, Cambridge, 2014.

[7] T.A. Boden, G. Marland, R.J. Andres, Global, Regional, and National Fossil-Fuel CO_2 Emissions, Carbon Dioxide Information Analysis Center, Oak Ridge National Laboratory, U.S. Department of Energy, Oak Ridge, TN, 2015.

[8] J. Andrews, N. Jelley, N.A. Jelley, Energy Science: Principles, Technologies, and Impacts, Oxford University Press, Oxford, 2013.
[9] Royal Academy of Engineering, Wind Energy—Implications of Large-Scale Deployment on the GB Electricity System, 2014.
[10] Parsons Brinckerhoff, Technical Assessment of the Operation of Coal & Gas Fired Plants: Prepared for DECC, 2014.
[11] P. Tielens, D. Van Hertem, Grid inertia and frequency control in power systems with high penetration of renewables, Young Researchers Symposium in Electrical Power Engineering, Delft, vol. 6, 2012.
[12] International Renewable Energy Agency, Renewable Energy Prospects: Germany. REmap 2030—a renewable energy roadmap, 2015.
[13] A.S. Iyer, S.J. Couch, G.P. Harrison, A.R. Wallace, Variability and phasing of tidal current energy around the United Kingdom, Renew. Energy 51 (2013) 343–357.
[14] S.P. Neill, M.R. Hashemi, M.J. Lewis, Tidal energy leasing and tidal phasing, Renew. Energy 85 (2016) 580–587.
[15] G. Allan, M. Gilmartin, P. McGregor, K. Swales, Levelised costs of wave and tidal energy in the UK: cost competitiveness and the importance of "banded" Renewables Obligation Certificates, Energy Policy 39 (1) (2011) 23–39.
[16] International Energy Agency, Nuclear Energy Agency, Projected Costs of Generating Electricity, 2010.
[17] Oxera Consulting Ltd., Discount rates for low-carbon and renewable generation technologies—a report prepared for the Committee on Climate Change, 2011.
[18] D.S. Jenne, Y.-H. Yu, V. Neary, Levelized cost of energy analysis of marine and hydrokinetic reference models, in: 3rd Marine Energy Technology Symposium, METS, Washington, DC, USA, 2015.
[19] B. Kinsman, Wind Waves: Their Generation and Propagation on the Ocean Surface, Prentice Hall, New Jeresey, 1965.
[20] Ocean Energy Forum, Ocean Energy Strategic Roadmap—Building Ocean Energy for Europe, 2016.
[21] C. Hendry, The role of tidal lagoons, Final Report to the UK Government, 2016.
[22] S.P. Neill, A. Vögler, A. Goward-Brown, S. Baston, M. Lewis, P. Gillibrand, S. Waldman, D. Woolf, The wave and tidal resource of Scotland, Renew. Energy 114 (2017) 3–17.
[23] E.C. Pielou, The Energy of Nature, University of Chicago Press, Chicago, IL, 2007.
[24] Wikipedia, Orders of magnitude (energy) 2017, Available from: https://en.wikipedia.org/wiki/Orders_of_magnitude_%28energy%29. Accessed 12 January 2017.

FURTHER READING

[1] International Energy Agency, Key World Energy Statistics, 2016.
[2] Enerdata, Global Energy Statistical Yearbook, 2016.

Chapter 2

Review of Hydrodynamic Theory

The dynamics of tides, waves, and winds can be understood/simulated by hydrodynamic-thermodynamic theory. In general, the conservation of mass, momentum, energy, and entropy can be formulated mathematically to describe the motion of fluids (water/air). Unfortunately, these equations, which are formulated by integral or partial differential equations, cannot be solved analytically (i.e. using mathematics), unless they are significantly simplified. Therefore, numerical models that are discussed in later chapters are employed to provide us with approximate numerical solutions to these equations.

In this chapter, we discuss the basic equations that are used to describe the dynamics of fluids. In addition, the simplified forms of these equations, including two-dimensional (2D) flow that is popular for tidal modelling, and one-dimensional (1D) equations that are used in the actuator disk theory of wind/tidal turbines, are discussed. This chapter provides a very brief overview of these equations, and more details can be found in other texts (e.g. see [1–5]).

As index/indicial notation is a popular and efficient method to present these equations, we first briefly explain the index notation.

2.1 VECTOR AND INDEX NOTATION

A three-dimensional (3D) vector, such as velocity, can be represented using several notations. Some notations are more efficient when writing out equations; and we will use various notations in this book depending on the complexity of the subject. Here, these notations are explained. Starting from a 3D velocity vector, which has three components in the x, y, and z directions, we can show the velocity ($\mathbf{u}$) in vector notation as follows,

$$\mathbf{u} = (u, v, w) = u\hat{i} + v\hat{j} + w\hat{k} \tag{2.1}$$

where u, v, and w are the components of velocity in the x, y, and z directions, respectively, and $\hat{i}$, $\hat{j}$, and $\hat{k}$ are the unit vectors in the x, y, and z directions,

Fundamentals of Ocean Renewable Energy. https://doi.org/10.1016/B978-0-12-810448-4.00002-1

respectively (e.g. $\hat{i} = (1, 0, 0)$). Occasionally, instead of boldfaced letters, arrow notation is used for vectors, especially in handwritten equations, as follows,

$$\vec{u} = \mathbf{u} = (u, v, w) \tag{2.2}$$

The inner (dot or scalar) product of two vectors (e.g. $\mathbf{u}$ and $\mathbf{S} = (S_x, S_y, S_z)$) in vector notation is written as

$$\mathbf{u} \cdot \mathbf{S} = uS_x + vS_y + wS_z \tag{2.3}$$

The cross (outer or vector) product of two vectors (e.g. $\mathbf{\Omega}$ and $\mathbf{R}$) is a vector quantity, and can be evaluated as follows

$$\mathbf{\Omega} \times \mathbf{R} = (\Omega_y R_z - \Omega_z R_y)\hat{i} + (\Omega_z R_x - \Omega_x R_z)\hat{j} + (\Omega_x R_y - \Omega_y R_x)\hat{k} \tag{2.4}$$

Alternatively, index or indicial notation is an efficient method of writing vectors and matrices. Indicial notation is based on indices. For instance, the velocity vector $\mathbf{u} = (u, v, w)$ can be represented by $\mathbf{u} = (u_1, u_2, u_3)$. Indices 1, 2, and 3 correspond to x, y, and z, respectively. If we use a 'free index' i which can take 1, 2, and 3, we can show a vector as

$$\mathbf{u} = u_i \tag{2.5}$$

Lowercase subscript (here i) in the index notation has a range $(1, 2, 3)$.

Similarly, a matrix can be represented by two free indices: i and j. For example, τ_{ij} $(i, j = 1, 2, 3)$ is a three-by-three matrix or array where $\tau_{11} = \tau_{xx}$, $\tau_{12} = \tau_{xy}, \tau_{13} = \tau_{xz}$, $\tau_{21} = \tau_{yx}, \ldots, \tau_{33} = \tau_{zz}$.

2.1.1 Einstein Convention

In the index notation, for summations, the rule is that repeated indices are summed over. For instance, for the inner product, we can write

$$u_i S_i = \sum_{i=1}^{3} u_i S_i = \mathbf{u} \cdot \mathbf{S} = u_1 S_1 + u_2 S_2 + u_3 S_3 \tag{2.6}$$

Based on this convention, when an index variable appears twice in a single term, it should be summed over. Here is another example,

$$b_i = c_k u_{ik} = \sum_{k=1}^{3} c_k u_{ik} \tag{2.7}$$

where k is repeated and should be summed over. The expanded version of the previous equation is

$$\begin{aligned} b_1 &= c_1 u_{11} + c_2 u_{12} + c_3 u_{13} \Rightarrow b_x = c_x u_{xx} + c_y u_{xy} + c_z u_{xz} \\ b_2 &= c_1 u_{21} + c_2 u_{22} + c_3 u_{23} \Rightarrow b_y = c_x u_{yx} + c_y u_{yy} + c_z u_{yz} \\ b_3 &= c_1 u_{31} + c_2 u_{32} + c_3 u_{33} \Rightarrow b_z = c_x u_{zx} + c_y u_{zy} + c_z u_{zz} \end{aligned} \tag{2.8}$$

Some functions are frequently used in indicial notation. The Kronecker delta δ, which is used later in this chapter, is defined as follows

$$\begin{aligned} \delta_{i,j} &= 1 \quad \text{if } i = j \\ \delta_{i,j} &= 0 \quad \text{if } i \neq j \end{aligned} \tag{2.9}$$

2.1.2 More Examples of Indicial Notation

Using the basic rules that we described previously, we are able to write some equations very efficiently. For instance, the mass balance equation (which is discussed later) can simply be written as

$$\frac{\partial u_i}{\partial x_i} = 0 \text{ or } \frac{\partial u_1}{\partial x_1} + \frac{\partial u_2}{\partial x_2} + \frac{\partial u_3}{\partial x_3} = 0 \text{ or } \frac{\partial u}{\partial x} + \frac{\partial v}{\partial y} + \frac{\partial w}{\partial z} = 0 \tag{2.10}$$

For the Laplace operator, which is used later to quantify shear stress, we have

$$\frac{\partial^2 u}{\partial x_k \partial x_k} = \sum_{k=1}^{3} \frac{\partial^2 u}{\partial x_k \partial x_k} = \frac{\partial^2 u}{\partial x_1^2} + \frac{\partial^2 u}{\partial x_2^2} + \frac{\partial^2 u}{\partial x_3^2} = \frac{\partial^2 u}{\partial x^2} + \frac{\partial^2 u}{\partial y^2} + \frac{\partial^2 u}{\partial z^2} = \nabla^2 u \tag{2.11}$$

2.2 REYNOLDS TRANSPORT THEOREM

In mechanics, the equations of motion can be described for either a system or a control volume. In particular, in fluid mechanics, a fixed volume of space (control volume) is, usually, more suitable than a definite mass of the fluid (system). Nevertheless, the basic laws of continuum mechanics, such as Newton's second law, are applied to mass (i.e. system). Therefore, we need to set the relation between system and control volume. Reynolds transport theorem describes the relationship between a system and a control volume, as follows (Fig. 2.1)

$$\frac{dN}{dt} = \frac{d}{dt} \int_{V_{\text{system}}} (\rho\beta) dV = \int_{V_{\text{CV}}} (\rho\beta) dV + \int_S (\rho\beta) \mathbf{u} \cdot d\mathbf{S} \tag{2.12}$$

where V_{CV} is the fixed control volume in space, which we hereafter call V; V_{system} is the system volume, which is moving as it has a constant mass; ρ is the fluid density; S is the control surface; $\mathbf{u}$ is the fluid velocity; $\mathbf{S} = S\mathbf{n}$ is the control surface; and $\mathbf{n}$ is the unit vector normal to the control surface. β is defined as dN/dm, where N is any fluid property (e.g. mass, momentum, energy), and m is mass. This theory simply indicates that the net flux of a fluid property (outflux minus influx) through a control volume plus changes in the control volume equals the system change of that property. For instance, consider a pond or a lagoon as a fixed control volume. If the inflow Q_{in} and outflow Q_{out} of water are equal, the amount of water in the pond (ρV_{pond}) does not change

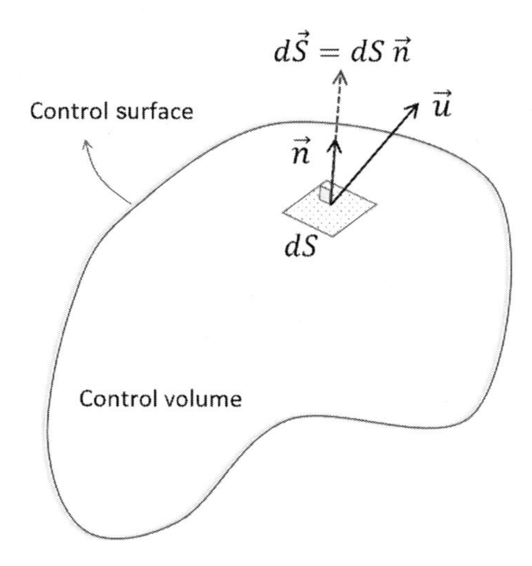

FIG. 2.1 Schematic of a control volume.

with time. However, if the inflow and outflow of water are not equal, they lead to a change in the amount of water in that pool/lagoon. For this simple case, we can write

$$\rho Q_{\text{out}} - \rho Q_{\text{in}} + \frac{\rho dV_{\text{pond}}}{dt} = 0 \tag{2.13}$$

Note that since the mass of a system does not change, $\frac{dN}{dt} = \frac{dm}{dt} = 0$ in the previous equation. The Reynolds transport theorem can be expressed in index notation as follows

$$\frac{dN}{dt} = \frac{d}{dt}\int_{V_{\text{system}}} (\rho\beta)dV = \int_V \frac{\partial(\rho\beta)}{\partial t}dV + \int_S \rho\beta u_i n_i dS \tag{2.14}$$

To change the integral form of an equation to a differential form, we need to use a theorem which relates the surface integral to the volume integral. Referring to vector algebra, using Gauss's theorem (or Divergence theorem), the integral over the control surface can be replaced with the integral over the volume as follows

$$\int_S \mathbf{u} \cdot d\mathbf{S} = \int_V (\nabla \cdot \mathbf{u})\, dV \tag{2.15}$$

where $\nabla \cdot \mathbf{u} = \frac{\partial u_i}{\partial x_i}$ is the divergent operator. In indicial notation, using Gauss's theorem, we can write

$$\int_S \rho\beta u_i n_i dS = \int_V \frac{\partial(\rho\beta u_i)}{\partial x_i} dV \tag{2.16}$$

Thus, the Reynolds transport theorem may also be written as

$$\frac{dN}{dt} = \int_V \left[\frac{\partial(\rho\beta)}{\partial t} + \frac{\partial(\rho\beta u_i)}{\partial x_i}\right] dV \tag{2.17}$$

2.3 NAVIER-STOKES EQUATIONS

Starting with the continuity equation by taking $N = m$ or $\beta = 1$, because the mass of a system does not change with time, $\frac{dN}{dt} = 0$. For an infinitesimal control volume, and using Eq. (2.17), we can write

$$\frac{\partial \rho}{\partial t} + \frac{\partial(\rho u_i)}{\partial x_i} = 0 \tag{2.18}$$

For the special case of incompressible fluids, ρ is constant, and so,

$$\frac{\partial u_i}{\partial x_i} = 0 \qquad \text{or} \qquad \nabla \cdot \mathbf{u} = 0 \tag{2.19}$$

For the momentum equation, the rate of change of linear momentum will be equal to the summation of the forces (on the surface and throughout the body). For momentum, $N = mu_i$ or $\beta = u_i$.

$$\sum F = \int_{V_{\text{system}}} \frac{d(\rho u_i)}{dt} dV = \int_V \left[\frac{\partial(\rho u_i)}{\partial t} + \frac{\partial(\rho u_i u_j)}{\partial x_j} \right] dV \tag{2.20}$$

Using Gauss's theorem, the surface (e.g. pressure and shear stress) and body forces (gravity) can be written as

$$\sum F = \int_V \left[\rho B_i + \frac{\partial \tau^*_{ji}}{\partial x_j} \right] dv \tag{2.21}$$

where B_i and τ^*_{ji} represent body force per unit mass and surface stresses, respectively. The differential form of the momentum equation may be written as

$$\frac{d(\rho u_i)}{dt} = \rho B_i + \frac{\partial \tau^*_{ji}}{\partial x_j} \tag{2.22}$$

Eq. (2.22) is called Cauchy's first law of motion. Commonly, the stress field is decomposed into pressure and friction fields, that is,

$$\tau^*_{ji} = -p\,\delta_{ji} + \tau_{ji} \tag{2.23}$$

Also, the left-hand side of Eq. (2.22) can be expanded using Eq. (2.20), therefore,

$$\frac{\partial \rho u_i}{\partial t} + \frac{\partial(\rho u_i u_j)}{\partial x_j} = \rho B_i - \frac{\partial p}{\partial x_i} + \frac{\partial \tau_{ji}}{\partial x_j} \tag{2.24}$$

For incompressible flows, $\frac{\partial u_j}{\partial x_j} = 0$ (using the continuity Eq. 2.19), therefore we can further simplify $\frac{\partial(\rho u_i u_j)}{\partial x_j}$ as follows

$$\begin{aligned} \frac{\partial(\rho u_i u_j)}{\partial x_j} &= \rho \left\{ u_i \frac{\partial u_j}{\partial x_j} + u_j \frac{\partial u_i}{\partial x_j} \right\} \\ &= \rho \left\{ 0 + u_j \frac{\partial u_i}{\partial x_j} \right\} = \rho u_j \frac{\partial u_i}{\partial x_j} \end{aligned} \tag{2.25}$$

Therefore, Eq. (2.24) for incompressible flow may be written as

$$\frac{\partial u_i}{\partial t} + u_j \frac{\partial u_i}{\partial x_j} = B_i - \frac{1}{\rho}\frac{\partial p}{\partial x_i} + \frac{1}{\rho}\frac{\partial \tau_{ji}}{\partial x_j} \tag{2.26}$$

Eqs (2.19), (2.26) represent the continuity and the momentum equations for incompressible flows. These equations can be expanded as follows

$$\frac{\partial u}{\partial x} + \frac{\partial v}{\partial y} + \frac{\partial w}{\partial z} = 0 \tag{2.27}$$

$$\frac{\partial u}{\partial t} + u\frac{\partial u}{\partial x} + v\frac{\partial u}{\partial y} + w\frac{\partial u}{\partial z} = B_x - \frac{1}{\rho}\frac{\partial p}{\partial x} + \frac{1}{\rho}\left(\frac{\partial \tau_{xx}}{\partial x} + \frac{\partial \tau_{yx}}{\partial y} + \frac{\partial \tau_{zx}}{\partial z}\right) \tag{2.28}$$

$$\frac{\partial v}{\partial t} + u\frac{\partial v}{\partial x} + v\frac{\partial v}{\partial y} + w\frac{\partial v}{\partial z} = B_y - \frac{1}{\rho}\frac{\partial p}{\partial y} + \frac{1}{\rho}\left(\frac{\partial \tau_{xy}}{\partial x} + \frac{\partial \tau_{yy}}{\partial y} + \frac{\partial \tau_{zy}}{\partial z}\right) \tag{2.29}$$

$$\frac{\partial w}{\partial t} + u\frac{\partial w}{\partial x} + v\frac{\partial w}{\partial y} + w\frac{\partial w}{\partial z} = B_z - \frac{1}{\rho}\frac{\partial p}{\partial z} + \frac{1}{\rho}\left(\frac{\partial \tau_{xz}}{\partial x} + \frac{\partial \tau_{yz}}{\partial y} + \frac{\partial \tau_{zz}}{\partial z}\right) \tag{2.30}$$

As the previous equations show, we have 4 equations but more unknowns (i.e. velocity (u, v, w), pressure, and stresses) than equations. Therefore, we need to either simplify the previous equations or include additional equations.

Let us look at a special case, where the shear stresses are ignored. This is the case for inviscid flow, where we ignore viscosity and, therefore, the shear stresses.

2.3.1 Euler Equations

If we assume that the flow is inviscid, then, $\tau_{ji} = 0$. Therefore, the hydrodynamic equations reduce to

$$\frac{\partial u_j}{\partial x_j} = 0 \tag{2.31}$$

$$\frac{\partial u_i}{\partial t} + u_j \frac{\partial u_i}{\partial x_j} = -\delta_{i3} g - \frac{1}{\rho}\frac{\partial p}{\partial x_i} \tag{2.32}$$

or, in vector format,

$$\nabla \cdot \mathbf{u} = 0 \tag{2.33}$$

$$\frac{d\mathbf{u}}{dt} = \frac{\partial \mathbf{u}}{\partial t} + \mathbf{u} \cdot \nabla \mathbf{u} = -\frac{1}{\rho}\nabla p - g\hat{k} \tag{2.34}$$

In the previous equations, it is assumed that the only body force is gravity, which acts in the z direction; therefore, $B_1 = B_2 = 0$, and $B_3 = -g$. Euler equations are the basis of linear wave theory and potential flow theory. Linear wave theory will be presented briefly in Chapter 5 that discusses wave energy.

2.3.2 Viscous and Turbulent Flows

The shear stresses in Eq. (2.29) are associated with the fluid viscosity and turbulence. For the case of laminar flow, where the shear stresses are caused by the viscosity, Newton's shear-stress relationship can help to evaluate the stresses. Newton's viscosity law in a simple form (horizontal flow) can be written as

$$\tau = \mu \frac{\partial u}{\partial z} \tag{2.35}$$

where μ is dynamic viscosity and is the property of a fluid. In general, Newton's law expresses the relationship between the stress field and the deformations in a fluid, and is called the constitutive relation. The simplest case of the constitutive relation is the linear one (e.g. Newton's viscosity law), in which the stress is a linear function of the strain rate. Newton's viscosity law, for incompressible 3D flow, can be generalized as

$$\tau_{ij} = \mu \left\{ \frac{\partial u_i}{\partial x_j} + \frac{\partial u_j}{\partial x_i} \right\} \tag{2.36}$$

Replacing the shear stresses by the previous equation, the momentum equation (2.26) becomes

$$\frac{\partial u_i}{\partial t} + u_j \frac{\partial u_i}{\partial x_j} = B_i - \frac{1}{\rho} \frac{\partial p}{\partial x_i} + \frac{\mu}{\rho} \frac{\partial^2 u_i}{\partial x_k \partial x_k} \tag{2.37}$$

where $\frac{\partial^2 u_i}{\partial x_k \partial x_k}$, as mentioned before, is the Laplacian of u_i and in the expanded form is $\frac{\partial^2 u_i}{\partial x^2} + \frac{\partial^2 u_i}{\partial y^2} + \frac{\partial^2 u_i}{\partial z^2}$. Eqs 2.37 and 2.19 (continuity) are called the incompressible Navier Stokes Equations.

For turbulent flows, the shear stresses would be dependent on the turbulency in the flow field. One approach is replacing the dynamic viscosity with an 'eddy viscosity' or 'turbulent viscosity'. However, unlike dynamic viscosity, which is constant and a property of fluid, the eddy viscosity is highly variable in a flow field, depending on the level of turbulency. Some turbulence models introduce additional equations to estimate the distribution of eddy viscosity in a domain. More details are provided in Chapter 8.

2.3.3 Shallow Water Equations

The Navier-Stokes equations, which were discussed in the previous section, cannot be directly solved by any mathematical method, unless many assumptions/simplifications are made. Numerical methods have been developed to solve the 3D form of the Navier-Stokes equations, for small-scale problems (e.g.

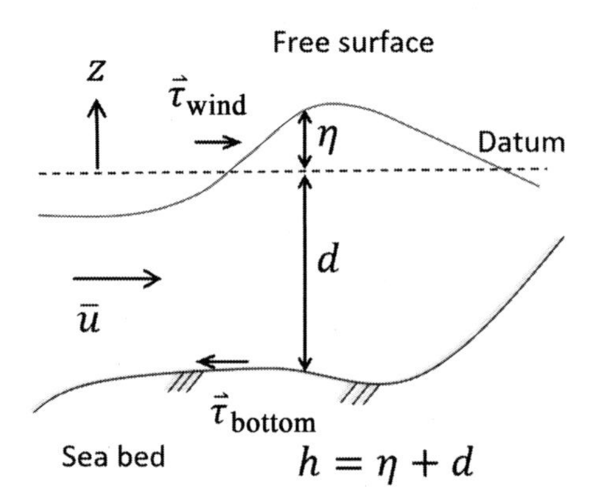

FIG. 2.2 Shallow water equation parameters.

flow around wind turbines). However, the computational cost of solving the Navier-Stokes equations for large scale, oceanic or atmospheric flow problems is extremely high. Therefore, scientists/engineers have tried to derive simpler forms of the Navier-Stokes equations, which can be realistically solved using available computing resources. One popular simplification is the Euler equation, which was introduced in Section 2.3.1, and is the basis of linear wave theory. The other form is the shallow water equations (SWEs), which will be discussed here.

In shallow water flow problems, the horizontal scale is much larger than the vertical scale, and therefore, the flow is 'nearly horizontal'. SWEs are very popular for modelling tidal flows and storm surges, and even for atmospheric flow simulations. The main simplification of the SWEs is that vertical variations in the velocity field are neglected. In other words, we assume just a single velocity for the entire column of a fluid. For cases such as stratified flows, baroclinic flows, or in general, in cases where the vertical acceleration of the fluid is important, SWE cannot be used.

Referring to Fig. 2.2, the idea is to integrate the Navier-Stokes equations over depth, and convert the 3D equation to 2D. Here, we discuss the SWEs for hydrodynamic problems.

The depth-averaged value of a state variable such as u can be computed as

$$\bar{u}(x, y, t) = \frac{\int_{-d}^{\eta} u(x, y, z, t)dz}{d + \eta} = \frac{\int_{-d}^{\eta} u(x, y, z, t)dz}{h} \tag{2.38}$$

where $\bar{u}$ is the depth-averaged velocity, and $h = d + \eta$ is the total water depth.

Leibnitz's Rule

This rule will be used frequently in the derivation of the depth averaged equations. Consider a function, $f(x, y, z)$. Leibnitz's rule can be used to take the derivative of an integral, as follows,

$$\frac{\partial}{\partial x}\int_{\alpha(x,y)}^{\beta(x,y)} f(x,y,z)dz = \int_{\alpha(x,y)}^{\beta(x,y)} \frac{\partial f}{\partial x}dz - \frac{\partial \alpha}{\partial x}f(x,y,\alpha(x,y)) + \frac{\partial \beta}{\partial x}f(x,y,\beta(x,y)) \tag{2.39}$$

or alternatively,

$$\int_{\alpha(x,y)}^{\beta(x,y)} \frac{\partial f}{\partial x}dz = \frac{\partial}{\partial x}\int_{\alpha(x,y)}^{\beta(x,y)} f(x,y,z)dz + \frac{\partial \alpha}{\partial x}f(x,y,\alpha(x,y)) - \frac{\partial \beta}{\partial x}f(x,y,\beta(x,y)) \tag{2.40}$$

where α and β are arbitrary functions.

Starting from the continuity equation (2.27), and integrating over the flow depth results in,

$$\int_{-d}^{\eta}\left(\frac{\partial u}{\partial x} + \frac{\partial v}{\partial y} + \frac{\partial w}{\partial z}\right)dz = 0 \tag{2.41}$$

where η is the water elevation above the datum (still water level), and d is the water depth (bathymetry). Eq. (2.41) leads to

$$\int_{-d}^{\eta} \frac{\partial u}{\partial x}dz + \int_{-d}^{\eta} \frac{\partial v}{\partial y}dz + w(\eta) - w(-d) = 0 \tag{2.42}$$

Referring to Leibnitz's rule, the derivatives inside integrals can be evaluated as follows

$$\int_{-d}^{\eta} \frac{\partial u}{\partial x}dz = \frac{\partial}{\partial x}\int_{-d}^{\eta} udz - u(\eta)\frac{\partial \eta}{\partial x} + u(-d)\frac{\partial(-d)}{\partial x} \tag{2.43}$$

$$\int_{-d}^{\eta} \frac{\partial v}{\partial y}dz = \frac{\partial}{\partial y}\int_{-d}^{\eta} vdz - v(\eta)\frac{\partial \eta}{\partial y} + v(-d)\frac{\partial(-d)}{\partial y} \tag{2.44}$$

The water free-surface is dependent on time and location (i.e. x, y), and can be expressed as,

$$z_{\mathrm{s}} = \eta(x,y,t) \tag{2.45}$$

The sea bed can similarly be expressed as (ignoring erosion and sedimentation)

$$z_{\mathrm{b}} = -d(x,y) \tag{2.46}$$

Therefore, the vertical velocity at the free surface, and the bed, may be expressed as

$$w(\eta) = \left.\frac{dz}{dt}\right|_{z=\eta} = \frac{\partial \eta}{\partial t} + u(\eta)\frac{\partial \eta}{\partial x} + v(\eta)\frac{\partial \eta}{\partial y} \tag{2.47}$$

$$w(-d) = \left.\frac{dz}{dt}\right|_{z=-d} = -\frac{\partial d}{\partial t} - u(-d)\frac{\partial d}{\partial x} - v(-d)\frac{\partial d}{\partial y} \tag{2.48}$$

In addition, if it is assumed that the sea bed does not change over time, then,

$$\left.\frac{\partial d}{\partial t}\right|_{z=-b} = 0 \tag{2.49}$$

Therefore,

$$w(-d) = \left.\frac{dz}{dt}\right|_{z=-d} = u(-d)\frac{\partial(-d)}{\partial x} + v(-d)\frac{\partial(-d)}{\partial y} \tag{2.50}$$

By substituting Eqs (2.43), (2.44), (2.48), (2.50) into Eq. (2.42), several terms cancel out; using the depth-averaged variables (Eq. 2.38), after rearranging the terms, the continuity equation becomes

$$\frac{\partial \eta}{\partial t} + \frac{\partial(\bar{u}h)}{\partial x} + \frac{\partial(\bar{v}h)}{\partial y} = 0 \tag{2.51}$$

The previous equation has three unknowns: depth-averaged velocities in the x and y directions, and the total water depth ($h(x, y, t)$). Additional differential equations are therefore necessary, which will be formed by integrating the momentum equations over the water depth.

A main assumption for SWE is that the vertical acceleration is small and can, therefore, be neglected. This means that the main components of velocities are horizontal. Using this assumption, the momentum equation in the vertical (z-direction) can be considerably simplified. We will show that by neglecting the terms that correspond to the vertical acceleration, the momentum equation in the z-direction reduces to the hydrostatic pressure law.

Starting from Eq. (2.37) for $i = 3$, these assumptions imply,

$$\frac{dw}{dt} \approx 0, \quad \mu\nabla^2 w \approx 0 \tag{2.52}$$

Therefore, the momentum equation in the z-direction becomes ($B_3 = -g$)

$$-g - \frac{1}{\rho}\frac{\partial p}{\partial z} = 0 \tag{2.53}$$

For hydrodynamic simulations, the pressure at the water surface is equal to the atmospheric pressure (p_a). Integrating Eq. (2.53) over depth results in

$$p_a - p(z) + \int_z^\eta \rho g dz = 0 \tag{2.54}$$

which is the hydrostatic equation law

$$p(z) = p_a + \rho g(\eta - z) \tag{2.55}$$

If the atmospheric pressure is considered constant, then the pressure terms in the momentum equation (at a constant z in the 3D case) can be replaced by

$$\frac{\partial p}{\partial x_i} = \rho g \frac{\partial \eta}{\partial x_i} \tag{2.56}$$

where i is 1 or 2, corresponding to the x or y direction. In some cases, such as storm surge modelling, changes in atmospheric pressure are important, as they can vary in the x or y direction. For these cases, we can write

$$\frac{\partial p}{\partial x_i} = \frac{\partial p_a}{\partial x_i} + \rho g \frac{\partial \eta}{\partial x_i} \tag{2.57}$$

Similar to the continuity equation, we can integrate the momentum equations in the x and y directions over depth. Integrating the momentum equation in the x direction (Eq. 2.28) produces

$$\int_{-d}^{\eta} \frac{\partial u}{\partial t} dz + \int_{-d}^{\eta} \left(u\frac{\partial u}{\partial x} + v\frac{\partial u}{\partial y} + w\frac{\partial u}{\partial z} \right) dz = -\int_{-d}^{\eta} g\frac{\partial \eta}{\partial x} dz + \cdots$$
$$+ \frac{1}{\rho} \int_{-d}^{\eta} \left(\frac{\partial \tau_{xx}}{\partial x} + \frac{\partial \tau_{yx}}{\partial y} + \frac{\partial \tau_{zx}}{\partial z} \right) dz \tag{2.58}$$

Again, we can evaluate the integrals using Leibnitz's law, which leads to the cancellation of several terms. Therefore, for the local acceleration term, we may write

$$\int_{-d}^{\eta} \frac{\partial u}{\partial t} dz = \frac{\partial}{\partial t} \int_{-h}^{\eta} u dz - u(\eta)\frac{\partial \eta}{\partial t} + u(-d)\frac{\partial(-d)}{\partial t} \tag{2.59}$$

Referring to Eq. (2.25), the convective acceleration term in the momentum equation can be written as

$$u\frac{\partial u}{\partial x} + v\frac{\partial u}{\partial y} + w\frac{\partial u}{\partial z} = \frac{\partial u^2}{\partial x} + \frac{\partial uv}{\partial y} + \frac{\partial uw}{\partial z} \tag{2.60}$$

Application of Leibnitz's rule for the convective term leads to

$$\int_{-d}^{\eta} \frac{\partial u^2}{\partial x} dz = \frac{\partial}{\partial x} \int_{-d}^{\eta} u^2 dz - u^2(\eta)\frac{\partial \eta}{\partial x} + u^2(-d)\frac{\partial(-d)}{\partial x} \tag{2.61}$$

$$\int_{-d}^{\eta} \frac{\partial uv}{\partial y} dz = \frac{\partial}{\partial y} \int_{-d}^{\eta} (uv) dz - u(\eta)v(\eta)\frac{\partial \eta}{\partial y} + u(-d)v(-d)\frac{\partial(-d)}{\partial y} \tag{2.62}$$

$$\int_{-d}^{\eta} \frac{\partial uv}{\partial x} dz = u(\eta)w(\eta) - u(-d)w(-d) \tag{2.63}$$

If we add the above equations (2.61–2.63), and use Eqs 2.47 and 2.48, several terms will be cancelled out. Integrating the hydrostatic pressure term generates

$$\int_{-d}^{\eta} \left(g\frac{\partial \eta}{\partial x} \right) dz = g\frac{\partial \eta}{\partial x} \int_{-d}^{\eta} dz$$
$$= g\frac{\partial \eta}{\partial x}(d + \eta) = g\frac{\partial \eta}{\partial x} h \tag{2.64}$$

Also, application of Leibnitz's rule to the shear-stress terms results in

$$\int_{-d}^{\eta} \frac{\partial \tau_{xx}}{\partial x} dz = \frac{\partial}{\partial x} \int_{-d}^{\eta} \tau_{xx} dz - \tau_{xx}(\eta)\frac{\partial \eta}{\partial x} + \tau_{xx}(-d)\frac{\partial(-d)}{\partial x} \tag{2.65}$$

$$\int_{-d}^{\eta} \frac{\partial \tau_{yx}}{\partial y} dz = \frac{\partial}{\partial y} \int_{-d}^{\eta} \tau_{yx} dz - \tau_{yx}(\eta)\frac{\partial \eta}{\partial y} + \tau_{yx}(-d)\frac{\partial(-d)}{\partial y} \tag{2.66}$$

$$\int_{-d}^{\eta} \frac{\partial \tau_{zx}}{\partial z} dz = \frac{\partial}{\partial z} \int_{-d}^{\eta} \tau_{zx} dz - \tau_{zx}(\eta)\frac{\partial \eta}{\partial z} + \tau_{zx}(-d)\frac{\partial(-d)}{\partial z} \tag{2.67}$$

The first terms on the RHS can be replaced by depth-averaged quantities as follows, for example

$$\bar{\tau}_{yx}h = \int_{-d}^{\eta} \tau_{yx} dz \tag{2.68}$$

Also, because the depth-averaged variables are independent of z, in Eq. (2.67), $\frac{\partial(\bar{\tau}_{zx}h)}{\partial z} = 0$.

By implementing the previous equations, replacing every term in Eq. (2.58), and rearranging the terms, the depth-averaged momentum equation in the x direction becomes

$$\begin{aligned}\frac{\partial \bar{u}h}{\partial t} + \frac{\partial \bar{u}^2 h}{\partial x} + \frac{\partial \bar{u}\bar{v}h}{\partial y} &= -g(d+\eta)\frac{\partial \eta}{\partial x} + \cdots \\ &+ \frac{1}{\rho}\left[\frac{\partial(\bar{\tau}_{xx}h)}{\partial x} + \frac{\partial(\bar{\tau}_{yx}h)}{\partial y}\right]_{\text{water column}} - \frac{1}{\rho}\left[\tau_{xx}\frac{\partial \eta}{\partial x} + \tau_{yx}\frac{\partial \eta}{\partial y}\right]_{z=\eta,\ \text{surface}} + \cdots \\ &+ \frac{1}{\rho}\left[\tau_{xx}\frac{\partial(-d)}{\partial x} + \tau_{yx}\frac{\partial(-d)}{\partial y}\right]_{z=-d,\text{bottom}}\end{aligned} \tag{2.69}$$

The shear-stress terms have been grouped into the bed shear stress, surface stress, and water column stress. The bed shear stress can be estimated by several empirical relationships. Using the Manning's equation,

$$\tau_b = \rho g h n^2 \frac{U|U|}{h^{4/3}} = \rho g n^2 \frac{U|U|}{h^{1/3}} \tag{2.70}$$

in which U is the total depth-averaged velocity (i.e. $U = \sqrt{\bar{u}^2 + \bar{v}^2}$) and n is the Manning coefficient. Eq. (2.70) indicates that the shear stress always acts in the opposite direction of the velocity (i.e. $U|U|$ instead of U^2). Therefore, for the component of the bed shear stress in the x direction, we have

$$\tau_{bx} = \tau_b \frac{\bar{u}}{\sqrt{\bar{u}^2 + \bar{v}^2}} = \rho g n^2 \frac{\bar{u}\sqrt{\bar{u}^2 + \bar{v}^2}}{h^{1/3}} \tag{2.71}$$

The wind shear stress can be computed using an empirical quadratic friction law as follows

$$\tau_w = \rho_a C_{wd} W|W| \tag{2.72}$$

where W is the wind velocity, usually specified at 10 m above the water surface, ρ_a is the air density, and C_{wd} is the wind drag coefficient. Consequently, the wind shear stress in the x direction can be evaluated as

$$\tau_{wx} = \frac{W_x}{\sqrt{W_x^2 + W_y^2}}\tau_w, \quad \tau_{wx} = \rho_a C_{wd} W_x \sqrt{W_x^2 + W_y^2} \tag{2.73}$$

where W_x and W_y are the components of the wind speed, W, in the x and y directions, respectively (i.e. $W = \sqrt{W_x^2 + W_y^2}$). By replacing the bed and surface shear-stress terms, the depth-averaged equations can finally be written as

$$\frac{\partial \eta}{\partial t}+\frac{\partial(\bar{u}h)}{\partial x}+\frac{\partial(\bar{v}h)}{\partial y}=0 \tag{2.74}$$

$$\begin{aligned}\frac{\partial(\bar{u}h)}{\partial t}+\frac{\partial(\bar{u}^2h)}{\partial x}+\frac{\partial(\bar{u}\bar{v}h)}{\partial y}&=-gh\frac{\partial \eta}{\partial x}+\frac{\rho_a}{\rho}CW_x\sqrt{W_x^2+W_y^2}+\cdots\\&-gn^2\frac{\bar{u}\sqrt{\bar{u}^2+\bar{v}^2}}{h^{1/3}}+\frac{1}{\rho}\left[\frac{\partial(\bar{\tau}_{xx}h)}{\partial x}+\frac{\partial(\bar{\tau}_{yx}h)}{\partial y}\right]\end{aligned} \tag{2.75}$$

$$\begin{aligned}\frac{\partial(\bar{v}h)}{\partial t}+\frac{\partial(\bar{u}\bar{v}h)}{\partial x}+\frac{\partial(\bar{v}^2h)}{\partial y}&=-gh\frac{\partial \eta}{\partial y}+\frac{\rho_a}{\rho}CW_y\sqrt{W_x^2+W_y^2}+\cdots\\&-gn^2\frac{\bar{v}\sqrt{\bar{u}^2+\bar{v}^2}}{h^{1/3}}+\frac{1}{\rho}\left[\frac{\partial(\bar{\tau}_{xy}h)}{\partial x}+\frac{\partial(\bar{\tau}_{yy}h)}{\partial y}\right]\end{aligned} \tag{2.76}$$

These equations are known as the Saint-Venant, or SWEs, and are the basis of tidal and surge modelling. Considering η, $\bar{u}$, and $\bar{v}$ as state variables, three equations can be used to find the solution of a hydrodynamic field; note that total water depth h can be computed based on η, and bathymetry (b).

We still have the problem of the water column shear-stress terms (i.e. $\frac{\partial(\bar{\tau}_{xx}h)}{\partial x}+\frac{\partial(\bar{\tau}_{yx}h)}{\partial y}$ and $\frac{\partial(\bar{\tau}_{xy}h)}{\partial x}+\frac{\partial(\bar{\tau}_{yy}h)}{\partial y}$) in the momentum equations, as they introduce additional unknowns. In many problems (e.g. open channel flow), shear stresses in the water column are either neglected or specified by very simple laws (e.g. constant eddy viscosity). These terms represent some forces such as turbulent shear stresses, wave radiation stresses, etc., which act in the water column. They are usually replaced by additional equations. More details about turbulence and wave radiation forces will be presented in later chapters.

2.4 HYDRODYNAMIC EQUATIONS IN 1D STEADY CASE

The equations of the fluid that was described in the previous sections are generally solved by numerical methods. On the other hand, for a very simple case in which the flow is incompressible, approximately 1D, and steady (i.e. no change in time), the conservation laws can be applied directly (without resort to numerical methods) to a problem. For instance, we will use these equations to study the basic hydrodynamics of wind/tidal turbines.

Starting from the continuity equation (2.14); $N = m$, because the flow is steady, the Reynolds transport equation becomes

$$\int_S \rho\mathbf{u}\cdot d\mathbf{S}=0 \Rightarrow \sum \mathbf{u}\cdot\ \mathbf{A}=0 \tag{2.77}$$

where $\mathbf{A} = A\mathbf{n}$ is the surface vector, and $\mathbf{n}$ is the normal vector which is perpendicular to the surface. Eq. (2.77) indicates that the summation of volumetric fluxes through the surfaces of a control volume should be zero in the steady-state case. For a special case of 1D flow, where the velocity is normal to the control surface, we have

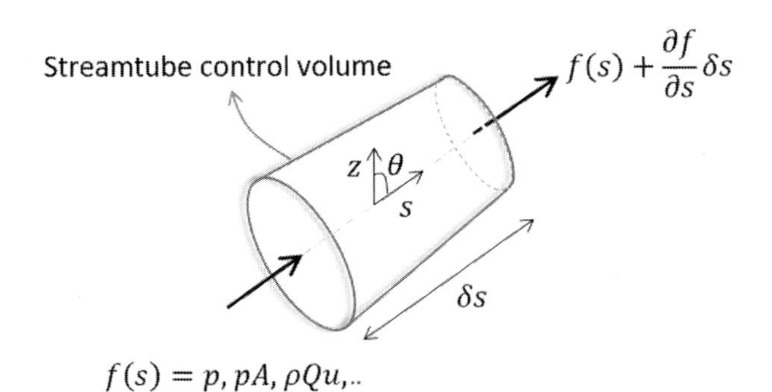

FIG. 2.3 Schematic of a streamtube control volume for the derivation of Bernoulli's equation.

$$- u_{\text{in}}A_{\text{in}} + u_{\text{out}}A_{\text{out}} = 0 \Rightarrow Q = uA = \text{cons.} \tag{2.78}$$

where 'in' and 'out' subscripts denote the inflow and outflow, respectively. Q is the volumetric flow rate or discharge, which is constant for this case. Note that the inner product of the velocity and the surface vector for inflow is negative, whilst for outflow it is positive.

For the 1D momentum equation, in the steady case, we have

$$\sum \mathbf{F} = \frac{d(m\mathbf{u})}{dt} = \int_S \rho\mathbf{u}\mathbf{u} \cdot d\mathbf{S} \Rightarrow \sum \mathbf{u}(\mathbf{u} \cdot \ \mathbf{A}) = \sum \rho u Q \tag{2.79}$$

where Q is positive for outflux and negative for the influx of momentum. The momentum equation can be simply written as

$$\sum F = \rho Q(u_{\text{out}} - u_{\text{in}}) \tag{2.80}$$

Another useful equation is Bernolli's equation, which is the energy equation for inviscid 1D flows in the steady-state case. Bernoulli's equation can be easily derived from the momentum equation. Fig. 2.3 shows an infinitesimal control volume of a streamtube. The forces acting on this control volume include weight and pressure. From calculus, we can write this relation for any function for small values of δs

$$f(s + \delta s) \approx f(s) + \frac{\partial f}{\partial s}\delta s \tag{2.81}$$

Therefore, the summation of forces acting on the control volume can be formulated as

$$\begin{aligned}\sum F &= -\rho g dV \cos\theta + pA|_s - pA|_{s+ds} \\ &= -\rho g \cos\theta A ds + pA + \left(pA + \frac{\partial(pA)}{\partial s} ds\right)\end{aligned} \tag{2.82}$$

where $ds = \delta s$ and A are the length and the area of the control volume, respectively. θ represents the angle of the streamtube to the vertical direction, as shown in the figure, and can be estimated as $\cos\theta = \frac{\partial z}{\partial s}$. The force terms reduce to

$$\sum F = -\rho g A \frac{\partial z}{\partial s} ds - \frac{\partial (pA)}{\partial s} ds \tag{2.83}$$

The flux of momentum through the control volume can be written as

$$\sum \rho u Q = -\rho Q u|_s + \rho Q u|_{s+\delta s} = -\rho Q u + \left(\rho Q u + \frac{\partial(\rho Q u)}{\partial s} ds\right) \tag{2.84}$$

We can further simplify the previous equation as follows

$$\sum \rho u Q = \frac{\partial(\rho Q u)}{\partial s} = \rho Q \frac{\partial u}{\partial s} = \rho A u \frac{\partial u}{\partial s} \tag{2.85}$$

Because the summation of forces (Eq. 2.83) should be equal to the net momentum flux, therefore,

$$-\rho g A \frac{\partial z}{\partial s} ds - \frac{\partial (pA)}{\partial s} ds = \rho A u \frac{\partial u}{\partial s} \tag{2.86}$$

which results in

$$\rho A \frac{\partial}{\partial s}\left[g \partial z + \frac{\partial p}{\rho} + u \partial u\right] = 0 \Rightarrow \frac{d}{ds}\left(gz + \frac{p}{\rho} + \frac{u^2}{2}\right) = 0 \tag{2.87}$$

which finally leads to Bernoulli's equation, as follows,

$$gz + \frac{p}{\rho} + \frac{u^2}{2} = \text{cons.} \tag{2.88}$$

This equation indicates that the work done by the pressure force $\frac{p}{\rho}$ will be balanced by the potential gz and kinetic energy $\frac{u^2}{2}$. The units of the terms in Eq. (2.88) is the energy per unit mass of a fluid.

REFERENCES

[1] F.M. White, Fluid Mechanics (2003), 7th ed., McGraw-Hill, NY, USA, 2011.

[2] R.G. Dean, R.A. Dalrymple, Water Wave Mechanics for Engineers and Scientists, World Scientific Publishing, Singapore, 1991.

[3] N.S. Heaps, Linearized veritically-integrated equations for residual circulation in coastal seas, Dt. Hydrogr. Z. 31 (1978) 147–169.

[4] J. Pedlosky, Geophysical Fluid Dynamics, second ed., Springer-Verlag, Berlin, 1992, 728 pp.

[5] J.H. Spurk, Fluid Mechanics, Springer-Verlag, Berlin, 1997, 513 pp.

Chapter 3

Tidal Energy

Tidal energy is presently one of the more favoured forms of marine renewable energy because, due to its origins in (astronomical) tide generating forces, it is predictable. This is in contrast to other more stochastic renewable energy sources such as wind and wave and, in part, solar. However, regardless of its predictability, tidal energy shares a key feature with the majority of renewable energy sources—it is intermittent, from diurnal (once per day) and semidiurnal (twice daily), to fortnightly (spring-neap) timescales. In this chapter, we explain the origin of the tides, and how tides evolve as they propagate over shelf sea regions. It covers methods of analysing and predicting the tides, and how tides can be used to generate electricity through arrays of tidal stream devices, and tidal range schemes (lagoons). The primary objective of this chapter is to equip those working within or researching marine renewable energy with an understanding of the fundamentals of tidal energy from both oceanographic and engineering perspectives.

3.1 TIDE GENERATING FORCES

Newton's Law of Universal Gravitation explains how every object in the universe attracts every other object with a force that is proportional to the product of their masses, and inversely proportional to the square of their distance apart

$$F = G\frac{m_1 m_2}{r^2} \tag{3.1}$$

where F is the force between two masses m_1 and m_2, G is the gravitational constant (6.674×10^{-11} $\mathrm{m^3/kg\ per\ s^2}$), and r is the distance between the centres of the masses (Fig. 3.1).

In the case of the Earth-Moon system, the Earth and Moon orbit each other about their common centre of gravity, and so the gravitational attraction is balanced by the outward directed centrifugal force[1] (Fig. 3.2). Because the mass of the Earth is two orders of magnitude greater than the mass of the Moon,

1. The outward directed centrifugal force is an apparent force, whilst centripetal force, the component of force acting on a body in curvilinear motion that is directed towards the centre of the axis of rotation, is an actual force.

Fundamentals of Ocean Renewable Energy. https://doi.org/10.1016/B978-0-12-810448-4.00003-3

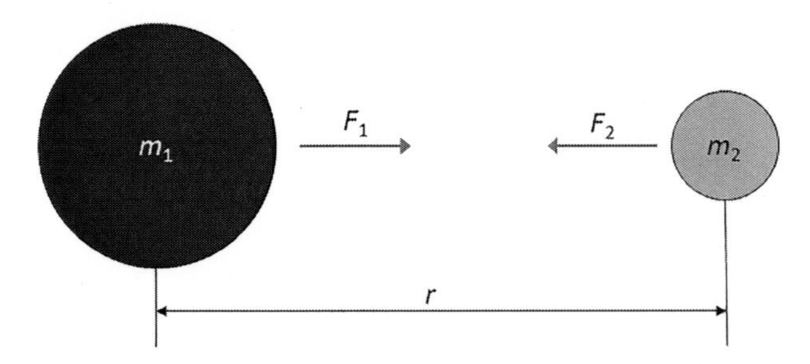

FIG. 3.1 Newton's Law of Universal Gravitation.

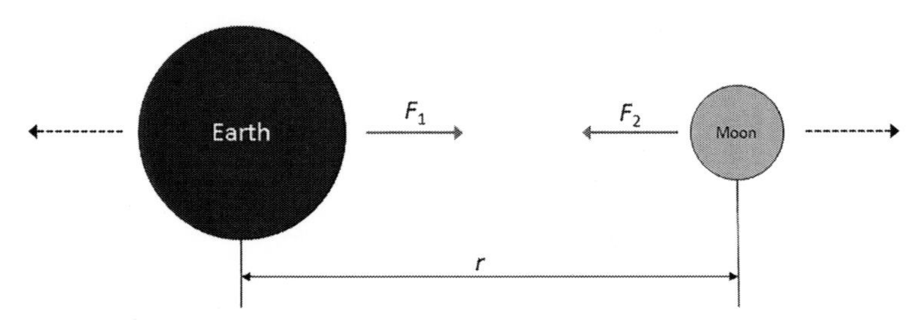

FIG. 3.2 Earth-Moon system. The Earth and Moon attract each other due to Newton's Law of Universal Gravitation (*solid arrows*), but this is balanced by the outward directed centrifugal force (*dashed arrows*).

the centre of gravity of the Earth-Moon system actually lies within the body of the Earth; indeed, it is around 1700 km below the surface of the Earth. The Earth-Moon system turns about this centre of gravity once per month, and so every object on Earth turns with it. The centrifugal force (the dashed arrow directed outward from the Earth on Fig. 3.2) is the same everywhere, because all points on Earth experience the same orbital motion. However, the gravitational attraction of the Moon is progressively weaker as we move away from the Moon, as stated by Newton's Law of Universal Gravitation (Eq. 3.1). The result is that the centrifugal force and the Moon's gravity cannot balance each other everywhere. The slight imbalances that occur are what cause the tides in the ocean. These imbalances are called the *tide generating forces*.

At the centre of the Earth (Fig. 3.3), the centrifugal force (dashed arrow) is equal and opposite to the Moon's gravitational pull (solid arrow). At point A (position on the Earth's surface closest to the Moon), the Moon's gravity exceeds the centrifugal force, and this results in a net force towards the Moon (block arrow at A). At point B (position on the Earth's surface furthest from the Moon), the Moon's gravity is weaker than the centrifugal force. This results in a net force directed away from the moon at B (block arrow at B).

The tide generating forces lead to a deformation in the shape of the ocean, stretching it out in both directions along the Earth-Moon axis (Fig. 3.4). Because

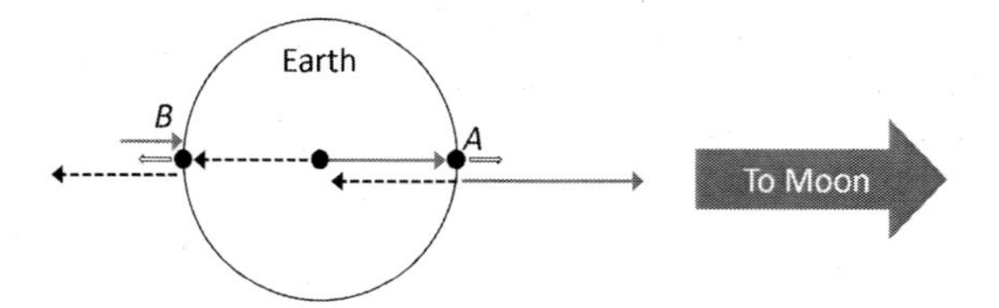

FIG. 3.3 Net forces in an Earth-Moon system. See text for explanation.

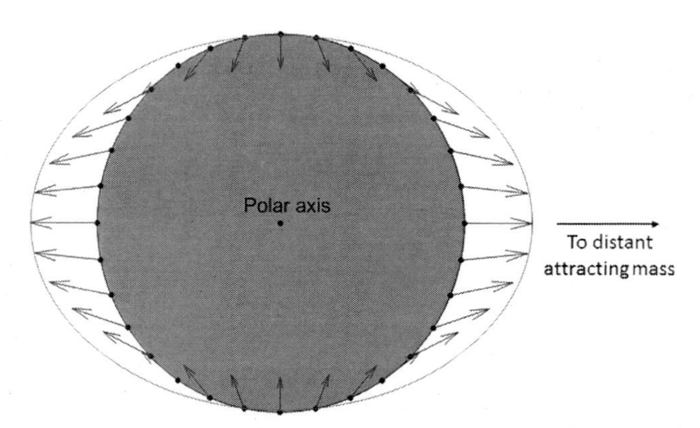

FIG. 3.4 Vectors on the Earth's surface indicate the difference between the gravitational force the Moon exerts at a given point on the Earth's surface, and the force it would exert at the Earth's centre. These resultant force vectors move water towards the Earth-Moon orbital plane, creating two bulges on opposite sides of the Earth.

the Earth rotates within this ellipsoid, two high waters (and two low waters) are experienced per day. This is known as the *equilibrium tidal model*. However, there are three main problems with this equilibrium tidal model:

- the tidal wave cannot move fast enough to keep up with the Earth's spin and maintain 'equilibrium';
- the equilibrium tide does not account for continents; and
- the equilibrium tide does not account for Earth's rotation.

As can be seen in Fig. 3.4, the equilibrium tide is actually a wave that is stretched around the circumference of the Earth. The wave has two crests: one directly under the Moon, and the other on the opposite side of the Earth. Between these two crests are two troughs. As the Earth spins, the equilibrium tidal model requires the wave to maintain itself with one crest directly under the moon. Tidal waves behave as shallow water waves, with speed c given as

$$c = \sqrt{gh} \tag{3.2}$$

where h is water depth. The mean depth of the world's oceans is around 4000 m, and so $c = 200$ m/s. At the equator, the Earth turns at around 460 m/s. Therefore, tidal waves cannot move as fast as the Earth spins.

The equilibrium tide does not account for continents, which hinder the progress of the tidal wave. In addition, water that moves over a rotating Earth is deflected by Coriolis forces, which arise as a result of the Earth's spin (see Section 3.6).

Wave speed is controlled by the local water depth (Eq. 3.2). For a typical shelf sea water depth of $h = 200$ m, the tidal wave propagates at 44 m/s. Given a tidal period T of 12.42 h (see Section 3.8), the wavelength $L = cT$ is around 2000 km.

3.2 PROGRESSIVE WAVES

The time of high water (HW) at a particular location depends on how fast the tidal wave can travel (Eq. 3.2), and so HW will not necessarily occur when the Moon is directly overhead. Currents in a progressive wave flow *with* the wave under the crest, and in the *opposite* direction under the trough (Fig. 3.5). Maximum currents in a progressive wave correspond with HW and low water (LW). The maximum current speed V beneath the wave is given by shallow water wave theory as

$$V = \frac{R}{2}\sqrt{\frac{g}{h}} \tag{3.3}$$

where $R = 2a$ is the tidal range.

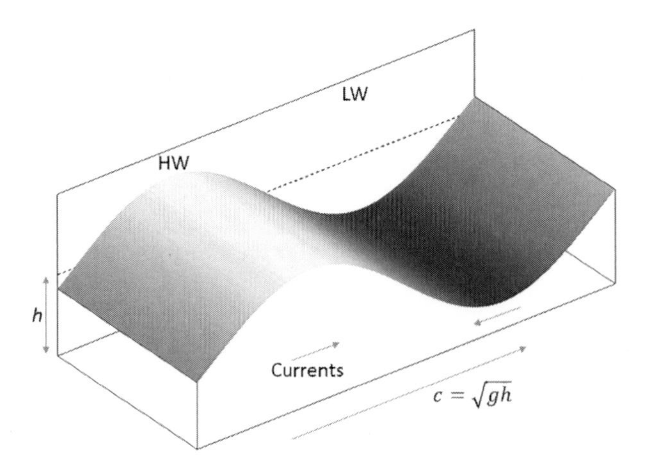

FIG. 3.5 An idealized progressive tidal wave. Peak (positive, or flood) currents occur under the crest (HW), and peak (negative, or ebb) currents under the trough (LW).

3.3 COTIDAL CHARTS

Cotidal charts of a region convey two important pieces of information: tidal amplitude and tidal phase. *Cotidal lines* join locations that are in phase with one another (e.g. HW would occur at the same time at these locations). *Coamplitude lines* join locations that have equal tidal amplitudes. Cotidal and coamplitude lines are shown on Fig. 3.6 for the M2 (principal semidiurnal lunar constituent) over the northwest European shelf seas. The M2 tidal constituent has a period of around 12.42 h (Section 3.8), and the cotidal contour interval on the chart is 30 degrees. Therefore, the contour interval can be converted into time as

$$\frac{30\,\text{degrees}}{360\,\text{degrees}} \times 12.42 \approx 1\,\text{h}\,2\,\text{min} \tag{3.4}$$

If we examine the Irish Sea in detail (Fig. 3.6) and assume that HW occurs at midday (12:00) at the southern entrance to the Irish Sea, HW at Holyhead (90 degrees phase difference) would be at 15:00. The distance between these two locations is 165 km. Therefore, the speed of the propagation of the tidal wave can be calculated as

$$\frac{165\,\text{km}}{3\,\text{h}} = 55\,\text{km/h} = 15\,\text{m/s} \tag{3.5}$$

For interest, and referring back to Eq. (3.2), the mean water depth h between these two locations could be estimated as approximately 23 m.

Two additional points to note from the cotidal chart are that the tidal amplitude in the northeastern part of the Irish Sea and in the Bristol Channel are high, due to wave reflection (Section 3.4), and that there are various locations, known as *amphidromic points*, where the tidal amplitude is theoretically zero, due to the Earth's rotation (Sections 3.6 and 3.7).

3.4 STANDING WAVES

In shelf seas, tides cannot propagate indefinitely as progressive waves, due to reflection at coastal boundaries and at sudden changes in water depth. Incident and reflected waves combine together to give the observed total tidal wave. In the case of a wave propagating along a channel that is relatively long in comparison to the tidal wavelength (Fig. 3.7), at a distance of $1/4$ of a wavelength from the coast, the crest of the incoming wave passes at the same time as the trough of the reflected wave (and vice versa). The two waves (incident and reflected) cancel at this point, known as a *node*, and so there is no tide at this location. Nodes also occur at $3L/4$, $5L/4$, etc. Note that since incoming and reflected waves combine, the amplitude a at the coast becomes $2a$.

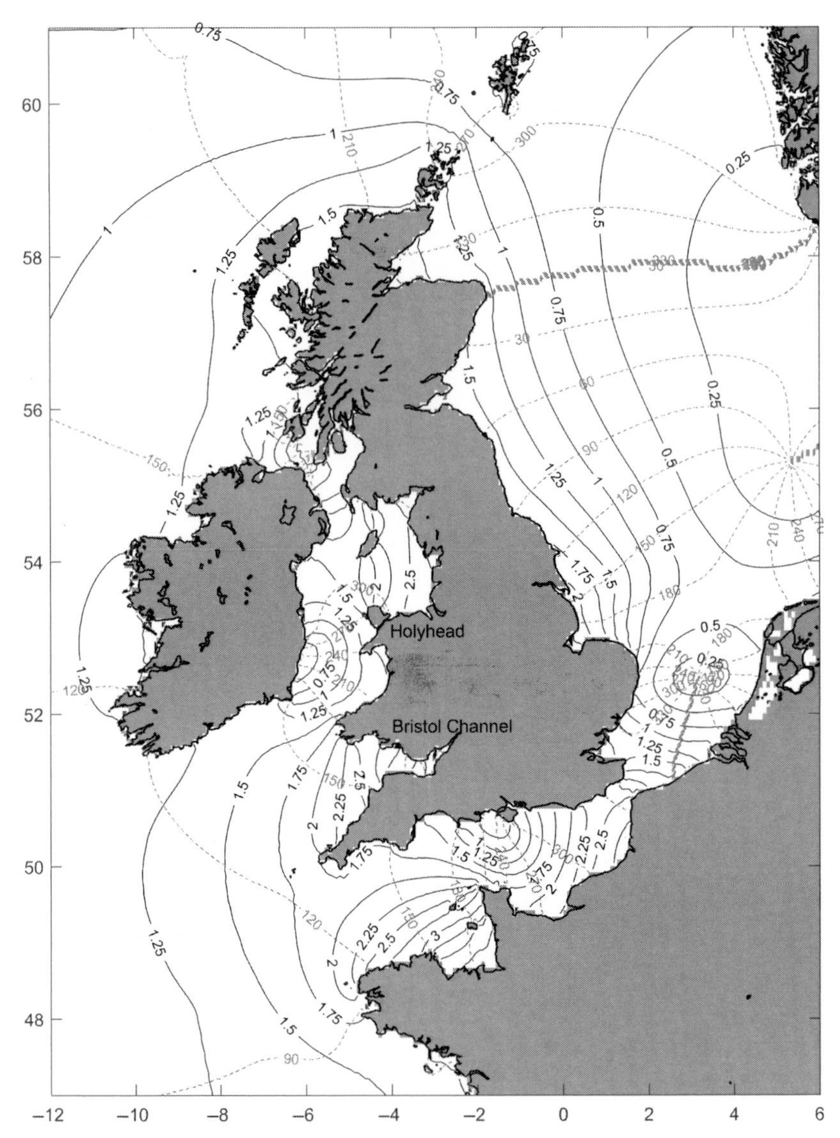

FIG. 3.6 M2 cotidal chart of the northwest European shelf seas. *Dashed contours* (plotted at intervals of 30 degrees) are cotidal lines, and *solid contours* (plotted at intervals of 0.25 m) are coamplitude lines. *(Adapted from S.P. Neill, J.D. Scourse, K. Uehara, Evolution of bed shear stress distribution over the northwest European shelf seas during the last 12,000 years, Ocean Dyn. 60 (2010) 1139–1156.)*

When it is HW at the coastline, the incident and reflected waves combine as shown in Fig. 3.8. Because the currents under the incident and reflected waves are equal and opposite along the channel, currents along the tide wave will be

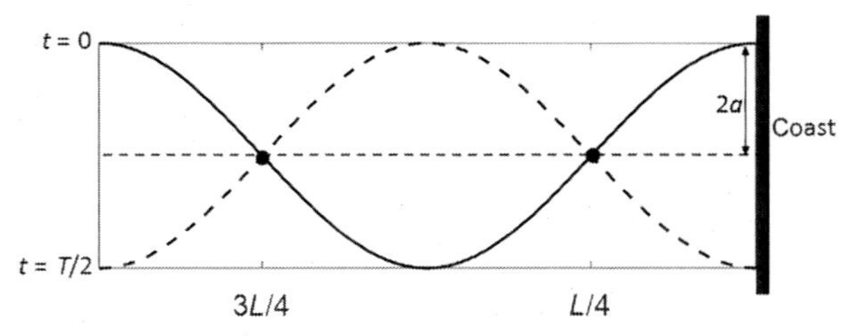

FIG. 3.7 Standing wave shown at two time intervals ($t = 0$ and $t = T/2$).

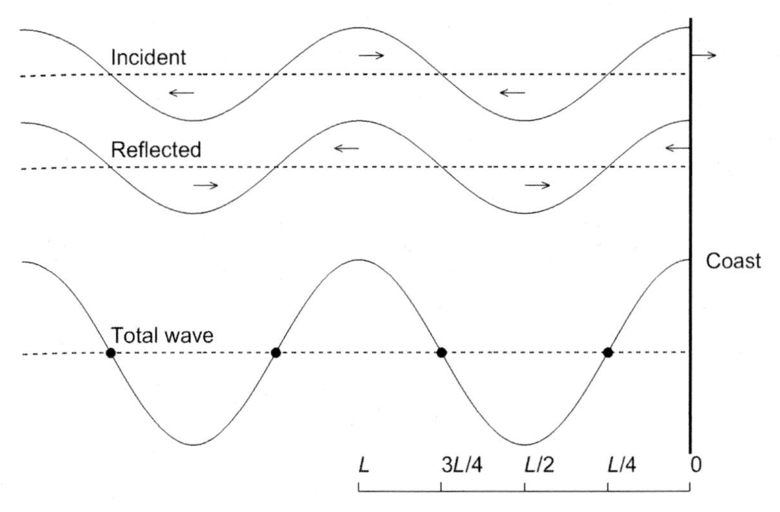

FIG. 3.8 Standing wave plotted along a channel at the time of HW at the coast. In this case, maximum elevations occur, but there are zero currents along the channel. Nodes are plotted as *filled circles* at $3L/4$, $L/4$, etc.

zero (slack water). By contrast, 1/4 of a wave period later (Fig. 3.9), combination of the incoming and reflected waves leads to zero elevations along the channel, but maximum amplified currents, which occur at the nodes.

Because the currents at the nodes of a standing wave are double those of a progressive wave (Eq. 3.6), the maximum current velocity at the nodes of a standing wave is

$$V = R\sqrt{\frac{g}{h}} \tag{3.6}$$

3.5 RESONANCE

Resonance occurs when a water body is forced at its natural period. If a sea has a natural period of 12 h, it will resonate with this forcing, resulting in amplified tides. Consider a channel which is exactly $L/4$ long; for example, this could be

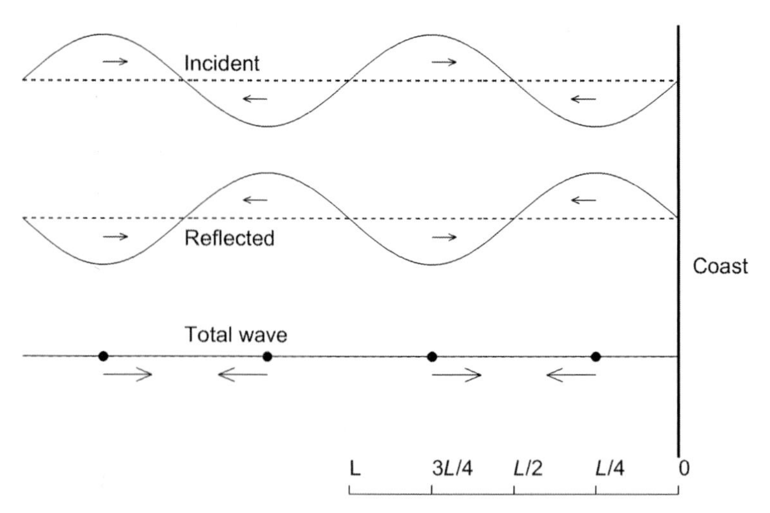

FIG. 3.9 Standing wave plotted along a channel at the time of HW+$T/4$ at the coast. In this case, zero elevations occur along the channel, and there are maximum (enhanced) currents at the nodes.

considered as the right-hand portion of Fig. 3.7, between the node at $L/4$ and the coast. The currents at the entrance to the channel ($L/4$) could lead to large changes in water surface elevation at the head of the channel. The natural period for such a forced oscillation is

$$\frac{4 \times \text{channel length}}{\sqrt{gh}} = \frac{L}{\sqrt{gh}} \tag{3.7}$$

For the M2 tidal constituent, which has a period of 12.42 h, if we assume a water depth of 50 m

$$L = T\sqrt{gh} \tag{3.8}$$

and $L \approx 1000$ km, and so the length of the channel, which would be at quarter wavelength resonance, is around 250 km. This is approximately the distance between the southern entrance to the Irish Sea and the coastline in the northeastern part of the Irish Sea (Fig. 3.6), and so the Irish Sea can be said to be close to resonance with the M2 tide.

In nature, forced resonant oscillations cannot grow indefinitely because energy losses due to friction increase more rapidly than the amplitudes of the oscillations. Regions of the world which experience resonance, and so are celebrated for their very large tidal ranges, are the Bay of Fundy (Canada) (spring tidal range of almost 15 m in the Minas Basin) and the Bristol Channel (UK) (spring tidal range of 13 m at Avonmouth).

3.6 CORIOLIS

The Earth rotates around its own axis from west to east. Therefore, a reference frame attached to a fixed position on the Earth rotates around the Earth axis. This leads to a complication when we try to apply Newton's law of motion on a coordinate system that is attached to the Earth, specially when studying large-scale ocean circulation and tidal dynamics. An object (or here, our reference frame), which rotates around an axis, has acceleration (i.e. centripetal acceleration), because its velocity is changing with time and Newton's law of motion is not valid in a frame, which has an acceleration. Whilst the speed (scaler) of an object rotating around an axis may be constant, its velocity (vector) changes due to the change in direction. To further clarify this concept, consider an object on the surface of the Earth which rotates around the Earth with an angular velocity of Ω (Fig. 3.10). The speed of the object will be $u = R_E\Omega$, where R_E is the radius of the Earth. The velocity of the object with respect to a nonrotating frame at the centre of the Earth is given by

$$\vec{u} = -u\sin\theta\hat{i} + u\cos\theta\hat{j} = R_E\Omega[-\sin\theta\hat{i} + \cos\theta\hat{j}] \tag{3.9}$$

Therefore, we can calculate the acceleration by taking the derivative of velocity as follows

$$\vec{a} = \frac{d\vec{u}}{dt} = R_E\Omega\left[-\cos\theta\frac{d\theta}{dt}\hat{i} - \sin\theta\frac{d\theta}{dt}\hat{j}\right] = -R_E\Omega\frac{d\theta}{dt}[\cos\theta\hat{i} + \sin\theta\hat{j}] \tag{3.10}$$

Because $\Omega = \frac{d\theta}{dt}$, then

$$\vec{a} = -R_E\Omega^2[\cos\theta\hat{i} + \sin\theta\hat{j}] \tag{3.11}$$

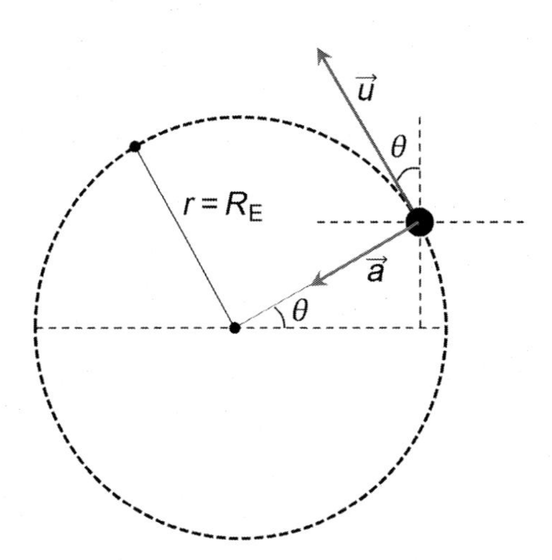

FIG. 3.10 Centrifugal acceleration for a rotating body.

Eq. (3.11) describes a vector, which is orientated towards the centre of the earth, with a magnitude of $R_E\Omega^2 = u^2/R_E$, which is called the centripetal acceleration.

In order to apply the equations of motion in a rotating frame attached to the Earth, we need to find the relationship between rotating and nonrotating frames. In general, an object can rotate around three axes in space; therefore, to deal with a rotational coordinate system in a more general way, we need to formulate a relationship between inertial and rotational frames of reference in 3D. Consider a frame of reference at the centre of the Earth (inertial), and a rotational frame of reference at a point in the ocean (rotates around the Earth's axis). The rotating frame may be described by three unit vectors, $\hat{i}_r$, $\hat{j}_r$, and $\hat{k}_r$. These unit vectors are rotating with angular velocity $\vec{\Omega}$ around the inertial frame of reference. If we represent the position of an object in 3D using the position vector $\mathbf{R}$, the rotational speed will be $R\Omega$, or angular velocity will be $\vec{\Omega} \times \mathbf{R}$. Based on the theory of relative motion in dynamics, for any vector variable $\mathbf{A}$, we can write

$$\left(\frac{d\mathbf{A}}{dt}\right)_{\text{iner}} = \left(\frac{d\mathbf{A}}{dt}\right)_{\text{rot}} + \vec{\Omega} \times \mathbf{A} \tag{3.12}$$

Consequently, the relationship between velocities in the inertial and rotating frame can be written as

$$\left(\frac{d\mathbf{R}}{dt}\right)_{\text{iner}} = \left(\frac{d\mathbf{R}}{dt}\right)_{\text{rot}} + \vec{\Omega} \times \mathbf{R} \Rightarrow \mathbf{u}_{\text{iner}} = \mathbf{u}_{\text{rot}} + \vec{\Omega} \times \mathbf{R} \tag{3.13}$$

Similarly, if we apply Eq. (3.12) for acceleration (as a vector), it leads to

$$\left(\frac{d\mathbf{u}_{\text{iner}}}{dt}\right)_{\text{iner}} = \left(\frac{d\mathbf{u}_{\text{iner}}}{dt}\right)_{\text{rot}} + \vec{\Omega} \times \mathbf{u}_{\text{iner}} \tag{3.14}$$

Because, $\mathbf{u}_{\text{iner}} = \mathbf{u}_{\text{rot}} + \vec{\Omega} \times \mathbf{R}$, therefore,

$$\left(\frac{d\mathbf{u}_{\text{iner}}}{dt}\right)_{\text{iner}} = \left(\frac{d(\mathbf{u}_{\text{rot}} + \vec{\Omega} \times \mathbf{R})}{dt}\right)_{\text{rot}} + \vec{\Omega} \times (\mathbf{u}_{\text{rot}} + \vec{\Omega} \times \mathbf{R}) \tag{3.15}$$

$$= \left(\frac{d\mathbf{u}_{\text{rot}}}{dt}\right)_{\text{rot}} + 2\vec{\Omega} \times \mathbf{u}_{\text{rot}} + \left\{\vec{\Omega} \times (\vec{\Omega} \times \mathbf{R}) + \frac{D\vec{\Omega}}{dt} \times \mathbf{R}\right\} \tag{3.16}$$

As can be seen, the acceleration, which is observed in a rotational frame (i.e. $\frac{d\mathbf{u}_{\text{rot}}}{dt}$), is different from the acceleration, which is observed in an inertial frame (i.e. $\frac{d\mathbf{u}_{\text{iner}}}{dt}$). The difference between these accelerations includes three terms: Coriolis acceleration ($-2\vec{\Omega} \times \mathbf{u}_{\text{rot}}$), the centripetal acceleration $-\vec{\Omega} \times (\vec{\Omega} \times \mathbf{r})$, and, Euler acceleration, the acceleration due to change in the angular velocity

$(-\frac{D\vec{\Omega}}{dt} \times \mathbf{R})$. These terms appear when we write the equation of motion in a rotational coordinate system. Therefore, we refer to them as apparent or fictitious accelerations or forces. Assuming that the angular velocity of the Earth is constant results in $\frac{d\vec{\Omega}}{dt} = 0$. The centripetal acceleration at the equator is about 0.3% of the Earth attraction, and in measurements it is difficult to differentiate between the Earth attraction and centripetal acceleration. The resultant force of Earth attraction and centrifugal force is called gravity, which acts normal to the surface of the Earth.

Now, consider a rotating frame attached to the Earth, in which x and y axes correspond to the west-east and the south-north directions, respectively. The z axis is oriented towards the centre of the Earth. An object moving northwards in the northern hemisphere is considered; using the cross product, the magnitude of Coriolis acceleration will be, $2\Omega \sin\phi v$, where ϕ is the latitude (or the angle between velocity and rotation vectors), and the direction of the Coriolis acceleration will be towards the east, or to the right of moving object (i.e. perpendicular to the plane containing rotation and velocity vectors). It can be seen that the magnitude of Coriolis is zero for an object moving northwards at the equator ($\phi = 0$).

Let us assume that an object is moving from the west towards the east in the northern hemisphere; the magnitude of Coriolis acceleration (using cross product) will be $2\Omega u$, as the rotation vector and velocity vector are perpendicular. For this scenario, the Coriolis acceleration has two components: one component towards the south, or negative y direction. With a magnitude of $2\Omega u \sin\phi$, and the other component away from the centre of the Earth (z axis), with a magnitude of $2\Omega u \cos\phi$. Therefore, the Coriolis effect turns objects rightward in the northern hemisphere. We can similarly show that Coriolis turns objects leftward in the southern hemisphere.

As we know, the Earth rotates 2π in 1 day. At the same time, it also rotates a little ($\approx 1/365 \times 360$ degrees) around the Sun. A sidereal day (86,164 s) is the exact time that it takes for the Earth to make one rotation, and is slightly shorter than a solar day (86,400 s). Therefore, the angular velocity of the Earth is $\Omega = \frac{2\pi}{86{,}164} = 7.29 \times 10^{-5}$ per s. Note that the Coriolis parameter or frequency, f, is defined as $2\Omega \sin(\phi)$.

3.7 KELVIN WAVES

The propagation of tidal waves in the oceans and continental shelves is primarily affected by the Earth's rotation. The dynamics of long waves on a rotating system was originally described by Lord Kelvin. In a rotating reference frame like the Earth, Newton's second law of motion—which is valid for an inertial frame of reference with zero acceleration—cannot be directly applied. By introducing Coriolis as a 'fictitious force', we can use Newton's second law in the Earth rotating system as discussed in Section 3.6. In the northern hemisphere, the Coriolis force causes a deflection of the currents towards the right of the direction of motion (e.g. poleward propagating currents are deflected

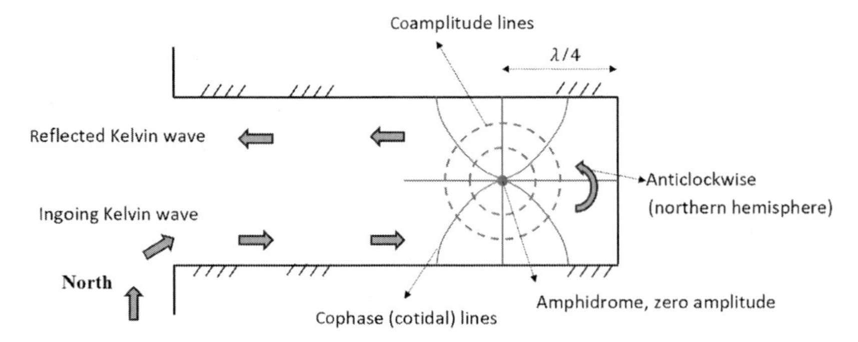

FIG. 3.11 Propagation of Kevin waves in an idealized rectangular basin in the northern hemisphere [1].

to the east). This deflection of flow continues until the flow reaches the right-hand boundary (i.e. coastline), where the build-up of water leads to a pressure gradient. A Kelvin wave propagates as a result of the balance between these two forces: Coriolis and pressure gradient. Standing oscillating waves are a special case of interest. When a Kelvin wave enters a basin, it will be reflected at the head of the basin; therefore, the tidal wave in a basin can be represented by two Kelvin waves travelling in opposite directions (Fig. 3.11). These waves rotate around a node, which is called an amphidrome. The cotidal contour lines radiate outwards from the amphidrome, and tidal amplitude is zero at this node. Considering an idealized rectangular basin and using the earlier concept, it is easy to demonstrate why, in the northern hemisphere, the direction of the rotation around an amphidrome is anticlockwise.

As an example, consider the propagation of tidal waves in the northwest European shelf seas (e.g. [2]). The Atlantic semidiurnal Kelvin wave travels towards the north and transfers the tidal energy into the Celtic Sea between Brittany and Southern Ireland (Fig. 3.6). Part of this energy propagates into the English Channel and the Irish Sea. The Atlantic Kelvin wave further progresses northwards and deflects towards the east (due to Coriolis), travels to the North of Shetland, and enters the North Sea. Tides in the North Sea are primarily semidiurnal, with two amphidromes in the southern North Sea and a third around the southern tip of Norway. The south travelling wave, which moves along the east coast of the United Kingdom, generates the largest amplitudes in the North Sea.

3.8 TIDAL ANALYSIS AND PREDICTION

Tidal analysis and prediction are based on the assumption that the tidal signal can be represented by a finite number N of harmonic terms of the form

$$H_n \cos(\omega_n t - g_n) \tag{3.17}$$

where H_n is the amplitude, g_n is the phase lag (e.g. relative to Greenwich), and ω_n is the angular speed of the constituent.

If we are interested in predicting how the surface elevation η varies over time, then

$$\eta_n = H_n \cos(\omega_n t - g_n) \tag{3.18}$$

Through observations and/or numerical simulations, we can obtain values for H and g, and we know the angular speeds of the various constituents (M2, S2, K1, O1, etc.), and so we have the basis for making a tidal prediction.

At the port of Holyhead (Fig. 3.6), information on the four principal tidal constituents can be obtained from Admiralty tide tables (Table 3.1). Taking a timestep of 15 min, we can easily predict the elevation time series at Holyhead for an arbitrary start time (Fig. 3.12). This time series shows a spring-neap cycle (due to combination of the M2 and S2 semidiurnal constituents), and

TABLE 3.1 Tidal Data for Holyhead

Constituent	Name	*H* (m)	*g* (degrees)	Speed (degrees per solar hour)
M2	Principal lunar semidiurnal	1.81	292	28.984
S2	Principal solar semidiurnal	0.59	329	30.000
K1	Lunisolar diurnal	0.11	177	15.041
O1	Lunar diurnal	0.10	029	13.943

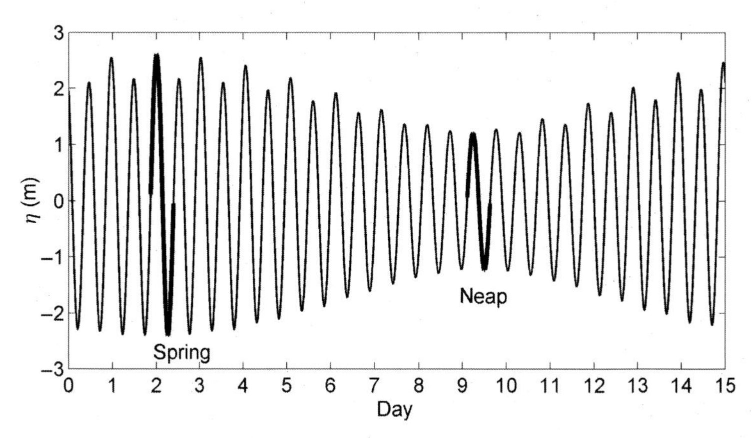

FIG. 3.12 Predicted elevation time series at the port of Holyhead, based on a limited number of tidal constituents (Table 3.1).

diurnal inequalities (due to O1 and K1, in combination with the semidiurnal constituents).

Although the previous example is for tidal elevations, a similar procedure can be applied to predict tidal currents, and by using tidal ellipses, which will be explained later in this chapter. The main difference between predictions of elevations and currents is that water current measurements that are suitable for analysis are more difficult to obtain. For tidal energy projects, these tend to be obtained (due to cost) at scoping or initial planning stage via numerical simulations (Chapter 8), and via ADCP deployments at later stages of project development (Chapter 7).

Tidal analysis can be conveniently performed by computer software, and one of the most popular choices for oceanographers and tidal energy practitioners is T_TIDE, which runs in the Matlab environment [3]. Although it is tempting to treat T_TIDE as a black box, it is useful to understand the choice of constituents used in the analysis. Critical to this is the Rayleigh criterion, which states that only constituents separated by at least a complete period from their neighbouring constituents should be included in the analysis [4]. For example, to analyse a time series for M2 and S2 requires a minimum record length of

$$360/(30.0 - 28.98)\,\text{h} = 14.77\,\text{days} \tag{3.19}$$

(speeds of the constituents taken from Table 3.1). An additional useful functionality of T_TIDE is that it estimates the error for each of the analysed constituents.

3.9 COMPOUND TIDES

In Section 3.1, we discussed tide generating forces within the context of a simplified Earth-Moon system. However, the effects of the Sun are also important. Although the Sun is considerably further away from the Earth than the Moon, its mass is 27 million times the mass of the Moon. As a result, the influence of the Sun on the tides is significant, but is around half of the Moon's influence, and it is manifested through the principal semidiurnal solar constituent, S2 which, because time on Earth is measured relative to the Sun, has a period of exactly 12 h. Relative positions of the Sun, Moon, and Earth are shown in Fig. 3.13 for two scenarios. When the Sun and the Moon are in line with the Earth, during either a full Moon or new Moon (Fig. 3.13A), the combined tide generating forces are at their greatest effect, and this leads to spring tides (mean high water spring (MHWS) and mean low water spring (MLWS), as shown on Fig. 3.14). By contrast, when the Sun and Moon are perpendicular to each other in relation to the Earth (Fig. 3.13B), they have their least effect. This occurs at the first or third quarter of the Moon's phase, and leads to neap tides, associated with mean high water neap (MHWN) and mean low water neap (MLWN), again shown in Fig. 3.14. The combination of M2 and S2 tidal constituents is known as a compound tide, and the combined signal (in addition to the diurnal constituents K1 and O1) can be seen in Fig. 3.12, with

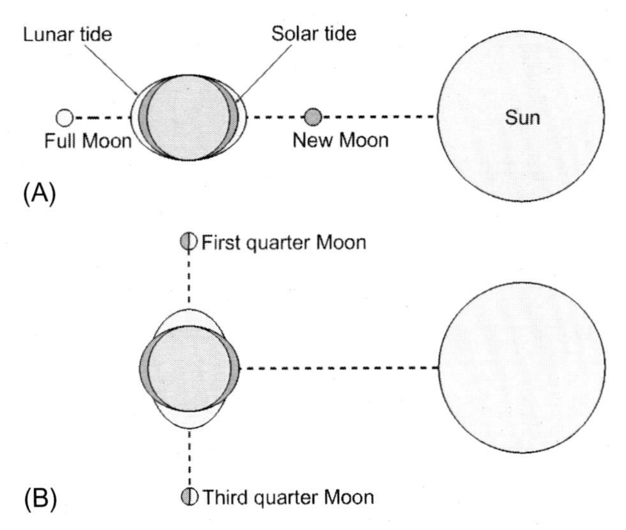

FIG. 3.13 Relative positions of the Sun, Moon, and Earth corresponding to (A) spring and (B) neap tides.

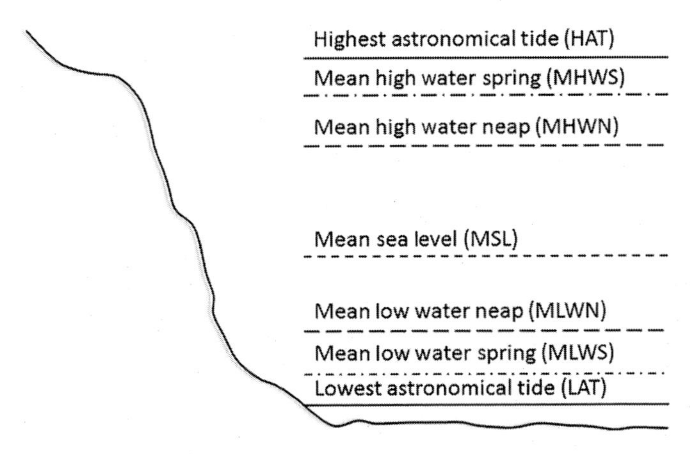

FIG. 3.14 Tidal datums.

a spring tide (highlighted) at day = 2, and a neap tide 7 days later (day = 9). Associated with spring and neap tides are the strongest and weakest peak tidal currents, respectively.

The type of tide varies considerably around the world. When the two high waters and two low waters of each tidal day are approximately equal in height, the tide is semidiurnal [5]. When there is a relatively large diurnal inequality in the high or low waters (or both), the tide is mixed. If only one high water and one low water occurs each tidal day, the tide is diurnal. Examples of tidal elevation time series for two contrasting locations (Boston: semidiurnal, and Mississippi Sound: diurnal) are given in Fig. 3.15. The form factor F can be used to classify tides [4]

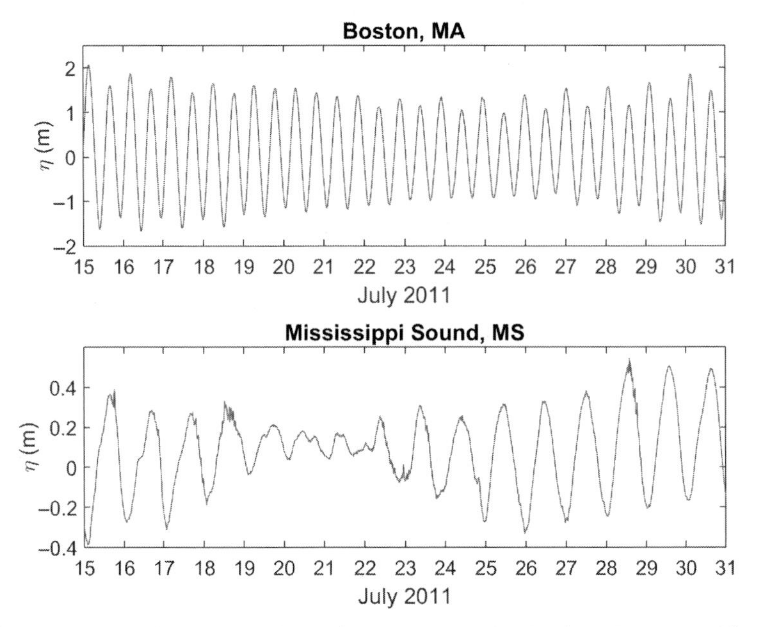

FIG. 3.15 Water surface elevation time series at two contrasting locations: Boston (semidiurnal) and Mississippi Sound (diurnal). *(Data from NOAA.)*

$$F = \frac{H_{\mathrm{K1}} + H_{\mathrm{O1}}}{H_{\mathrm{M2}} + H_{\mathrm{S2}}} \tag{3.20}$$

where H_{M2}, etc. is the amplitude of the M2 tidal constituent, etc. F can be used to quantify the type of tide, using the interpretation shown in Table 3.2. Form factor is plotted globally from the FES2012 global tidal atlas in Fig. 3.16. Although the majority of high tidal stream regions throughout the world are semidiurnal (e.g. the northwest European shelf seas or the northeast of the United States), many important regions are diurnal, such as the South China Sea, East China Sea, New Guinea, and Northern Australia. It is therefore important that semidiurnal and diurnal constituents be included in any resource assessment.

3.10 OVERTIDES AND TIDAL ASYMMETRY

As described in previous sections, astronomical tides are generated by the gravitational forces of the Sun and the Moon on the oceans; therefore, the frequencies of tidal signals (water elevation and velocity) can be directly related to lunar or solar days in the deep oceans. The propagation of barotropic tides in the deep ocean can be assumed to be mainly governed by linear processes. The interaction of tidal components in the deep ocean leads to subharmonic

TABLE 3.2 Type of Tide, Based on Form Factor F (Eq. 3.20)

F	Type of Tide
0–0.25	Semidiurnal form
0.25–1.50	Mixed, mainly semidiurnal
1.50–3.00	Mixed, mainly diurnal
>3.0	Diurnal form

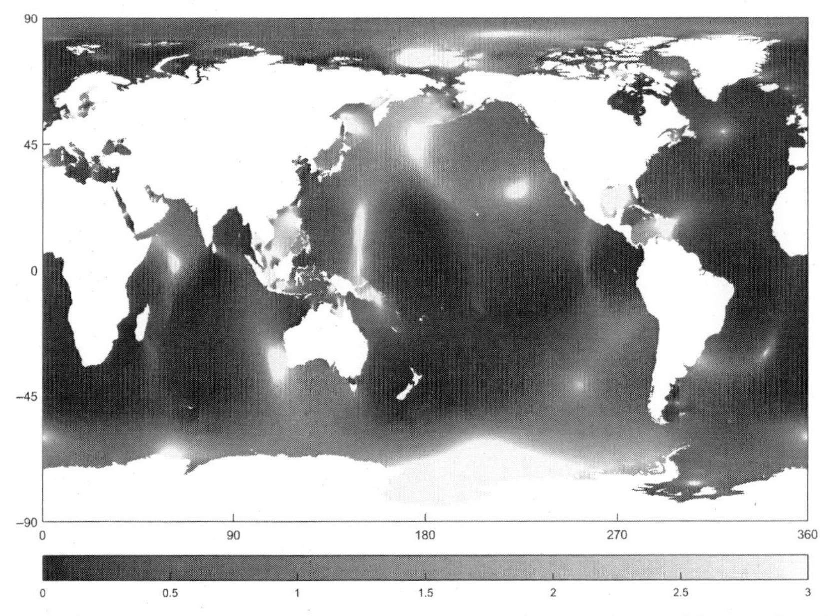

FIG. 3.16 Form factor F, showing global distribution of semidiurnal ($F < 0.25$) and diurnal ($F > 3$) tides.

tides [6]. For instance, as described in Section 3.9, the combination of semidiurnal M2 and S2 tidal constituents results in the spring-neap cycle, with a frequency of

$$\omega_{S2} - \omega_{M2} = \frac{1}{12\text{ h}} - \frac{1}{12.42\text{ h}} = \frac{1}{14.79\text{ days}} \tag{3.21}$$

As tides propagate onto the continental shelves and into shallow coastal waters, nonlinear forces and processes become significant. These forces include friction, advective inertia force, and diffusion due to turbulence. As a result of these nonlinear forces, the tidal signal in shallow water regions is more complex than its deep water counterpart, and sometimes asymmetrical. Therefore, the tidal signal can no longer be reconstructed by combining/superimposing astronomical (e.g. diurnal and semidiurnal) components.

To understand the generation of overtides, consider a simple case: an estuary in which the open (seaward) boundary is forced by an M2 signal (Fig. 3.17). Considering the 1D equations of motion for the estuary, the water elevation and averaged velocity (over estuarine cross-sections) can be evaluated by solving the continuity equation

$$\frac{\partial \eta}{\partial t} + \frac{\partial[(h+\eta)\overline{V}]}{\partial x} = 0 \tag{3.22}$$

and the momentum equation

$$\frac{\partial \overline{V}}{\partial t} + \overline{V}\frac{\partial \overline{V}}{\partial x} + g\frac{\partial \eta}{\partial x} + g\frac{n^2\overline{V}|\overline{V}|}{(h+\eta)^{4/3}} = 0 \tag{3.23}$$

where h is the depth relative to MSL, η is water elevation relative to this datum, and n is the Manning coefficient (which depends on the bed roughness). The momentum equation shows the balance of linear and nonlinear forces: the pressure force (linear; $g\frac{\partial \eta}{\partial x}$), friction force (nonlinear; $g\frac{n^2\overline{V}|\overline{V}|}{(h+\eta)^{4/3}}$), and the inertia force, which has linear ($\frac{\partial \overline{V}}{\partial t}$) and nonlinear ($\overline{V}\frac{\partial \overline{V}}{\partial x}$) terms. In deep water, the frictional force approaches zero, because water depth is in the denominator of the friction term. Also, the spatial variation of velocity $\frac{\partial \overline{V}}{\partial x}$ is negligible for tides in deep waters, so we can neglect convective acceleration. Therefore, tidal propagation in deep water is governed by the linear momentum equation

$$\frac{\partial \overline{V}}{\partial t} + g\frac{\partial \eta}{\partial x} = 0 \tag{3.24}$$

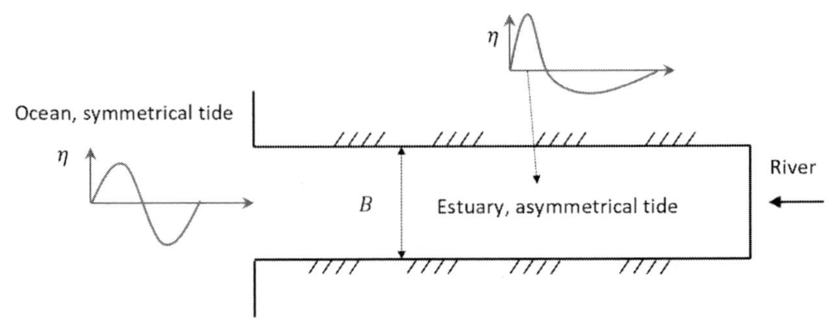

FIG. 3.17 Tidal asymmetry in an idealized estuary.

The solution of Eq. (3.24) can be based on a simple harmonic wave (i.e. M2 tidal signal)

$$\eta_t = a_{\mathrm{M2}} \cos(\omega_{\mathrm{M2}} t - \phi) \tag{3.25}$$

or combination of two or more astronomical components (e.g. M2 and S2), depending on the forcing at the boundary of the domain. However, the solution of Eq. (3.23) cannot be only based on the M2 signal, due to the presence of nonlinear terms. As we know from Fourier analysis, any function, regardless of complexity, can be reconstructed by harmonic components if we add other higher-frequency terms. In other words, the solution of Eq. (3.23) can be based on a series like this

$$\begin{aligned}\eta_t = {} & [a_{\mathrm{M2}} \cos(\omega_{\mathrm{M2}} t - \phi)]_{\text{astronomical tide}} \\ & + \cdots + [a_2 \cos(2\omega_{\mathrm{M2}} t - \phi_2) + a_3 \cos(3\omega_{\mathrm{M2}} t - \phi_3) \\ & + a_4 \cos(4\omega_{\mathrm{M2}} t - \phi_2) + \cdots]_{\text{overtide}}\end{aligned} \tag{3.26}$$

The additional terms, which appear in the previous equation, are called overtides. In particular, M4 is the second term, with a frequency of $2\omega_{\mathrm{M2}}$, M6 has a frequency of $3\omega_{\mathrm{M2}}$, M8 has the frequency of $4\omega_{\mathrm{M2}}$, and so on.

Fig. 3.18 shows an M2 tidal signal and an M4 tidal signal (with a phase lag of 0 and $\pi/2$), and the combined signal. As this plot shows, the resulting signal, depending on the phase relationship between M2 and M4, can be symmetrical or asymmetrical. For the asymmetrical case ($\pi/2$), the peak value between flood and ebb are very different; also, the duration of the flood is less than ebb.

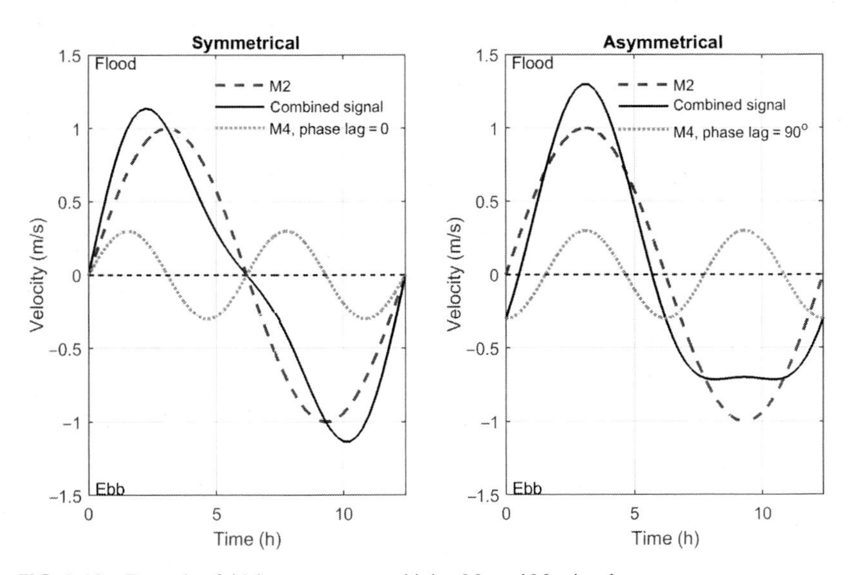

FIG. 3.18 Example of tidal asymmetry, combining M_2 and M_4 signals.

Overtides and compound tides are the main causes of tidal asymmetry, and their role in processes such as sediment transport and tidal energy conversion are important in some regions [7–9]. It can easily be shown that tidal asymmetry is controlled by the phase difference of semidiurnal and quarter-diurnal tidal constituents. Therefore, tidal asymmetry can be described by comparing the phases of M_2 and M_4 tidal constituents, as well as the ratio of M_4/M_2 amplitudes.

3.11 CHARACTERIZING TIDES AT A SITE

3.11.1 Velocity Profile

The bottom boundary layer is defined as the region of flow in which the dynamics are influenced by frictional effects due to the sea bed. In relatively shallow waters, the boundary layer may occupy the entire water column, but in deeper water it will occupy only the lower part of the water column. The way in which the current increases with height above the sea bed is known as the velocity profile. Many engineers adopt what is known as the one-seventh power law to characterize the velocity profile

$$U_z = \overline{U}\left(\frac{z}{0.32h}\right)^{1/7} \tag{3.27}$$

where $\overline{U}$ is the depth-averaged current speed, z is the height above the sea bed, h is the water depth, and 0.32 is the bed roughness chosen for this example [10]. The resulting velocity profile (Fig. 3.19) demonstrates that, theoretically, the current speed is zero at the sea bed, and increases with height above the sea bed. In addition, shear (du/dz) is greater in the lower part of the water column. Therefore, it is advantageous from both resource and design perspectives to place the rotor as high in the water column as possible, subject to navigational and economic constraints.

3.11.2 Power Density

Instantaneous 'theoretical' power density (per unit area) can be calculated as

$$\frac{P}{A} = \frac{1}{2}\rho u^3 \tag{3.28}$$

where P is the power output in W, A is the swept area of the rotor, ρ is the water density, and u is the depth-averaged current speed. Clearly, since power output depends on the cube of current speed, a small increase in current speed leads to a large increase in power output (Fig. 3.20). This is why developers generally seek sites with high current speeds to maximize net power output. For example, at a peak current speed of 1 m/s, net power output over a tidal cycle in this example is <1 MWh. By increasing peak current speed to 3 m/s, net power output increases to >23 MWh, that is over 20 times the power output! Eq. (3.28) neglects device efficiency, and this is considered in detail in Section 3.13.

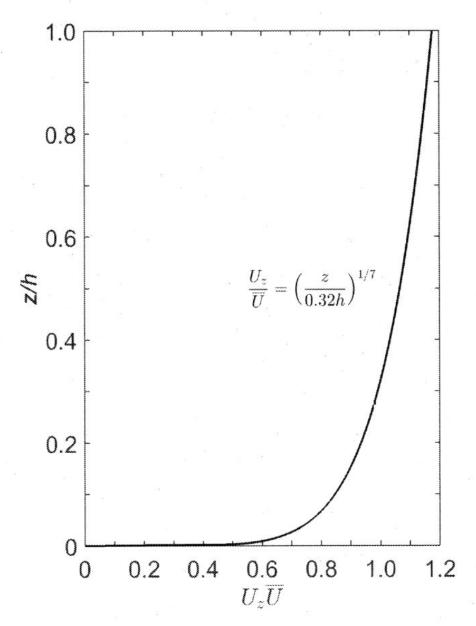

FIG. 3.19 Variation of current speed with height above the sea bed.

3.11.3 Tidal Ellipses

Consider the 2D current velocity, which is generated by a tidal component (e.g. M2). We can represent the x and y components of the tidal current—due to the M2 tide—at a point by u_x and u_y, respectively; the current velocity can be written as

$$\vec{V} = u_x + iu_y \rightarrow \vec{V} = \left[a_{u_x}\cos(\omega t - g_x) + ia_{u_y}\cos(\omega t - g_y)\right] \tag{3.29}$$

where $i = \sqrt{-1}$, $\vec{V}$ is the current velocity due to a tidal component; ω is the angular frequency of the tidal component; a_{u_x} and a_{u_y} are amplitudes of a tidal component in the x and y directions, respectively; g_x and g_y are phases of a tidal component in x and y directions, respectively; and t is time. Referring to the calculus of complex variables, and applying Euler's formula ($e^{i\theta} = \cos\theta + i\sin\theta$), Eq. (3.29) leads to

$$\begin{aligned}\vec{V} &= a_{u_x}\cos(\omega t - g_x) + ia_{u_y}\cos(\omega t - g_y)\\ &= a_{u_x}\frac{e^{i(\omega t - g_x)} + e^{-i(\omega t - g_x)}}{2} + ia_{u_y}\frac{e^{i(\omega t - g_y)} + e^{-i(\omega t - g_y)}}{2}\end{aligned} \tag{3.30}$$

After rearranging the terms, we have

$$\vec{V} = W_{\mathrm{p}}e^{i(\omega t + \theta_{\mathrm{p}})} + W_{\mathrm{q}}e^{-i(\omega t - \theta_{\mathrm{q}})} \tag{3.31}$$

$$\vec{V} = \vec{W}_{\mathrm{I}} + \vec{W}_{\mathrm{II}} \tag{3.32}$$

where W_{p}, θ_{p}, W_{q}, and θ_{q} can be directly related to the amplitudes and phases of tidal velocity components. Eq. 3.32 shows that the velocity for a tidal component

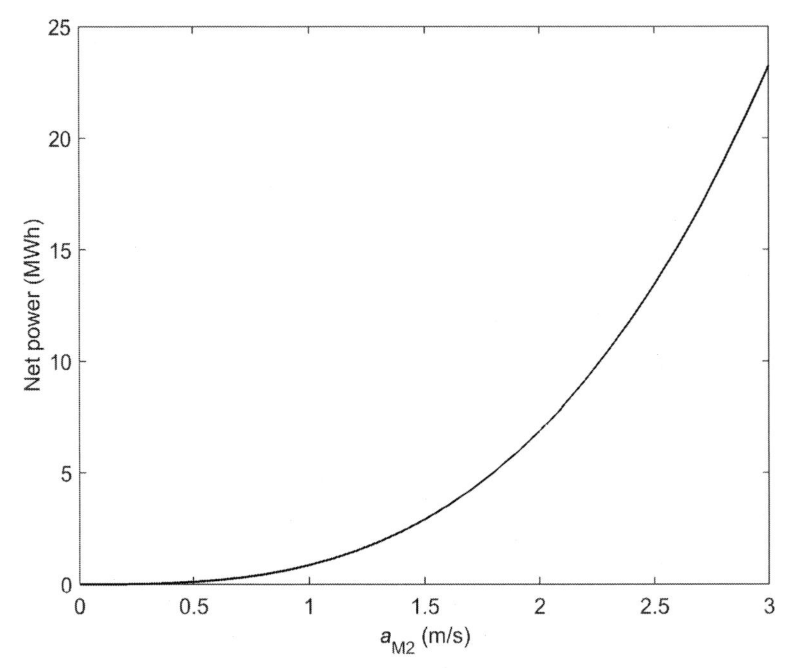

FIG. 3.20 Net power over a tidal cycle for a range of M2 current amplitudes, assuming a 20-m diameter rotor and neglecting device efficiency.

can decomposed into two radial rotating vectors, $\vec{W}_{\mathrm{I}}$ and $\vec{W}_{\mathrm{II}}$, which rotate in opposite directions. The summation of the two vectors, $\vec{V}$, is a vector rotating in the direction of the larger circle, and forms an ellipse, as shown in Fig. 3.21. A tidal ellipse has the following parameters:

- Semimajor axis: maximum current velocity
- Semiminor axis: minimum current velocity
- Eccentricity: ratio of minor to major axis, negative showing CW and positive CCW direction
- Inclination: angle between x and major axis
- Phase: the angle between the two circular rotating vectors at initial position

Tidal analysis programs such as T_TIDE use time series of current velocity as input and evaluate the tidal ellipse parameters for each tidal constituent.

3.12 TIDAL-STREAM DEVICES

Various technology types have been proposed for generating electricity from tidal streams, and these can generally be categorized as the following five types:

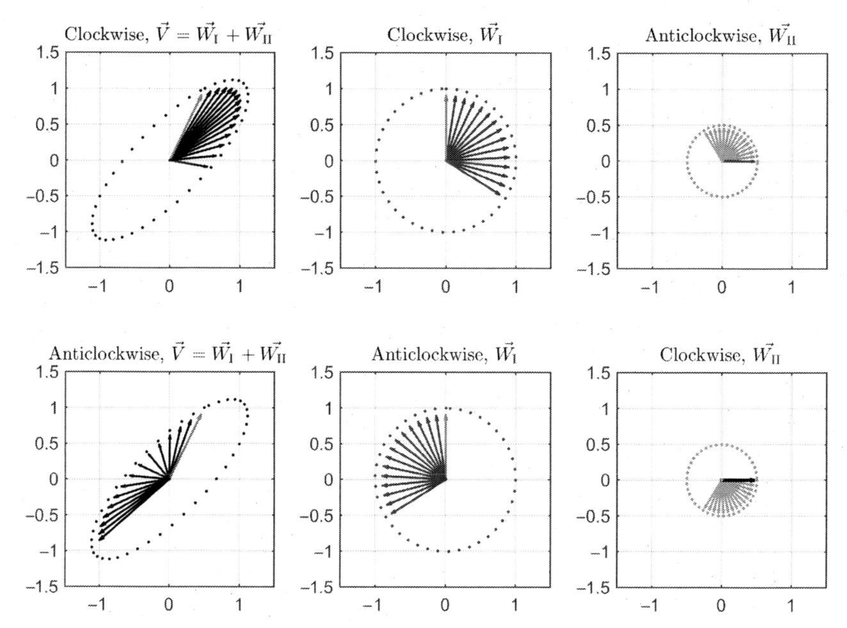

FIG. 3.21 Constructing a tidal ellipse by adding two rotating vectors in opposite directions. Note that the sense of rotation of velocity is the same as the larger vector (i.e. W_I).

- horizontal axis turbines
- vertical axis turbines
- oscillating hydrofoils
- venturi effect devices
- tidal kites

3.12.1 Horizontal Axis Turbines

These devices are similar in principal to the operation of conventional wind turbines. A turbine is placed in a tidal stream, which causes the turbine to rotate about a horizontal axis. In the horizontal plane, horizontal axis turbines can either be fixed, or include a yawing mechanism. However, in contrast to wind, because tidal flows are highly predictable and highly energetic rectilinear flows are much sought sites for development, most devices being developed do not include a yawing mechanism. This has the added advantage of reducing complexity in an environment where simplicity is clearly beneficial. Examples of horizontal axis turbines are the 1.2 MW twin rotor Marine Current Turbine (MCT) SeaGen device, which was deployed in Strangford Narrows, Northern Ireland, and the Tidal Energy Limited DeltaStream device installed in Ramsey Sound, Wales (Fig. 3.22). Recent developments have seen companies looking

FIG. 3.22 Tidal Energy Limited DeltaStream device prior to installation in Ramsey Sound, Wales. *(From Tidal Energy Limited.)*

into the possibility of floating tidal devices, potentially reducing capital and installation costs. The hydrodynamic theory of horizontal axis turbines is covered in Section 3.13.

3.12.2 Vertical Axis Turbines

Vertical axis turbines are similar in principal to horizontal axis turbines, but with a different axis of rotation (Fig. 3.23). These devices are more suited to regions where there is a large variation in the direction of tidal streams (e.g. rotary currents) in contrast to the rectilinear flows exploited by horizontal axis turbines. An example vertical axis turbine is the Gorlov Helical Turbine, which utilizes three twisted blades in the shape of a helix [e.g. 11].

3.12.3 Oscillating Hydrofoils

Oscillating hydrofoils have one or more hydrofoils attached to an oscillating arm (Fig. 3.24). The oscillating motion used to produce power is due to the lift generated by the tidal stream flowing either side of the wing. The motion can be used to drive a hydraulic system to produce useful power. Oscillating hydrofoils are generally suited to shallower water environments such as estuaries, because competing technologies such as horizontal axis turbines require significant clearance for the rotor to operate (in addition to navigational constraints).

FIG. 3.23 Conceptual diagram of vertical axis turbines. *(From www.Aquaret.com.)*

3.12.4 Venturi Effect Devices

These are funnel-like devices, which direct the water flow through a constricted section of a duct (Fig. 3.25). This increases the water velocity and decreases the pressure. The resultant flow can be used to directly drive a turbine, or the induced pressure differential in the system can drive an air turbine. Venturi effect devices are useful to exploit lower tidal streams and could potentially be deployed in much shallower water than horizontal axis turbines.

3.12.5 Tidal Kites

A tidal kite turbine is an underwater kite system that converts tidal energy into electricity by moving through the tidal stream. The water current creates a hydrodynamic lift force on the wing of a 'kite' (Fig. 3.26), which pushes the kite forward. The kite is steered in a figure-of-eight trajectory by a rudder, and reaches a speed many times the water current speed. As the kite moves, water flows through the turbine to generate electricity. The electricity is transmitted through a cable in the tether attached to the wing. The Minesto 1/4 scale Deep Green tidal kite has been successfully trialled in Northern Ireland, and plans are well underway for Minesto to install a small array of full-scale devices in the Irish Sea.

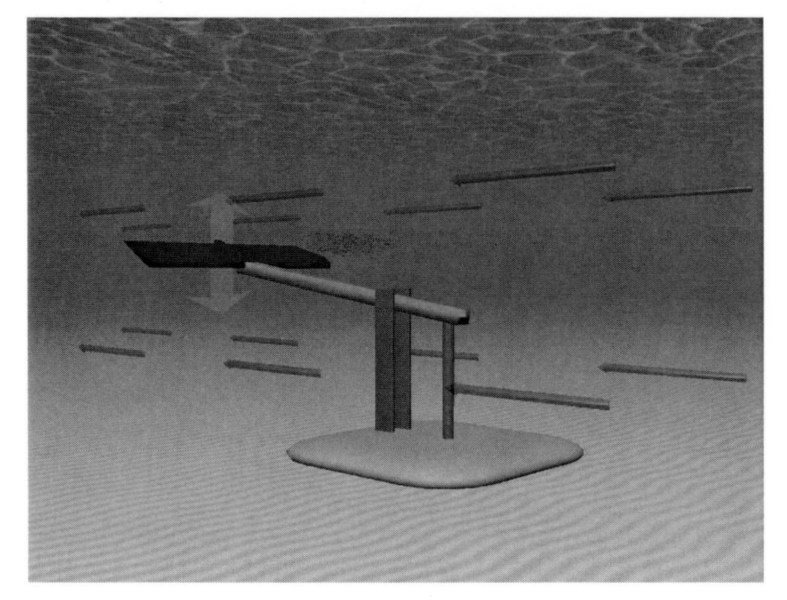

FIG. 3.24 Conceptual diagram of an oscillating hydrofoil. *(From www.Aquaret.com.)*

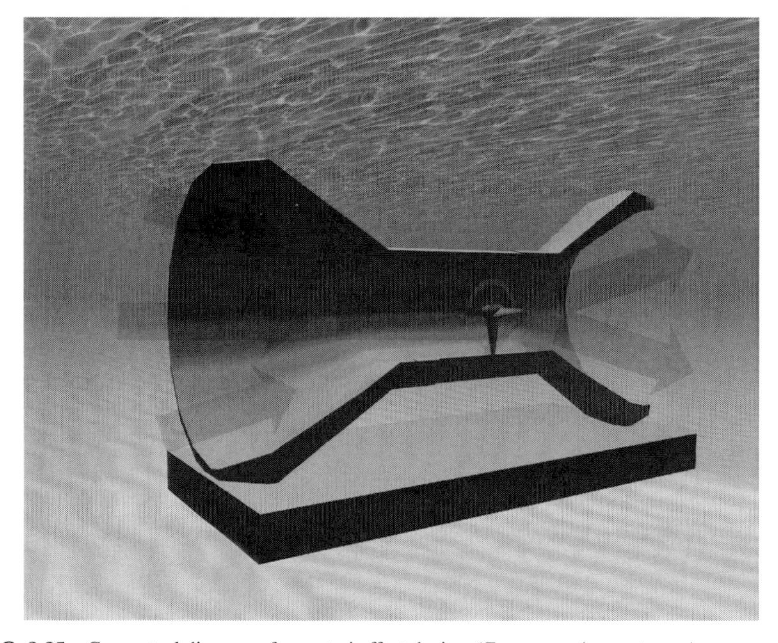

FIG. 3.25 Conceptual diagram of a venturi effect device. *(From www.Aquaret.com.)*

FIG. 3.26 Minesto Deep Green tidal kite. *(From: Minesto.)*

3.12.6 Arrays

Clearly, it is only through installing devices within arrays that significant levels of electricity generation can be achieved. This also has the advantage of shared infrastructure and maintenance costs. Many considerations must be taken into account for device placement within arrays (e.g. wakes influencing neighbouring devices), but many of these issues are technology-dependent. These issues come under the topic of optimization, and are covered in Chapter 9.

3.13 BASIC HYDRODYNAMICS OF HORIZONTAL AXIS TURBINES

Simulation of the hydrodynamic field around tidal turbines (or wind turbines) is a challenging task, and requires advanced numerical modelling to take into account turbulence, and fluid structure interaction (e.g. [12]). It also requires the collection of experimental data for model validation. However, simplified analytical approaches such as *actuator disk theory* can provide a good understanding of the flow field around horizontal axis turbines, as well as some useful preliminary results at initial stages of design [13]. The analytical methods are usually based on many assumptions, which should be considered in application, as discussed here.

Consider a control volume around a horizontal axis turbine. Due to axial symmetry, the control volume would be a stream tube with variable cross-sectional area (Fig. 3.27). Assume that the flow field inside this control volume does not interact with the fluid outside, and the flow is incompressible and steady state. Therefore, the continuity equation results in $Q = uA =$ constant. A turbine extracts energy from the flow and retards the current; therefore, the upstream flow velocity (u_o) would reduce to u_{dis} at the turbine. The velocity in the wake section (S3) would be even smaller; however, we can assume that the pressure far from the turbine is equal to the ambient pressure in the fluid (i.e. p_o at S1 and S3). The velocity at the disk can also be written in terms of axial flow induction factor, a, as follows

$$u_{dis} = u_o(1 - a) \tag{3.33}$$

Larger values of a indicate more drop in the velocity (i.e. more effect on the flow from a turbine). Also, based on the continuity equation, we can write

$$Q = A_o u_o = A_{dis} u_{dis} = A_w u_w \tag{3.34}$$

where u_w is the wake velocity (Fig. 3.27). The presence of a turbine in the flow field causes a pressure drop at the disk (or turbine) whilst—assuming the steady-state case—no sudden change in velocity ($Q = u_{dis} A_{dis}$) is expected at the disk. The pressure drop at the disk generates a force, which can be expressed as

$$\sum F = (\delta p) A_{dis} = [p_d^+ - p_d^-] A_{dis} \tag{3.35}$$

where p_d^+ and p_d^- are pressures upstream and downstream of the disk, respectively. Referring to momentum theory or Newton's law of motion (see Chapter 2), this force should be equal to the change of momentum in the control volume. Consider the input and output fluxes of momentum (i.e. $\rho u Q$) at Sections S1 and S3, respectively

$$\sum F = \rho Q(u_o - u_w) = \rho\,[A_{dis} u_{dis}]\,(u_o - u_w) = [p_d^+ - p_d^-] A_{dis} \Rightarrow [p_d^+ - p_d^-] = \rho\,[u_{dis}]\,(u_o - u_w) \tag{3.36}$$

So far, we have only employed the continuity and momentum equations. To further simplify the previous equations, we can also use the energy or Bernoulli's equation. As the amount of energy extracted by the turbine is

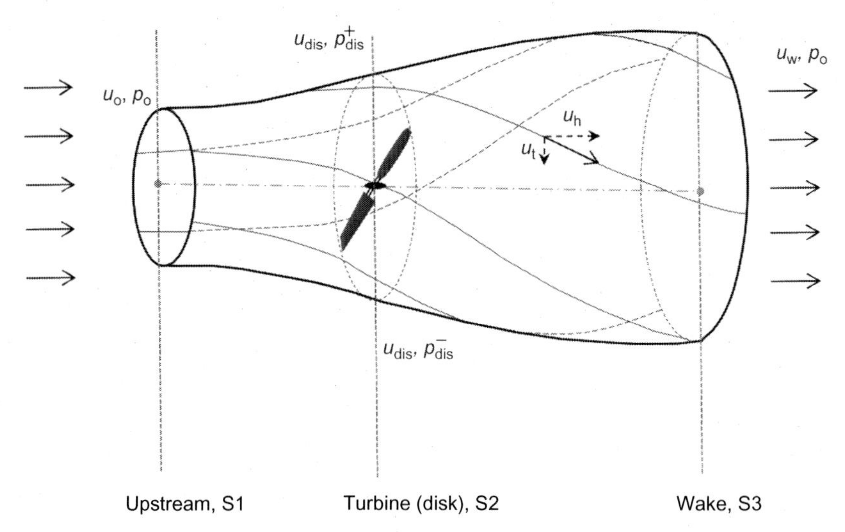

FIG. 3.27 Stream tube around a horizontal axis tidal turbine. The cross-sections for implementation of the momentum and energy equations have been also shown (S1, S2, and S3).

unknown, we can apply the energy equation in the upstream (between S1 and S2) or the downstream part (between S2 and S3) of the disk.

Applying the energy equation between S1 and S2, leads to

$$\frac{u_o^2}{2g} + \frac{p_o}{\rho g} = \frac{u_{dis}^2}{2g} + \frac{p_{dis}^+}{\rho g} \Rightarrow \frac{p_o}{\rho g} = \frac{u_{dis}^2}{2g} + \frac{p_{dis}^+}{\rho g} - \frac{u_o^2}{2g} \tag{3.37}$$

Similarly, for the downstream sections (S2 and S3), we have

$$\frac{u_{dis}^2}{2g} + \frac{p_{dis}^-}{\rho g} = \frac{u_w^2}{2g} + \frac{p_o}{\rho g} \Rightarrow \frac{p_o}{\rho g} = \frac{u_{dis}^2}{2g} + \frac{p_{dis}^-}{\rho g} - \frac{u_w^2}{2g} \tag{3.38}$$

therefore

$$\frac{u_{dis}^2}{2g} + \frac{p_{dis}^+}{\rho g} - \frac{u_o^2}{2g} = \frac{u_{dis}^2}{2g} + \frac{p_{dis}^-}{\rho g} - \frac{u_w^2}{2g} \tag{3.39}$$

which leads to

$$p_{dis}^+ - p_{dis}^- = \frac{1}{2}\rho(u_o^2 - u_w^2) \tag{3.40}$$

By equating the RHS of Eqs (3.36), (3.40), we can write

$$\frac{1}{2}\rho(u_o^2 - u_w^2) = \rho u_{dis}(u_o - u_w) \tag{3.41}$$

or

$$\frac{1}{2}(u_o + u_w) = u_{dis} = u_o(1 - a) \tag{3.42}$$

In other words, the axial velocity at the disk is simply the average of the velocities at the upstream and downstream sections. Eq. (3.42) also leads to

$$u_{\mathrm{w}} = (1 - 2a)u_{\mathrm{o}} \tag{3.43}$$

3.13.1 Power Coefficient and the Betz Limit

Now that we have computed the forces and velocity at the actuator disk, we can evaluate the extracted power using Eq. (3.36) as follows

$$P = Fu_{\mathrm{dis}} = [p_d^+ - p_d^-]A_{\mathrm{dis}}u_{\mathrm{dis}} = A_{\mathrm{dis}}\rho[(u_{\mathrm{o}} - u_{\mathrm{w}})]u_{\mathrm{dis}}^2 \tag{3.44}$$

Referring to Eqs (3.42), (3.43) and replacing u_{w} and u_{dis}, leads to

$$P = \rho A_{\mathrm{dis}}(u_{\mathrm{o}} - [(1 - 2a)u_{\mathrm{o}}])[u_{\mathrm{o}}(1 - a)]^2 = 2\rho A_{\mathrm{dis}}u_{\mathrm{o}}^3 a(1 - a)^2 \tag{3.45}$$

Considering the flux of kinetic energy through the swept area of the turbine, the total available power (before the turbine influences the flow field) is $P_{\mathrm{ava}} = \frac{1}{2}\rho A_{\mathrm{dis}}u_{\mathrm{o}}^3$, whilst the extracted power is

$$P = \rho A_{\mathrm{dis}}u_{\mathrm{o}}^3 a(1 - a)^2 = \left\{\frac{1}{2}\rho A_{\mathrm{dis}}u_{\mathrm{o}}^3\right\}^3 [4a(1 - a)^2] = [4a(1 - a)^2]P_{\mathrm{ava}} \tag{3.46}$$

The power coefficient, C_{p}, may be defined as the ratio of extracted power to available power, and is given by

$$C_{\mathrm{p}} = \frac{P}{P_{\mathrm{ava}}} = 4a(1 - a)^2 \tag{3.47}$$

Fig. 3.28 plots the power coefficient as a function of axial flow induction factor. As this figure shows, the maximum power coefficient is 0.593, corresponding to $a = 1/3$. Alternatively we can write

$$\frac{dC_{\mathrm{p}}}{dp} = 4(1 - a)^2 - 8a(1 - a) = 0 \Rightarrow a = \frac{1}{3} \tag{3.48}$$

which leads to $C_p = \frac{16}{27} = 0.593$. This limit is called the Betz limit after the German physicist Albert Betz. Therefore, theoretically, no turbine can convert more than 59.3% of the kinetic energy in a tidal current into mechanical energy to turn a rotor.

3.14 TIDAL RANGE: LAGOONS AND BARRAGES

In contrast to tidal stream technology (Section 3.12), the extraction of tidal energy using tidal lagoons (or tidal barrages) is based on the simple concept of converting potential energy to kinetic energy. This concept is quite old and has been largely implemented in hydropower projects for centuries.

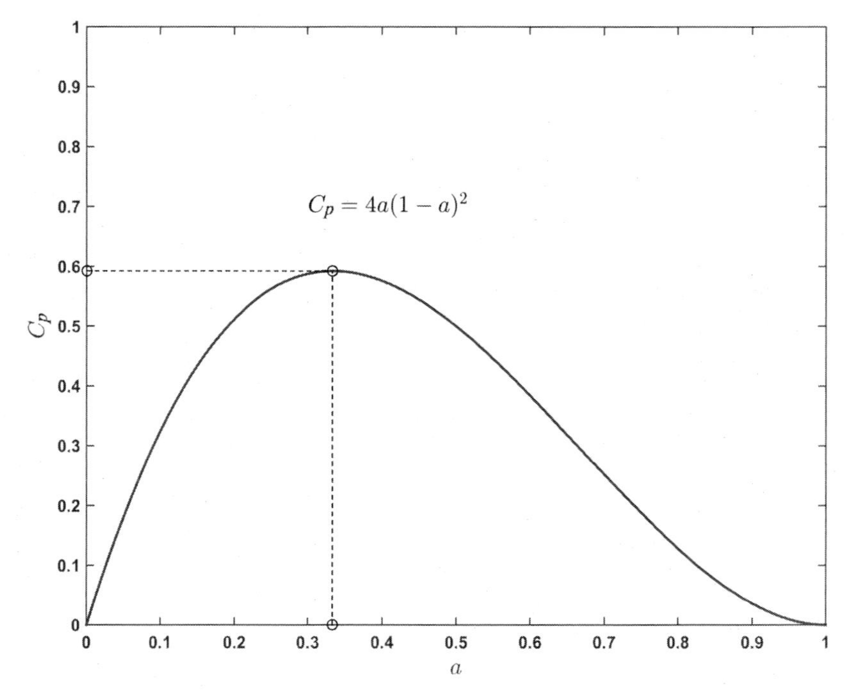

FIG. 3.28 Power coefficient and the Betz limit.

La Rance tidal power station in France (Fig. 3.29), with a peak output capacity of 240 MW, is a good example and is one of the oldest operational tidal energy projects [14].

As we know, water elevation changes due to astronomical tides. If we build a barrier in a sea or an estuary, we can create an artificial basin and control the inflow and outflow of water by gates (see Fig. 3.30). As a result of the barrier and control of flow through gates we can *impose* a time lag between the water elevation inside and outside of basin/lagoon. The difference of water elevation gives rise to a potential energy that can be harvested.

Here, we explain and formulate the equations for an ebb-generation scenario. The theory is quite similar for flood-generation or combined flood and ebb generation scenario (double-effect; [15]). Referring to Fig. 3.31, let us assume that the sluice gates are open during the flood tide; therefore, the water elevation both inside and outside of a lagoon rise to the peak (Point 1) or high tide (filling period; between Points 4 and 5). There would be small time lag between the water level inside and outside of the lagoon because of flow constriction in the gates. At high tide, the sluice gates would be closed, and consequently water level inside the lagoon will initially remain constant. However, the water elevation in the ocean falls, which will lead to a head on the turbines. The sluice gates will remain closed until a suitable head for turbine operation is achieved. This period is called standing (between Points 1 and 2), and during this time

FIG. 3.29 La Rance tidal barrage in Brittany—generating electricity since 1966. *(From S. Neill.)*

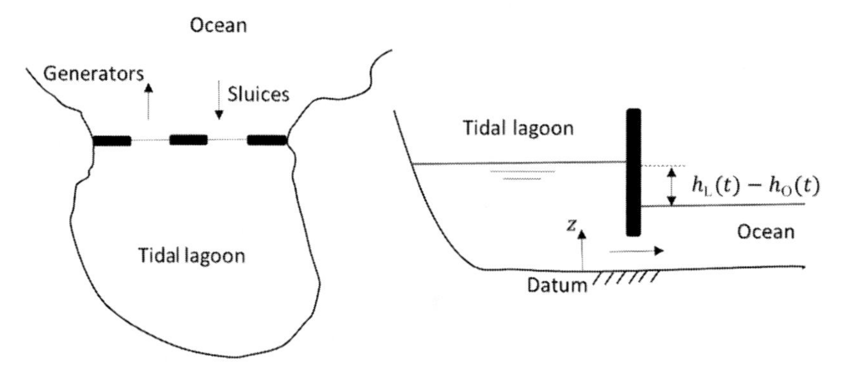

FIG. 3.30 Schematic of a tidal lagoon; ebb generation case.

no electricity would be generated. The turbine gates will open at Point 2 and energy conversion would continue as long as there is sufficient head on the turbines. During this period (generation), the water elevation inside the lagoon decreases. At Point 3, which usually happens after low tide, the difference of water elevation inside and outside the lagoon is no longer sufficient for turbine operation. At this point, the turbine gates will be closed. There will be another standing period (all gates closed), until the tide returns and sluice gates are opened for filling. This process would be repeated for each tidal cycle. Here, we will show that the amount of energy converted is approximately proportional to the area between water elevation curves—inside and outside the lagoon—when the turbines operate.

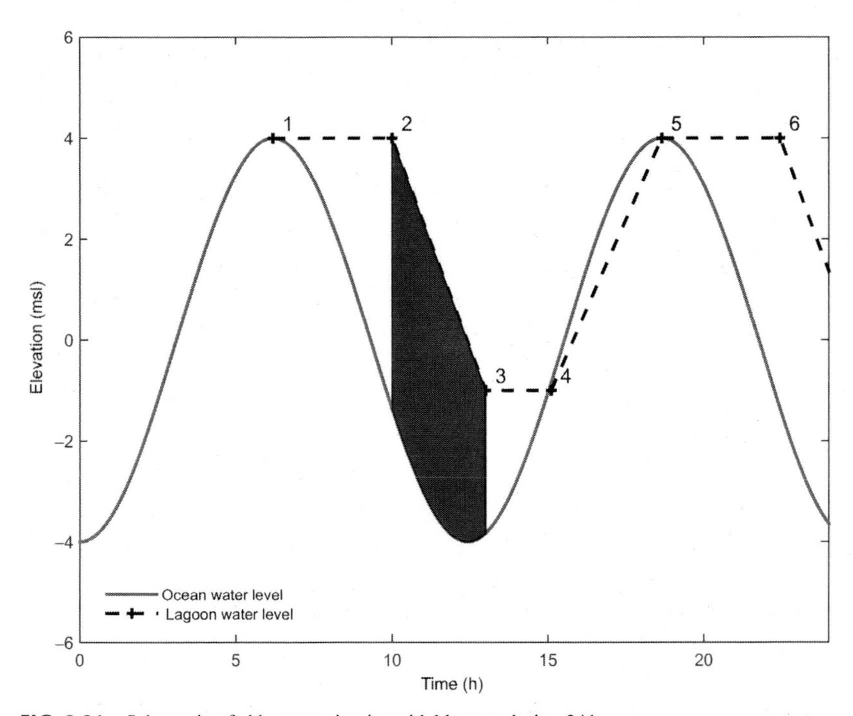

FIG. 3.31 Schematic of ebb generation in a tidal lagoon during 24 h.

Referring to Fig. 3.30, the total potential energy in a tidal lagoon can be evaluated by summing the potential energy of each individual particle in the reservoir with respect to a datum

$$E_{\text{potential}} = \sum (\delta m)gz = \sum \rho(\delta \mathcal{V})gz = \int_{\mathcal{V}} \rho z d\mathcal{V} \tag{3.49}$$

where z is the elevation above the datum and $\mathcal{V}$ is volume. First, consider a tidal lagoon in which the surface area of the lagoon does not vary with water elevation (which is not the case in Fig. 3.30); for this special case, $d\mathcal{V} = A_{\text{L}}dz$, where A_{L} is the surface area of the tidal lagoon. Therefore

$$E_{\text{potential}} = \int_0^{h_{\text{L}}} \rho z A_{\text{L}} dz = \rho A_{\text{L}} \int_0^{h_{\text{L}}} z dz = \rho A_{\text{L}} \frac{h_{\text{L}}^2}{2} \tag{3.50}$$

where h_{L} is the water surface elevation above a datum. Eq. (3.50) simply indicates that the energy which can be stored in a tidal lagoon is proportional to its *surface area*, and the *square of water elevation*.

Now we consider a more general case. Suppose that water surface elevations inside and outside of a tidal lagoon are represented by h_{L} and h_{O}, respectively; assuming no energy loss, the kinetic energy and the current velocity through the tidal turbines can be evaluated using Bernoulli's equation

$$\frac{v^2}{2g} = h_L(t) - h_O(t) \Rightarrow v = \sqrt{2g\,[h_L(t) - h_O(t)]} \tag{3.51}$$

where v is the water velocity at the turbine section. Therefore, the flow rate or discharge through the tidal turbines will be

$$Q_t = av = a\sqrt{2g\,[h_L(t) - h_O(t)]} \tag{3.52}$$

where a is the swept area of all turbines. As water flows through the turbines to the ocean, the water elevation inside the tidal lagoon gradually drops. Using the continuity equation, we can find a relationship between the change in water elevation inside the lagoon and the outflow rate. Consider the volume of water which leaves the lagoon during δt

$$\delta \forall = Q(t)\delta t = -A_L \delta h_L \Rightarrow Q = -\frac{A_L dh_L}{dt} \tag{3.53}$$

Now, we can derive an equation to estimate the theoretical power through the turbines. For a turbine, we know that the power at any time t will be

$$P = \frac{1}{2}\rho a v^3 = \frac{1}{2}\rho Q v^2 \tag{3.54}$$

Referring to Eq. (3.51) and replacing v, the power can also be evaluated by water elevations as follows

$$P = \rho Q(t) g\,[h_L(t) - h_O(t)] \tag{3.55}$$

replacing Q (Eq. 3.53), leads to

$$P = -\rho A_L \frac{dh_L}{dt} g\,[h_L(t) - h_O(t)] \tag{3.56}$$

Eq. (3.56) indicates that the theoretical power at each point in time is proportional to the difference of water elevation inside and outside a lagoon, and the rate of change of water surface elevation inside the lagoon (or flow discharge). Finally, if we have the time series of water elevation inside and outside a tidal lagoon (i.e. ocean), the total theoretical energy, which passes through turbines from t_1 to t_2, can be calculated as

$$E = \int_{t_1}^{t_2} \left\{ -\rho A_L \frac{dh_L}{dt} g\,[h_L(t) - h_O(t)] \right\} dt \tag{3.57}$$

Consider a special case where the discharge and the surface area of a lagoon remain constant during a period (i.e. $Q = -A_L dh_L/dt = \text{const}$), then

$$E = \rho g Q \int_{t_1}^{t_2} [h_L(t) - h_O(t)]\, dt \tag{3.58}$$

which simply shows that the power generated is proportional to the area between the time series of water elevation inside and outside a tidal lagoon. Eq. (5.58) can also be applied to the study of hydropower, in a similar way.

REFERENCES

[1] G.I. Taylor, Tidal oscillations in Gulfs and rectangular basins, Proc. Lond. Math. Soc. 2 (1) (1922) 148–181.
[2] D.A. Huntley, Tides on the north-west European continental shelf, in: Elsevier Oceanography Series, vol. 24, 1980, pp. 301–351.
[3] R. Pawlowicz, B. Beardsley, S. Lentz, Classical tidal harmonic analysis including error estimates in MATLAB using T_TIDE, Comput. Geosci. 28 (2002) 929–937.
[4] D.T. Pugh, Tides, Surges and Mean Sea-Level: A Handbook for Engineers and Scientists, John Wiley, Chichester, UK, 1987, 472 pp.
[5] S.D. Hicks, R.L. Sillcox, C.R. Nichols, B. Via, E.C. McCray, Tide and current glossary, in: Center for Operational Oceanographic Products and Services, NOAA National Ocean Service, Silver Spring, MD, 2000, pp. 1–29.
[6] C. Le Provost, Generation of over-tides and compound tides (review), in: Tidal Hydrodynamics, John Wiley & Sons, Inc., New York, NY, 1991, pp. 269–295.
[7] S.P. Neill, E.J. Litt, S.J. Couch, A.G. Davies, The impact of tidal stream turbines on large-scale sediment dynamics, Renew. Energy 34 (2009) 2803–2812.
[8] S.P. Neill, M.R. Hashemi, M.J. Lewis, The role of tidal asymmetry in characterizing the tidal energy resource of Orkney, Renew. Energy 68 (2014) 337–350.
[9] M.N. Gallo, S.B. Vinzon, Generation of overtides and compound tides in Amazon estuary, Ocean Dyn. 55 (5–6) (2005) 441–448.
[10] R. Soulsby, Dynamics of Marine Sands: A Manual for Practical Applications, Thomas Telford, London, 1997.
[11] F.O. Rourke, F. Boyle, A. Reynolds, Tidal energy update 2009, Appl. Energy 87 (2) (2010) 398–409.
[12] Y.L. Young, M.R. Motley, R.W. Yeung, Three-dimensional numerical modeling of the transient fluid-structural interaction response of tidal turbines, J. Offshore Mech. Arct. Eng. 132 (1) (2010) 011101.
[13] M.E. Harrison, W.M.J. Batten, L.E. Myers, A.S. Bahaj, Comparison between CFD simulations and experiments for predicting the far wake of horizontal axis tidal turbines, IET Renew. Power Gen. 4 (6) (2010) 613–627.
[14] J.P. Frau, Tidal energy: promising projects: La Rance, a successful industrial-scale experiment, IEEE Trans. Energy Convers. 8 (3) (1993) 552–558.
[15] R.H. Clark, Elements of Tidal-Electric Engineering, vol. 33, John Wiley & Sons, New York, NY, 2007.

FURTHER READING

[1] S.P. Neill, J.D. Scourse, K. Uehara, Evolution of bed shear stress distribution over the northwest European shelf seas during the last 12,000 years, Ocean Dyn. 60 (2010) 1139–1156.

Chapter 4

Offshore Wind

Wind energy is one of the fastest growing renewable energy sectors, with levelized costs of wind energy now comparable to the cost of electricity generated by thermal power stations such as coal and gas. However, due to lower surface roughness, the wind resource is higher offshore, and has further advantages such as lower visual impact. In this chapter, following a general introduction, the components of a wind turbine are introduced. Wind energy resource assessment, which is a crucial part of macro-siting of wind energy projects, is discussed in detail, and applied to a case study of the first offshore wind energy project in the United States—Block Island Wind Farm. Finally, a brief introduction to marine spatial planning is provided—the method of siting wind energy projects. The objective of this chapter is to introduce the basic technical aspects of offshore wind energy, and the reader is referred to more specific books for further information.

4.1 INTRODUCTION

Wind can be defined as the movement of air over the surface of the Earth. Because air is a fluid, the movement of air and water (in the ocean) follow the same principles. A pressure gradient in the air (or ocean) leads to a flow from regions of high pressure to regions of low pressure. Variations in pressure that lead to pressure gradients are due to uneven heating of the Earth's land and sea surfaces—particularly the differences in heating between the tropics and high latitude regions. Wind is, therefore, an indirect form of solar energy. Because of the Earth's rotation, these large scale wind patterns will be affected by the Coriolis force (Section 3.6). Near the surface of the Earth, land or ocean, friction slows down the wind, and surface roughness (e.g. topography, forests, waves) is the main cause of this friction force. Therefore, the spatial and temporal variations in wind are controlled by solar radiation, Coriolis (i.e. Earth's rotation), and the Earth's surface (mountains, buildings, ocean, etc.). These variations can be studied at various scales: global wind patterns, regional climates, variations within a wind farm, and around the blades of individual turbines.

The use of wind energy (e.g. for sailing) has a history extending back thousands of years. The oldest wind energy device is possibly the Persian windmill (Fig. 4.1) at Nashtifan, Iran, that was built around the 9th century [1,2]. The Nashtifan village is located in the northeast of Iran, in Khorasan Razavi

Fundamentals of Ocean Renewable Energy. https://doi.org/10.1016/B978-0-12-810448-4.00004-5

FIG. 4.1 Persian windmills (vertical axis turbines) at Nashtifan Village, constructed around the 9th century. *(Reproduced from [2], with permission from Elsevier.)*

Province. Recently, wind energy is a rapidly growing industry, and can compete with fossil fuels at similar levelized costs of electricity. Fig. 4.2 shows the global trend of wind power capacity since 2002. Wind energy, along with solar energy, can be regarded as the fastest growing renewable energy sectors [3]. So far, the majority of wind farms have been installed on land (over 90%), whilst the offshore wind industry is also very popular in many countries such as the United Kingdom and Germany.

In general, offshore wind farms are more expensive and challenging to build compared with their onshore counterparts due to several factors: the foundation and supporting structure of offshore wind turbines, grid connection, installation of turbines, and the operation and maintenance of offshore wind projects are all more expensive and complicated. However, due to lower surface roughness (absence of building, mountains, trees), the wind energy resource is better offshore. Further, the visual impact and noise of offshore wind projects are

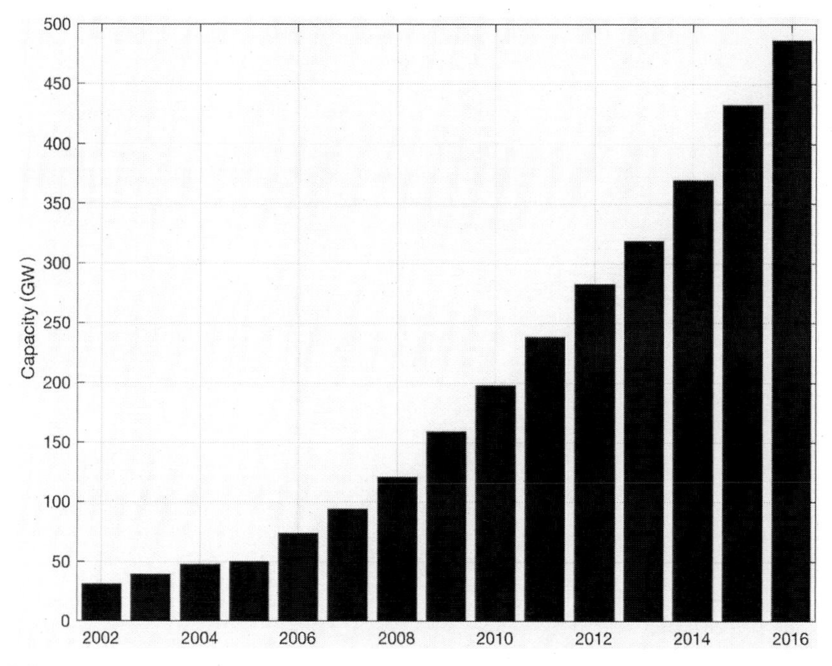

FIG. 4.2 Global installed capacity of wind power, 2002–16 [3].

much less than onshore projects. Also, offshore wind farms are a good source of electricity for coastal cities; around 40% of the global population lives in the coastal zone. Some analysts predict that much of the future development of the wind industry will move towards offshore wind farms [4], because there will be much less capacity to install wind farms on land (e.g. lack of suitable land near coastal cities). Therefore, the offshore wind industry is gradually becoming an important part of the global energy mix. In some regions of the word (e.g. the northeast region of the United States), offshore wind is the most promising renewable energy resource, compared with other types of renewables (e.g. wave and solar).

A major consideration in the design and construction of offshore wind farms is water depth. Up to now, the majority of wind farms have been built in relatively shallow waters—typically less than 30 m water depth. The substructure of offshore wind turbines can be considered as the most limiting factor in the development of offshore wind projects. Fig. 4.3 shows the trend in the design of substructure for offshore wind turbines. In shallow water zones, simple monopile support structures can be used. As water depth increases to transition zones (less than 60 m in depth), more complex supporting structures such as Jacket or Tripod are required (Fig. 4.4). In deep waters (more than 60 m), floating structures are used to support wind turbines. Whilst building a substructure for offshore wind turbines in shallow water is a proven technology,

FIG. 4.3 Offshore wind turbine substructure designs for use in various water depths. *(Illustration courtesy of Josh Bauer, NREL.)*

FIG. 4.4 Horizontal axis turbines in the Block Island Wind Farm (RI, USA) during installation in 2016. The installed turbines are Haliade-150, each with a capacity of 6 MW, diameter of 150 m, and hub height of 100 m. *(Photo courtesy of Deepwater Wind.)*

floating wind turbine technology is at the demonstration stage (similar to the tidal energy industry). Much research is ongoing to test and assess the performance of offshore floating wind turbines.

4.2 AN INTRODUCTION TO OFFSHORE WIND TURBINES

Horizontal axis wind turbines are commonly used for generating electricity offshore. Fig. 4.4 shows an example of horizontal axis offshore wind turbines (Haliade-150, with a capacity of 6 MW and diameter of 150 m) during installation. Fig. 4.5 shows a schematic of a horizontal axis wind turbine. In simple terms, the wind exerts forces on the blades (lift and drag) and turns the rotor. A low-speed shaft is connected to the rotor. The gearbox transmits the power (rotation and torque) of the low-speed shaft to a high-speed shaft that spins a generator. Functions of the individual components of a wind turbine are explained briefly as follows: [1]

Anemometer: Used to measure the wind speed. The collected data are transmitted to the controller unit, for instance, to stop the wind turbine at very high speeds (e.g. during hurricanes).

Blades: Harvest the wind energy. Wind exerts forces (lift and drag) on the blades, causing the rotor to spin.

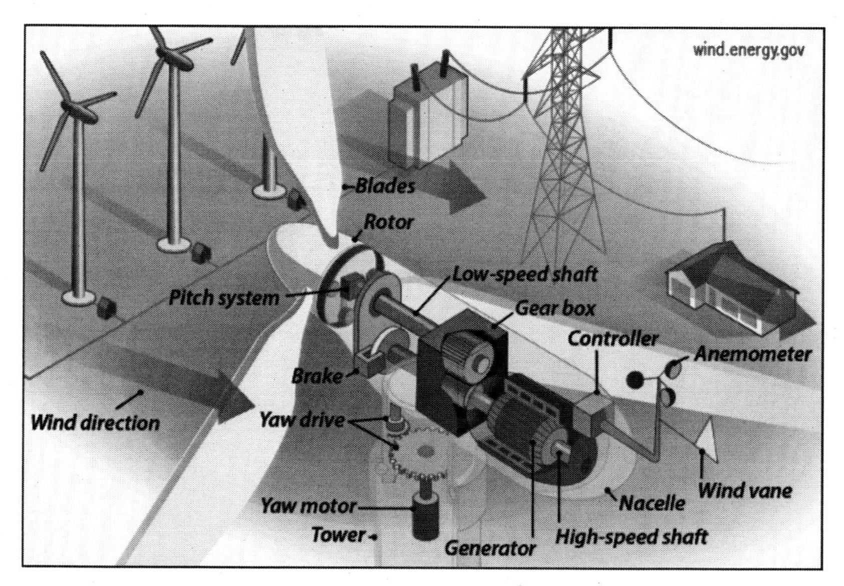

FIG. 4.5 Schematic of a wind turbine. *(Courtesy of the US Department of Energy.)*

1. See https://energy.gov/eere/wind/wind-energy-technologies-office.

Brake: Stops the rotor in case of emergency.

Controller: Starts the wind turbine at cut-in speeds (generally around 3.5 m/s) and shuts off the turbine at very high wind speeds (over 25 m/s, cut-out speed) to protect the device.

Gear box: Transmits the power (torque times angular speed) from the low-speed shaft to the high-speed shaft. The rotational speed required for a turbine is very high (≈1000 rpm) compared with the rotational speed of the rotor (≈20 rpm).

Generator: Produces the AC electricity.

High-speed shaft: Transmits the power from the gearbox to generator, and rotates/drives the generator.

Low-speed shaft: Transmits the power from the rotor to the gearbox.

Nacelle: Houses the generating components of a wind turbine including the gearbox, generator, controller, and brake.

Pitch system: Adjusts the angle of attack of wind by turning the blades. The rotational speed and the generated power can be controlled/optimized by the pitch system. The pitch system can stop the turbine rotating at cut-in or cut-out speeds.

Rotor: Consists of blades and the hub.

Tower: Supports the wind turbine. Towers are mainly built using conical tubular steel.

Wind direction: Upwind wind turbines (shown in Fig. 4.5) face into the wind, whilst downwind turbines face away from the wind.

Yaw drive: Aligns the turbines towards the wind. This keeps the wind turbine facing the wind when the wind direction changes for upwind turbines.

Yaw motor: Used to power the yaw drive.

4.2.1 Aerodynamics of Wind Turbines

The aerodynamics of horizontal axis wind turbines is very similar to the hydrodynamics of horizontal axis tidal turbines, reviewed in Chapter 3. Here, a brief overview is presented, whilst more details are provided in Chapter 3 and elsewhere (e.g. [5]).

Consider a wind turbine that has a diameter D and a swept area A. If the wind speed is u, the available or 'theoretical' wind kinetic energy per unit time (i.e. wind power, P_t) can be evaluated as

$$P_t = \frac{d}{dt}\left(\frac{1}{2}mu^2\right) = \frac{1}{2}\frac{dm}{dt}u^2 = \frac{1}{2}\rho Q u^2 \tag{4.1}$$

where Q is the volumetric flow rate of the wind and ρ is the air density. Replacing Q with Au, leads to

$$P_t = \frac{1}{2}\rho A u^3 \tag{4.2}$$

Only part of the available/theoretical energy can be harvested by a wind turbine. The power coefficient (C_p; or overall efficiency of a turbine) can be defined as the ratio of the extracted power (generated electricity) to the available wind power as follows:

$$C_p = \frac{P_e}{P_t} \tag{4.3}$$

where P_e is the extracted power. The power coefficient depends on a combination of aerodynamic, mechanical, and electrical efficiencies. It can be shown theoretically that the maximum power coefficient of a horizontal axis wind turbine is 59%, which is called the Betz limit/law.

Betz Limit

Fig. 4.6 shows a schematic of a streamtube that is classically used to analyse wind turbines based on actuator disk theory. Because turbine blades partially

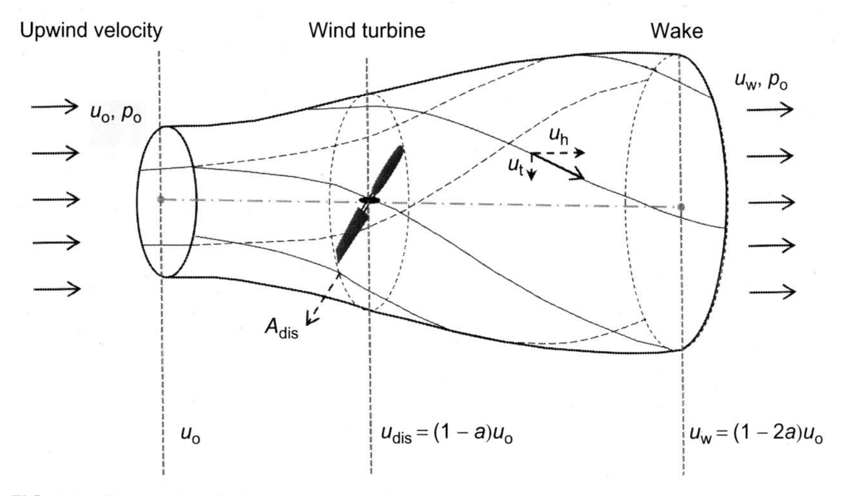

FIG. 4.6 Streamtube of a horizontal axis wind turbine.

block and also extract energy from the flow, they cause a reduction in the wind velocity at the turbine (or disk), and also in the wake of the turbine. Considering the continuity equation in the steady state (i.e. $Q = Au$), this flow reduction means that the flow area will increase at the turbine, and even more at the wake as shown in the figure. The average wind velocity at the turbine can be written in terms of the axial flow induction factor (a). An induction factor of 1 means that the wind turbine does not affect the upwind velocity at all ($u_{\text{dis}} = u_{\text{o}}$), whereas $a = 0$ means that the turbine completely blocks/stops the wind ($u_{\text{dis}} = 0$). Practically, the induction factor is between 0 and 1. It can be shown that the power coefficient is a function of the flow induction factor as follows (see Chapter 3):

$$C_{\text{p}} = \frac{P_{\text{e}}}{P_{\text{t}}} = 4a(1-a)^2 \tag{4.4}$$

The maximum power coefficient can be evaluated by taking the derivative

$$\frac{dC_{\text{p}}}{dp} = 4(1-a)^2 - 8a(1-a) = 0 \Rightarrow a = \frac{1}{3} \text{ and } C_{\text{p}} = \frac{16}{27} \tag{4.5}$$

Therefore, the maximum power coefficient is $\frac{16}{27} = 0.593$, corresponding to $a = 1/3$. The above limit is called the Betz limit after Albert Betz, a German scientist, who pioneered wind turbine technology.

Power Curve

The efficiency, or power coefficient, of wind turbines is not constant, and depends on the wind speed. A power curve shows how the power of a wind turbine varies with the wind speed. A typical power curve is depicted in Fig. 4.7.

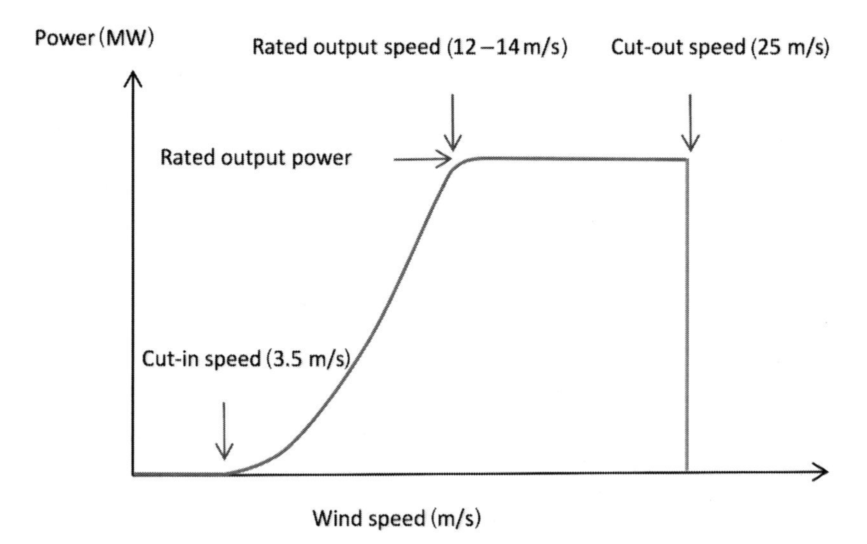

FIG. 4.7 A typical power curve for a wind turbine.

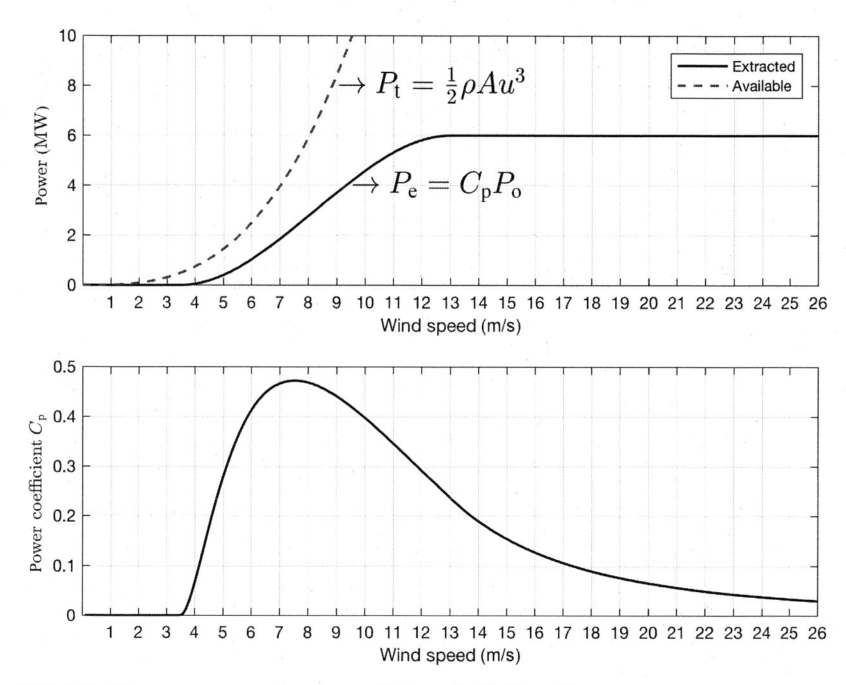

FIG. 4.8 The power curve and power coefficient of a 6 MW turbine.

Wind turbines cannot generate electricity for very low (insufficient torque to overcome friction) or extremely high velocities (they will be damaged). Therefore, cut-in (around 3.5 m/s) and cut-out (around 25 m/s) speeds are specified on power curves. As the wind exceeds the cut-in speed, the power output increases rapidly. However, around certain speeds, known as rated wind speed (12–14 m/s), the power output reaches a limit that is called the rated power output. This is the maximum level that the electrical generator can still work. When wind speed is higher than rated output wind speed, the power is kept almost constant by a method, for instance by adjusting the blade angles (pitch system).

Using a power curve, it is also possible to plot the power coefficient as a function of wind speed. Fig. 4.8 shows the power curve and power coefficient of a 6 MW turbine with a rated output wind speed of 12.5 m/s. As this figure shows, the maximum efficiency of the turbine (i.e. around 47%) is achieved at wind speeds between 7 and 8 m/s.

4.3 ASSESSMENT OF WIND ENERGY AT A SITE

As discussed in the previous sections, the power generated by a wind turbine, and consequently the capacity factor of a wind farm, depends on the wind speed, which varies in time and space. Therefore, to select a suitable site, maximizing the power output from the wind at a selected location, and choosing/designing the best wind turbine that can produce maximum power at minimum cost, it is necessary to understand and examine variations in the wind speed.

There are large variations in wind speed, both geographically and temporally. The spatial and temporal variations of wind can be studied at a number of scales. The spatial variations of the wind resource can be examined for various climates (e.g. across the globe), and also within a climatic region. The spatial variations within a region (e.g. northeast of the United States), in general, depend on the climate of the region, Coriolis force (due to rotation of the Earth), and physical geography. For instance, surface roughness (land vs. sea, type of vegetation, size of buildings and structures), and the presence of mountains greatly influences the variation of wind speed in a region. Fortunately, climate models and wind data collected over the past decades give us sufficient data to understand these variations. On a smaller scale, wind energy varies within a wind farm, as the wind speed significantly reduces in the wake of wind turbines. This is the subject of micrositing, and will be discussed in Chapter 9.

4.3.1 Atmospheric Boundary Layer

Above certain heights, it can be assumed that the wind speed is not influenced by the Earth's surface (ocean or land). At these high altitudes, the wind is mainly driven by synoptic air pressure gradient and Coriolis (Earth's rotation) forces. Synoptic air pressure gradients are associated with weather systems that have the scale of days (e.g. high- and low-pressure systems), and can be predicted by numerical weather models. The wind (air flow) that is free from surface effects is called 'geostrophic wind'. At lower altitudes, the roughness of the surface and thermal effects such as air-sea interactions significantly affect the wind speed. This zone is called the atmospheric boundary layer. Assuming that the wind is strong enough to sufficiently mix the boundary layer, the wind speed is mainly controlled by surface roughness; in so-called neutral stability the surface heating/cooling does not affect the wind profile. The distribution of the wind near the surface of the ocean or land can be simply represented by a logarithmic profile:

$$u(z) = \frac{u^*}{\kappa}\left[\ln(z/z_o) + \psi\right]; \quad u^* = \sqrt{\frac{\tau_o}{\rho}} \tag{4.6}$$

in which u^* is the shear velocity, τ_o is surface shear stress, κ is von Karman's constant (about 0.41), and z_o is the surface roughness. ψ is a stability term and is considered zero in neutral stability condition. Typical values for roughness range from 0.7 m for forests to 0.0001 m in the ocean. For more accurate estimation of surface roughness in the ocean, waves should be included. Assuming that the velocity of wind (u_r) at a reference height above the Earth (z_r) is given (e.g. by a meteorological station), Eq. (4.6) leads to

$$u(z) = u(z_r)\frac{\ln(z/z_o)}{\ln(z_r/z_o)} \tag{4.7}$$

which can be used to estimate the wind velocity at other heights, such as the hub-height of a wind turbine. Alternatively, the power law has also been used to characterize the vertical distribution of wind velocity,

$$u(z) = u(z_r)\left(\frac{z}{z_r}\right)^{1/7} \tag{4.8}$$

If the wind velocity is measured/modelled at a particular height above the ocean/land surface, the velocity of wind at the turbine height can be estimated using the previous equations. For instance, Fig. 4.9 shows the distribution of wind velocity using a logarithmic distribution (assuming two values for the bed roughness) and power law. If the average measured velocity 10 m above the ground is 11 m/s, the estimated velocity at 90 m above the ground is about 13.5–15 m/s.

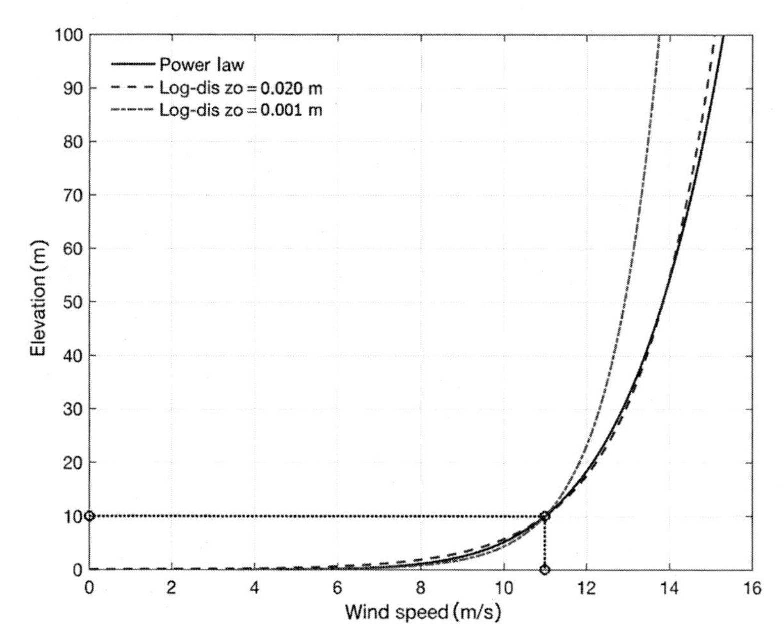

FIG. 4.9 Vertical distribution of mean wind velocity based on power law and logarithmic distributions. For this example, the average measured velocity 10 m above the ground is 11 m/s, and the estimated velocity at 90 m above the ground (e.g. turbine hub-height) is 13.6–15 m/s, depending on the surface roughness.

4.3.2 Temporal Distribution: Probability Density Function of Wind Speed

Here, we focus more on the temporal variations of wind speed at a site. As mentioned, the wind speed varies over a range of timescales: decades, years, seasons, days, hours, and seconds. There is still much uncertainty to understand and predict these variations. For instance, it is not possible to predict how the wind speed varies between years. However, seasonal variations are easier to understand and predict. Very small fluctuations in wind speed at the scale of seconds give rise to turbulence, which is not important for resource assessment; however, they should be considered in the design of blades. Turbulence causes cyclic loading on turbine blades and can lead to damage.

Temporal variations in the wind speed at scales of hours, seasons, and years that are important in wind energy resource assessments can be represented by a probability density function (PDF). PDFs and their properties are usually discussed in probability and statistics books, so only a short review is provided here.

If we consider the wind speed as a continuous random variable, the probability of wind speed falling within a specified range can be evaluated using a PDF. The PDF of wind speed at a site can be constructed using historical wind data, as we will discuss later. By definition, if we denote $f(u)$ as the PDF of wind speed, the probability of having a wind speed within u_1 and u_2 can be evaluated as

$$Pr\,[u_1 \leq u \leq u_2] = \int_{u_1}^{u_2} f(u)du \tag{4.9}$$

Fig. 4.10 shows a sample PDF that represents the variation in wind speed at a site. As we can see, the probability of wind speed falling between 5 and 10 m/s (the area between 5 m/s and 10 m/s; A1) is relatively high (42%). In other words, 42% of the time, the wind speed is between 5 and 10 m/s. However, the probability of having a wind speed greater than 15 m/s (i.e. the area under the curve for wind speeds greater than 15 m/s; A2) is relatively low (10%). Further, it is clear that the chance of having wind speeds greater than 0 is 100%. Therefore, the total area under a PDF is 1.

$$Pr\,[0 \leq u \leq \infty] = \int_{0}^{\infty} f(u)du = 1 \tag{4.10}$$

The average (or expected value) of wind speed is the first moment of the PDF, that is,

$$E(u) = \bar{u} = \int_{0}^{\infty} uf(u)du \tag{4.11}$$

Another useful function that can be derived from a PDF is the cumulative distribution function. It simply evaluates the probability of wind speed being greater than a specified value. Therefore,

$$F(u) = Pr\,[x \geq u] = \int_{u}^{\infty} f(x)dx \tag{4.12}$$

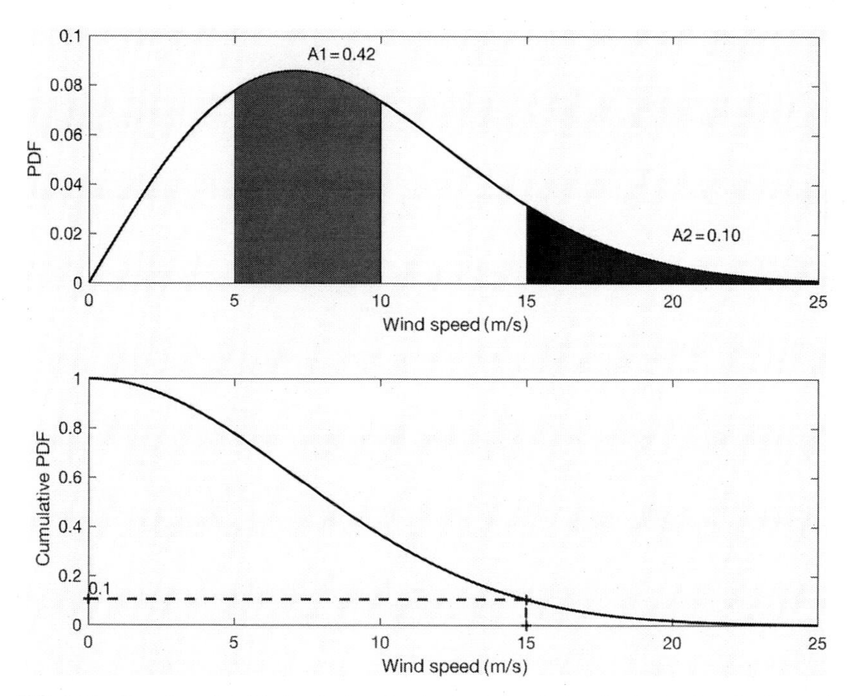

FIG. 4.10 Sample probability density function (*top*) and cumulative distribution function (*bottom*) for wind speed. For instance, the area under the curve for wind speed values greater than 15 m/s (A2) in the top plot is the probability of wind speed being ≥15 m/s, which is 10% for this case. The same information can be extracted from the cumulative distribution function.

Referring again to Fig. 4.10, the cumulative distribution function starts at 1 (because the chance of having a wind speed greater than 0 m/s is 100%). It gradually decreases to zero. For instance, the probability of wind speed being greater than 15 m/s is 10%—shown as dashed lines in the bottom panel. This is exactly the area under the PDF (A2; top panel). Based on Eq. (4.12), we can conclude that the PDF is the derivative of the cumulative distribution

$$f(u) = -\frac{dF(u)}{du} \tag{4.13}$$

A PDF or a cumulative distribution function can alternatively be used to demonstrate the probability of wind speed at a site. However, cumulative distributions may be preferred, as they directly show the probabilities, whilst it is necessary to integrate a PDF to compute the probability.

Previous studies have shown that the Weibull distribution is a good representation of hourly variations of wind speed at a location. The cumulative Weibull distribution is given by

$$F(u) = \exp\left[-\left(\frac{u}{c}\right)^k\right] \tag{4.14}$$

where c and k are parameters for this distribution. In particular, c is the scale parameter and k is the shape parameter. Both parameters are estimated based on historical wind speed data. The scale parameter is related to the average wind speed as follows

$$c = \frac{\bar{u}}{\Gamma(1 + 1/k)} \quad \text{or} \quad E(u) = c\Gamma(1 + 1/k) \tag{4.15}$$

in which Γ is the Gamma function. The shape parameter reflects the shape of the PDF. If we take the derivative of the cumulative distribution, the PDF of the Weibull distribution becomes

$$f(u) = -\frac{dF(u)}{du} = k\frac{u^{k-1}}{c^k} \exp\left[-\left(\frac{u}{c}\right)^k\right] \tag{4.16}$$

Fig. 4.11 shows examples of the Weibull PDF, assuming various shape factors. A special case of the Weibull distribution is called the Rayleigh distribution for $k = 2$. This is the most common value that can be used to describe the hourly variation of wind speed at many locations.

As mentioned before, given the historical wind speed data at a site, the best values for k and c are selected. This will be demonstrated for a sample site in the next section. Assuming that the PDF of wind speed at a site is evaluated, the average theoretical power intensity can be also estimated. Referring to

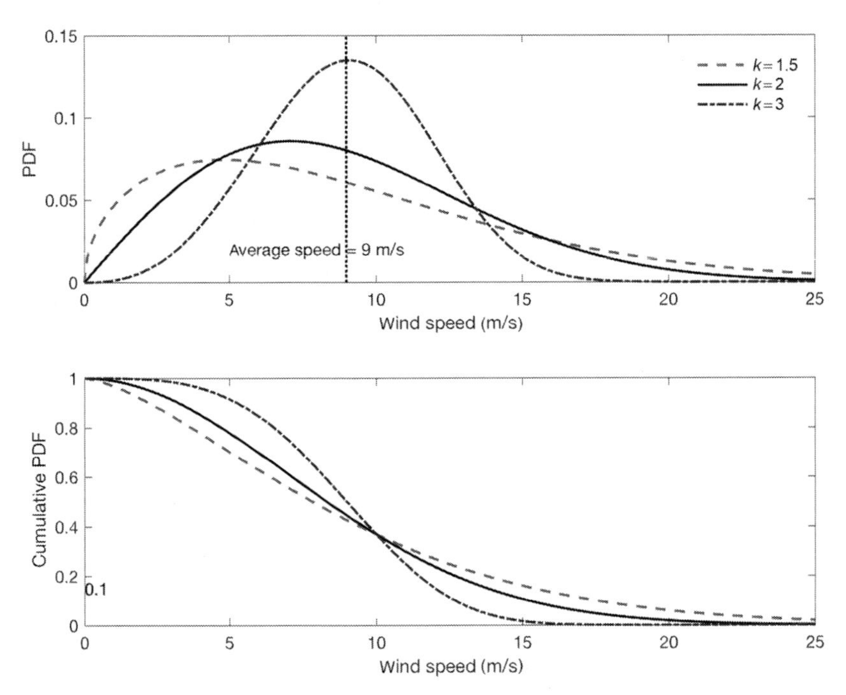

FIG. 4.11 Weibull distribution corresponding to various shape parameters (k). The average wind speed for all cases is 9 m/s. The corresponding values for scale parameter, c, are calculated based on Eq. (4.15).

Eq. (4.15), and using the properties of the Gamma function, we can write

$$\bar{u} = E(u) = c\Gamma(1 + 1/k) \rightarrow E(u^3) = c^3\Gamma(1 + 3/k) \tag{4.17}$$

Consequently, the expected value or mean power density is given by

$$E(P) = E\left\{\frac{1}{2}\rho u^3\right\} = \frac{1}{2}\rho c^3\Gamma(1 + 3/k) \tag{4.18}$$

4.3.3 Block Island Wind Farm

In this section, the resource assessment for a real offshore wind farm in the United States is performed using the methodology that was described in the previous section. The Block Island Wind Farm (see Fig. 4.12) is the first offshore wind farm in the United States and was constructed in 2016 by Deepwater Wind as a demonstration project. The farm includes five turbines, each with a rated power of 6 MW (Haliade-150 6 MW), and therefore the total capacity of the project is 30 MW.

Long-time series (e.g. covering a decade) of hourly wind data should be used for wind resource assessment to capture interannual and seasonal variability of wind energy for a location. In some cases, wind data are collected at the project site, but if the period of data collection is not long, other datasets should be combined. For instance, collected data at a site can be used to find the correlation of wind speed with other nearby stations that have longer records. Then, long-time series of wind data can be generated using those stations.

In order to conduct the resource assessment for the Block Island Wind Farm, the hindcast wind data provided by Wave Information Studies (WIS)

FIG. 4.12 An aerial photograph of Block Island, RI, United States, and the five wind turbines (each 6 MW) that have been installed about 6 km from this island. The turbine hub-height is about 100 m. *(Photograph courtesy of Deepwater Wind.)*

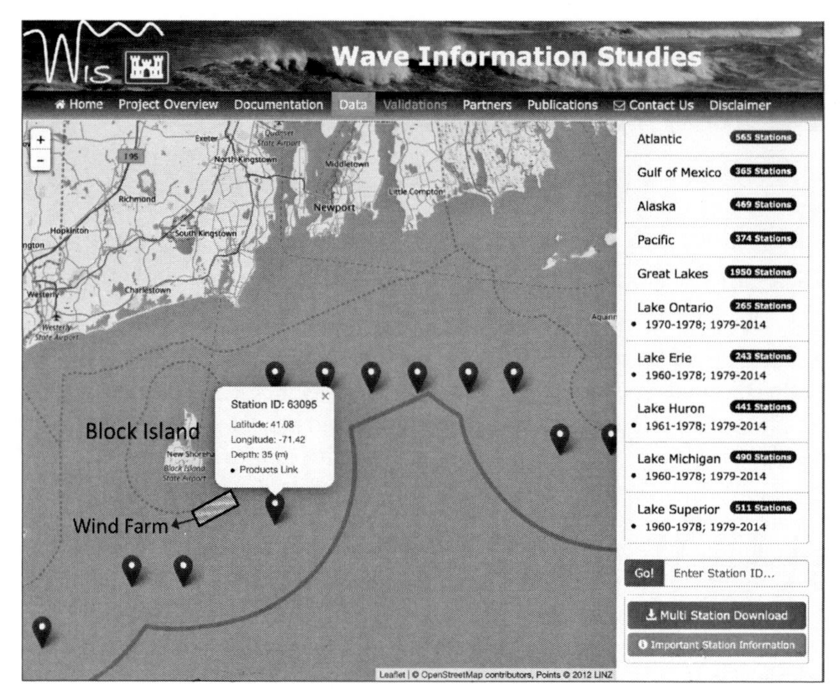

FIG. 4.13 Location of the Block Island Offshore Wind Farm and a number of neighbouring WIS stations. Station 63095 was used for resource assessment here.

were used. WIS is a project sponsored by the US Army Corps of Engineers (see wis.usace.army.mil). It provides more than 30 years of hindcast wind and wave information at thousands of nodes along the US coast, based on numerical models. These models have been validated using observed data. Fig. 4.13 shows a number of WIS stations, along with the location of the Block Island wind project. Here, we assume that the wind speed that is estimated at Station 63095 is representative of the wind speed at the project site. In studies that were carried out by Deepwater Wind, other meteorological stations at Block Island and Buzzard Bay were used. However, we will show that the final results are similar. The WIS wind data are the estimated wind velocity 10 m above the water surface. The wind speed at the hub-height (100 m) using the logarithmic distribution and assuming a roughness of 0.0001 m [6] is given by

$$u_{100} = u_{10} \frac{\ln(100/z_o)}{\ln(10/z_o)}; \quad z_o = 0.0001\,\text{m} \rightarrow u_{100} = 1.2u_{10} \tag{4.19}$$

Note that the actual sea surface roughness also depends on the waves, ignored here for simplicity. Alternatively, using the power law (Eq. 4.8), the wind speed should be increased by 32% instead of 20%. Here we applied the power law.

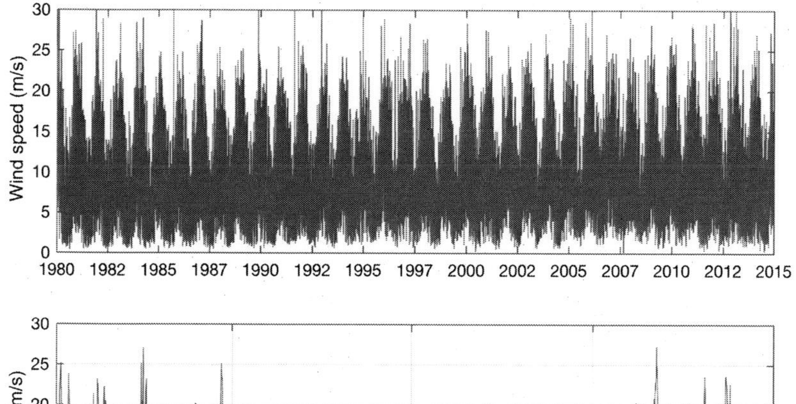

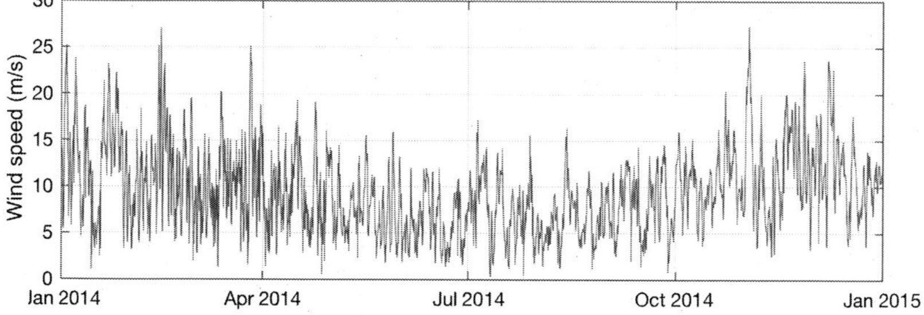

FIG. 4.14 Time series of hourly wind speed data at 100 m hub height at WIS Station 6309 for the period 1980–2014, including 306,815 data points. The *bottom plot* shows a magnified view for a sample year, 2014.

Fig. 4.14 shows the time series of hourly wind speed data at Station 6309 for the period 1980–2014, covering 35 years. Because a leap year occurs every 4 years, the average number of data points for a year is $365.25 \times 24 = 8766$, and the total number of data points would be $35 \times 8766 = 306{,}810$.[2]

In order to fit a PDF to the wind speed, the frequency distribution of the wind speed data should be evaluated. A histogram is a good graphical method to represent the distribution of the wind speed. The minimum and maximum wind speeds in the dataset are 0 and 36 m/s. Therefore, we can divide the wind speed data into a series of intervals (or bins) from 0 to 36 m/s. Let us choose bin sizes of 1 m/s. For example, the 12th bin will contain all recorded wind speeds that are greater than or equal to 11.5 m/s and less than 12.5 m/s (regardless of when recorded). The 13th bin includes all data points that have a wind speed greater than or equal to 12.5 m/s and less than 13.5 m/s, and so forth. If we put every single hourly wind speed observation/hindcast into a bin, we can sum the number of data points in each bin. If the frequency (number of data points in each bin) is plotted against the bins, a histogram will be produced. All of these steps can be conveniently performed by computer (e.g. in MATLAB or Excel).

Fig. 4.15 shows the histogram of the wind data. In the top panel of this figure, the number of cases (frequency of each bin) is shown. For instance, in 29,985

2. The actual number of data points is slightly different (306,815) in the downloaded dataset.

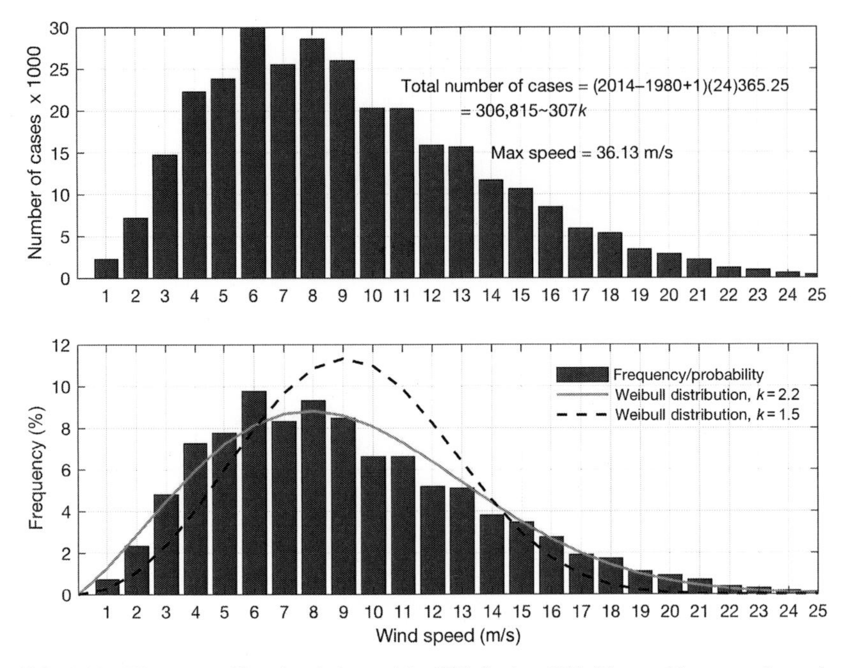

FIG. 4.15 Histogram of hourly wind speed for WIS Station 6309. The *top* histogram shows the number of cases (i.e. frequency) in each bin. The *bottom* histogram shows the relative frequency, along with the fitted PDF (the Weibull distribution).

(nearly 30,000) cases out of 306,815, the wind speed is around 6 m/s (more precisely, $5.5 \leq u < 6.5$). The relative frequency (in percent) is plotted in the bottom panel, which can be interpreted as the probability of wind. The relative frequency is defined as

$$\text{fr}_i = \frac{N_i}{\sum(N_i)} = \frac{N_i}{N} \tag{4.20}$$

where fr_i is relative frequency in Bin i, N_i is the frequency (number of cases) in that bin, and N is the total number of cases (i.e. 306,815 in our example). For example, the relative frequency is 9.8% for the 6 m/s Bin. We can say that 9.8% of the time, the wind speed is around 6 m/s. Or (looking at Fig. 4.15) 2% of the time, it is around 17 m/s. This is a classical way to define the probability. More advanced approaches also exist (e.g. Bayesian probability), which is beyond the scope of this book.

Referring back to the previous section, if a PDF represents the wind speed probability densities, the probability of wind speed in each bin will be $f(u_i)\,\delta u$, where δu is the bin size. This is the area under the PDF in each bin. Therefore,

$$f(u_i) = \frac{\text{fr}_i}{\delta u} \tag{4.21}$$

The previous equation gives the values of the PDF at the centre of the bins. It is usually not possible to find a perfect PDF that matches all these values. However, using a curve fitting approach (e.g. trial and error), we can find the best shape parameter (k) that can fit these values. The bottom plot in Fig. 4.15 shows the Weibull distribution for two shape parameters: 2.2 and 3. Based on this plot, $k = 2.2$ is a better fit. Note that once k is specified, the scaling parameter (c) will be determined by the average wind speed and k (Eq. 4.15).

Calculation of Power Output and Capacity Factor

So far, the frequency distribution of the hourly wind speed has been determined for this site. This distribution can be represented by a relative frequency histogram or the Weibull distribution. By combining the power curve for the wind turbines and wind probability distribution, the power output can be estimated.

Consider a typical year that has 365 days or 8760 h. Based on our calculations, we can predict how many hours the wind blows with a certain speed, and compute the power using the power curve. Then, we can do calculations for all wind speeds and add the total power. Fig. 4.8 shows the power curve of a 6 MW turbine representing (approximately) the Haliade-150 wind turbine.

Table 4.1 summarizes the details of the calculations for the electricity production during a year for this project. Based on this table, the capacity factor of the wind farm is 47.5%, and the annual production of electricity is about 125 GWh, if we directly use the histogram and relative frequencies. If we use the PDF, the capacity factor and annual energy production will be 49.5% and 130 GW, respectively. The calculation steps are explained as follows:

1. Evaluate the power for each wind speed interval/bin based on the power curve (Columns I and II)
2. Estimate the percentage of time (probability) that the wind blows for each wind speed in bins (Columns III and IV). This probability can be estimated based on the Weibull distribution (method 1; Column III) or relative frequency (method 2; Column IV).
3. Determine how many hours in a year the wind blows for each wind speed. This is calculated by multiplying the probabilities and the number of hours in a year (8760): Columns V and VI.
4. Calculate energy for each bin by multiplying the number of hours that wind blows with that speed (Column V or VI) and corresponding power (Column II).
5. Sum all energy productions to calculate energy for a year.
6. Consider a reduction due to operation and maintenance. We assumed that 10% of the time the turbines are not available and reduced the energy production by 10%, on average.
7. Multiply the energy production by number of turbines. According to the table, the total energy production using method 1 (the Weibull

TABLE 4.1 Computation of the Capacity Factor for the Block Island Wind Farm Project, Based on the Power Curve and Probability Distribution of Wind Speed During 1980–2015 (Fig. 4.14). The Energy is Calculated for a Year, or 8760 h (i.e. Average Annual Output)

I	II	III	IV	V (8760 x III)	VI (8760 x IV)	II x V	II x VI
Wind Speed (m/s)	Power (MW)	Weibull (1) (%)	Freq. (2) (%)	Duration (1) (h)	Duration (2) (h)	Energy (1) (MWh)	Energy (2) (MWh)
1	0	1.3	0.7	114	61	0.0	0.0
2	0	2.9	2.3	254	201	0.0	0.0
3	0	4.5	4.8	394	420	0.0	0.0
4	0.05	6	7.3	526	639	25	31
5	0.40	7.2	7.8	631	683	253	274
6	1.03	8.1	9.8	710	858	729	882
7	1.84	8.7	8.3	762	727	1405	1340
8	2.76	8.8	9.3	771	815	2130	2251
9	3.70	8.6	8.5	753	745	2791	2758
10	4.58	8.1	6.6	710	578	3252	2650
11	5.31	7.3	6.6	639	578	3398	3072
12	5.81	6.4	5.2	561	456	3260	2649
13	6.00	5.4	5.1	473	447	2838	2681
14	6.00	4.4	3.8	385	333	2313	1997
15	6.00	3.5	3.5	307	307	1840	1840

16	6.00	2.7	2.8	237	245	1419	1472
17	6.00	2	1.9	175	166	1051	999
18	6.00	1.4	1.7	123	149	736	894
19	6.00	1	1.1	88	96	526	578
20	6.00	0.7	0.9	61	79	368	473
21	6.00	0.4	0.7	35	61	210	368
22	6.00	0.3	0.4	26	35	158	210
23	6.00	0.2	0.3	18	26	105	158
24	6.00	0.1	0.2	9	18	53	105
25	6.00	0.1	0.1	9	9	53	53
>25	0	~0	~0	~0	~0		
Sum		~100	~100	~8760	~8760	28,912	27,734
Sum	(Availability)	~90	~90			26,021	24,961
Sum	(5 turbines)					130,105	**124,805**
Capacity	(Total)				30 MW×8760 h	262,800	262,800
Capacity Factor						49.5%	**47.5%**

Notes: Computations are based on 365 days (8760 h). Computed average wind speed is 9.2 m/s. It is also assumed that, on average, the wind farm is not available for 10% of the time due to maintenance. The boldfaced values indicate the total energy production and capacity factor, which are close to what is predicted in Deepwater Wind reports for the average year scenario.

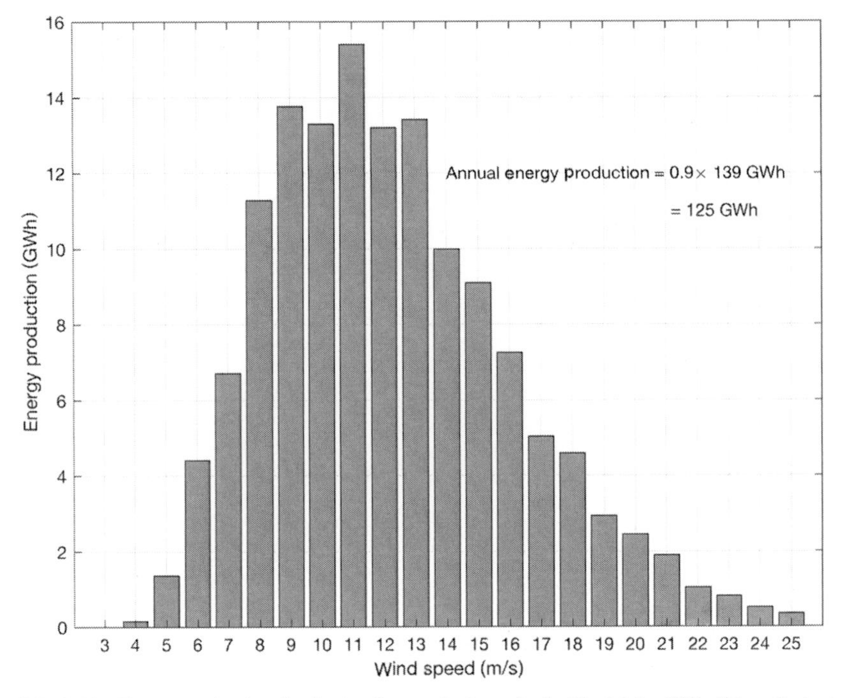

FIG. 4.16 Energy production distribution for a typical year in the Block Island Wind Farm Project (30 MW capacity). The computational details have been provided in Table 4.1.

PDF) is 130,105 MWh. Using method 2, the total energy production is 124,805 MWh.

8. Compute the capacity factor by dividing the energy output by the total capacity of the project. Here, the total capacity of the project is 30 MW times the number of hours in a year: 262,800 MWh.

Fig. 4.16 shows the distribution of the energy output. As we can see, the majority of power is produced when the wind blows with a speed of 11 m/s at this site.

According to a report published by Deepwater Wind [7], the overall energy production (on average) is 124,799 MWh, and the capacity factor is 47.5%. The average wind speed is reported 9.69 m/s. These numbers are very close to what we estimated in Table 4.1, given the assumptions that we made, and relative simplicity of our approach.

4.4 MARINE SPATIAL PLANNING

Siting an offshore wind farm is a complicated process, in which the feasibility of a project in several aspects such as technical (wind energy resource, foundation), economical (cost-benefit analysis, levelized cost of energy), environmental/

ecological (impacts on marine mammals, fish, birds), societal (tourism, visual impact), conflict with other users (fishing, sea and air navigation, military), and permit/legal aspects should be carefully investigated.

Marine spatial planning is a process in which all users/stake holders of the ocean from industries, government, coastal communities, conservation, tourism, and recreation collaborate in a unified framework to make informed decisions about using an area for a particular purpose. The objective of marine spatial planning is to identify areas that are most suitable for a specific usage (e.g. wind farm) with minimum conflicts amongst other users, minimum environmental impacts, and also preserving ecosystem services in a sustainable way.

Therefore, various types of data should be collected, analysed, and synthesized by a team of experts in collaboration with authorities and other stake holders to identify potential feasible sites for offshore wind projects. One of the best ways to present and combine geospatial data is in the geographic information science (GIS) environment. Here is a sample list of maps/data that needs to be prepared for marine spatial planning in relation to offshore wind projects [8].

- Technical-related maps/data
  - ◇ Wind resource (e.g. average wind speed at hub-height)
  - ◇ Water depth (bathymetry)
  - ◇ Geology and foundation
- Other users of the ocean
  - ◇ Navigation areas; shipping lanes
  - ◇ Vessel tracks, ferry routes
  - ◇ Fishing areas
  - ◇ Recreational/tourism (e.g. sailing race courses, diving sites, bird/shark watching areas)
  - ◇ Airport buffer zones
  - ◇ Aquaculture
  - ◇ Existing leases
  - ◇ Military
  - ◇ Cable routes
- Marine protected areas
- Environmental-related maps/data
  - ◇ Marine mammals and turtles
  - ◇ Birds
  - ◇ Sediment and benthic habitats

- ◇ Acoustic effects of a wind farm
- ◇ Other ecological data (fish, phytoplankton)
- Historical and cultural resources maps/data
- Statutes, regulations, and policies
  - ◇ Territorial (state/federal) and international waters
  - ◇ Coastal buffer zones

Because marine spatial planning is a long and comprehensive process, governmental agencies (e.g. The Bureau of Ocean Energy Management in the United States or The Crown Estate in the United Kingdom) take the lead and identify potential lease sites for developers, after conducting necessary studies.

REFERENCES

[1] A.Y. Al-Hassan, D.R. Hill, et al., Islamic Technology: An Illustrated History, Cambridge University Press, Cambridge, 1986.

[2] G. Müller, M. Chavushoglu, M. Kerri, T. Tsuzaki, A resistance type vertical axis wind turbine for building integration, Renew. Energy 111 (2017) 803–814.

[3] S. Sawyer, M. Dyrholm, Global Wind Report; Annual Market Update, Global Wind Energy Council, Brussels, Belgium, 2016, Available from: www.gwec.net (accessed November 1, 2017).

[4] M. Bilgili, A. Yasar, E. Simsek, Offshore wind power development in Europe and its comparison with onshore counterpart, Renew. Sustain. Energy Rev. 15 (2) (2011) 905–915.

[5] T. Burton, N. Jenkins, D. Sharpe, E. Bossanyi, Wind Energy Handbook, John Wiley & Sons, Hoboken, NJ, 2011.

[6] S. Emeis, M. Turk, Comparison of logarithmic wind profiles and power law wind profiles and their applicability for offshore wind profiles, in: Wind Energy: Proceedings of the Euromech Colloquium, Springer Science & Business Media, New York, NY, 2007, p. 61.

[7] AWS-Truepower, Block Island wind farm energy production estimate, Deepwater Wind, Rhode Island, USA, 2010, Available from: www.dwwind.com (accessed November 10, 2017).

[8] CRMC, Ocean Special Area Management Plan, vol. 1, Coastal Resources Management Council, RI, US, 2010, Available from: http://seagrant.gso.uri.edu (accessed October 15, 2017).

Chapter 5

Wave Energy

Similar to wind energy, and in contrast to tidal energy, wave energy is a stochastic form of electricity generation. Although modern forecasts make it possible to predict waves with certainty over relatively short timescales (e.g. 24–48 h), any longer-term planning must rely on statistical trends such as seasonal variability. This is one of the challenges of wave energy conversion, another being the extreme nature of the locations where, by their very nature, the wave climate could be suitable for electricity generation. However, wave energy has huge global potential, and its geographical distribution is generally more diverse than tidal energy, which tends to be confined to a relatively low number of 'hot spots' such as flow around headlands or through straits. Therefore, there is much global interest, R&D, and investment in wave energy projects and technologies.

In this chapter, we investigate the nature of wind waves, through a consideration of linear wave theory, examining fundamental properties of waves such as dispersion wave power, and wave transformation in shoaling water. We introduce the various wave energy converter (WEC) technologies and examine the theory of heaving point absorbers in some detail. Finally, we consider wave resource assessment and characterization, examining timescales of variability, and the theoretical versus technical resource.

5.1 WAVE PROCESSES

Looking back at Fig. 1.14, 'waves' in the ocean occur over a vast range of scales, from long-period (tidal) waves, with a period of several hours, to capillary waves that have periods of less than 0.1 s. However, the waves that are suitable for electricity generation within the context of 'wave energy' are wind waves and swell waves, which generally have periods in the range 2–25 s. Wind waves are generated due to transfer of wind energy and momentum into the wave field. Initially, when a sea surface is calm, small pressure fluctuations associated with turbulence in the airflow above the water surface are sufficient to induce small ripples, or dimples, on the sea surface. Once these ripples have formed, the small slopes provide a mechanism for horizontal winds to act further upon the sea surface, leading to the development of sizeable waves. These locally generated waves are known as 'wind waves', and waves that propagate far from their source of generation are known as 'swell waves'.

Fundamentals of Ocean Renewable Energy. https://doi.org/10.1016/B978-0-12-810448-4.00005-7

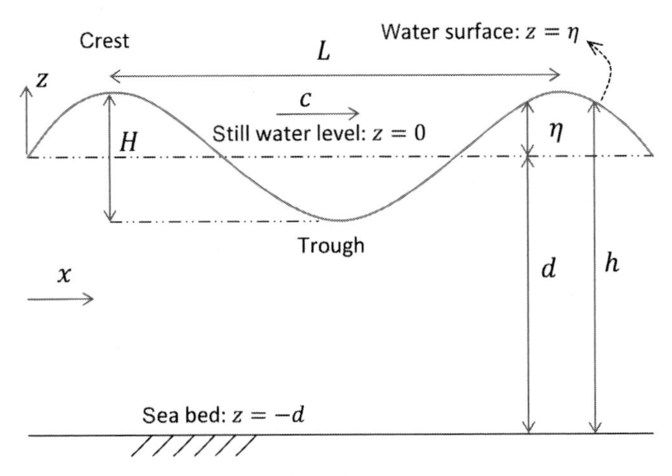

FIG. 5.1 Schematic of a monochromatic sinusoidal wave and important wave parameters.

There are various properties of ocean waves that are common to almost all waves that occur in the natural environment. Consider a 'linear' wave in the space domain that can be represented by a sinusoidal profile (Fig. 5.1). The maximum displacement of the wave from still water level ($z = 0$) is the wave amplitude, a. The vertical distance between the crest and trough (i.e. $2a$) is defined as the wave height, H, and the distance between two successive crests (or troughs) is the wave length, L. Although not shown in Fig. 5.1, a corresponding sketch in the time domain would show that the time between two successive wave crests (or troughs) is the wave period, T. The reciprocal of the wave period ($1/T$) is the wave frequency, with units of s^{-1} or Hz.

As all trigonometric functions repeat over the interval 2π, the mathematical description of waves is much simplified if we introduce the concepts of wavenumber ($k = 2\pi/L$) and angular frequency ($\sigma = 2\pi/T$). To give the wavenumber physical meaning, k is the number of complete wave cycles that exist in 1 m of linear space. A commonly reported property of waves is the significant wave height (H_s). This is defined as the mean height of the highest one-third of the waves in a record. This approximately corresponds to the wave height that can be estimated visually by a trained observer.

5.1.1 Linear Wave Theory

Waves in the ocean are nonlinear, and do not exactly follow linear wave theory; nevertheless, linear wave theory is very helpful in understanding and assessing more complicated waves such as nonlinear, irregular, and random waves, which will be discussed further later in this chapter. Linear wave theory can also be applied to many engineering and science problems with reasonable accuracy. A detailed derivation of linear wave theory can be found in many standard wave mechanic textbooks. Here, we present the main assumptions, parameters, equations, and wave properties associated with linear wave theory.

The main assumptions of linear wave theory can be summarized as [1]:

- The fluid is incompressible and inviscid.
- The flow is irrotational.[1]
- The bed is horizontal with a constant depth; the bed is impermeable.
- The wave amplitude is small.
- A regular wave with a constant wave period is considered.
- Coriolis effects are neglected.
- Waves are long-crested: the hydrodynamic solution is provided for a 2D vertical plane.
- The pressure at the water surface is constant.

Referring back to Chapter 2, we mentioned that the Euler equations represent an incompressible and inviscid flow as follows:

$$\nabla \cdot \mathbf{u} = 0 \tag{5.1}$$

$$\frac{d\mathbf{u}}{dt} = \frac{\partial \mathbf{u}}{\partial t} + \mathbf{u} \cdot \nabla \mathbf{u} = -\frac{1}{\rho}\nabla p - g\hat{k} \tag{5.2}$$

The Euler equations are the governing equations of *an ideal fluid*. A particular case of interest is potential flow, in which the flow *vorticity* is also zero. The other term, which is commonly used in *potential flow theory*, is irrotational flow (i.e. zero vorticity). Irrotational flow is simply a flow in which the particles do not rotate. It can be shown that the rate of rotation of particles in a flow field is directly proportional to the curl (i.e. $\nabla \times$) of velocity

$$\vec{\omega} = \frac{1}{2}\nabla \times \mathbf{u} \tag{5.3}$$

where $\vec{\omega}$ is called vorticity (i.e. rotation vector); hence, irrotational flows have zero vorticity. Referring to Eq. (5.2), if we take the curl of this equation, the RHS will become zero, and this leads to

$$\frac{d(\nabla \times \mathbf{u})}{dt} = \frac{d\vec{\omega}}{dt} = 0 \tag{5.4}$$

In other words, the flow vorticity is constant for ideal flow. Therefore, by assuming that the flow is inviscid and incompressible, we cannot necessarily conclude that the flow is irrotational. Nevertheless, irrotational flow is a special case in this context, where vorticity is not only constant, but it is also zero (i.e. $\vec{\omega} = \nabla \times \mathbf{u} = 0$). In hydrodynamics and wave mechanics, it is much easier to deal with incompressible, inviscid and irrotational flows, because a scalar

---

1. The flow in an inviscid and incompressible fluid is not necessarily irrotational, as will be discussed.

potential function can satisfy the governing equations and provide us with the hydrodynamic flow field. This is referred to as *potential flow theory*. Through some mathematical manipulation, you can easily show that the curl of a gradient of any scalar function is always zero. In other words,

$$\nabla \times \nabla \Phi = 0 \quad \text{and} \quad \nabla \times \mathbf{u} = 0 \Rightarrow \mathbf{u} = \nabla \Phi \tag{5.5}$$

where Φ is called the potential function, which as mentioned is a scalar function. In potential flow theory, the flow velocity is the gradient of this potential function. The momentum equation is already satisfied because $\frac{d\vec{\omega}}{dt} = \frac{d0}{dt} = 0$. For the continuity equation, we have

$$\nabla \cdot \mathbf{u} = 0 \Rightarrow \nabla \cdot \mathbf{u} = \nabla \cdot (\nabla \Phi) = 0 \Rightarrow \nabla^2 \Phi = 0 \tag{5.6}$$

This is called the Laplace equation, and for a 2D wave field can be expanded as follows:

$$\nabla^2 \Phi = \frac{\partial^2 \Phi}{\partial x^2} + \frac{\partial^2 \Phi}{\partial z^2} = 0 \tag{5.7}$$

In linear wave theory, it is assumed that the flow is inviscid, incompressible, and irrotational; therefore, a potential function, which satisfies the periodic boundary conditions, will be the solution.

Consider a sinusoidal progressive wave (Fig. 5.1). The basic parameters such as wave length (L), wave height (H), still water depth (d), and water surface elevation (η) are presented in this figure. For a progressive wave, it can be shown that the potential function is given by

$$\Phi = \frac{H}{2}\frac{g}{\sigma}\frac{\cosh k(d+z)}{\cosh kd}\sin(kx - \sigma t) \tag{5.8}$$

where k is the wave number and σ is the wave angular frequency. Although we skipped the derivation of this equation, let us just show that the above solution satisfies the Laplace equation. Taking the derivatives of the potential function leads to

$$\frac{\partial^2 \Phi}{\partial x^2} = -k^2 \left[\frac{H}{2}\frac{g}{\sigma}\frac{\cosh k(d+z)}{\cosh kd}\sin(kx - \sigma t)\right] \tag{5.9}$$

$$\frac{\partial^2 \Phi}{\partial z^2} = k^2 \left[\frac{H}{2}\frac{g}{\sigma}\frac{\cosh k(d+z)}{\cosh kd}\sin(kx - \sigma t)\right] \tag{5.10}$$

$$\Rightarrow \frac{\partial^2 \Phi}{\partial x^2} + \frac{\partial^2 \Phi}{\partial z^2} = 0 \tag{5.11}$$

Also, the potential function (Eq. 5.8) is periodic. This is because $\Phi(x, z, t) = \Phi(x, z, t + T) = \Phi(x, z, t + \frac{2\pi}{\sigma})$, which is demonstrated as follows

$$\sin\left(kx - \sigma\left[t + \frac{2\pi}{\sigma}\right]\right) = \sin\left(kx - \sigma t - 2\pi\right) = \sin(kx - \sigma t) \tag{5.12}$$

TABLE 5.1 Wave Properties Based on Linear Wave Theory

Wave Property	Equation
Velocity potential	$\Phi = \frac{H}{2}\frac{g}{\sigma}\frac{\cosh k(d+z)}{\cosh kd}\sin(kx-\sigma t)$
Wave celerity	$C = \frac{L}{T} = \frac{\sigma}{k}$
Horizontal component of velocity	$u = \frac{\partial\Phi}{\partial x} = \frac{H}{2}\frac{gk}{\sigma}\frac{\cosh k(d+z)}{\cosh kd}\cos(kx-\sigma t)$
Vertical component of velocity	$v = \frac{\partial\Phi}{\partial z} = \frac{H}{2}\frac{gk}{\sigma}\frac{\sinh k(d+z)}{\cosh kd}\sin(kx-\sigma t)$
Wave surface displacement	$\eta = \frac{H}{2}\cos(kx-\sigma t)$
Pressure (hydrostatic + dynamic)	$p = -\rho gz + \rho g\frac{\cosh k(d+z)}{\cosh kd}\left(\frac{H}{2}\cos(kx-\sigma t)\right) = p_O + p_D$
Horizontal acceleration	$a_x = \frac{\partial u}{\partial t} = \frac{gkH}{2}\frac{\cosh k(d+z)}{\cosh kd}\sin(kx-\sigma t)$
Vertical acceleration	$a_z = \frac{\partial w}{\partial t} = -\frac{gkH}{2}\frac{\sinh k(d+z)}{\cosh kd}\cos(kx-\sigma t)$
Dispersion	$C^2 = \frac{g}{k}\tanh kd$

Using linear wave theory, all of the hydrodynamic variables, including velocity, pressure, water surface elevation, and wave celerity, can be derived from the potential function and implementation of the boundary conditions. Table 5.1 summarizes the wave properties based on linear wave theory.

5.1.2 Relationship Between Wave Celerity, Wave Number, and Water Depth: The Dispersion Equation

Wave celerity is dependent on water depth and wave length, and important wave processes such as wave refraction can be explained by the dependence of wave celerity on depth (Section 5.2.2). The relation of wave celerity and water depth can be derived by implementing the (kinematic) free surface boundary condition, which implies that the vertical speed of water particles at the water surface is equal to the speed of the free surface (i.e. $v = \frac{d\eta}{dt} = \frac{\partial\eta}{\partial t} + u\frac{\partial\eta}{\partial t}$ at the water surface). Assuming a small amplitude wave, it results in

$$C^2 = \frac{g}{k}\tanh kd \Rightarrow \sigma^2 = gk\tanh kd \quad (5.13)$$

which is called the *dispersion equation*, because waves of different celerities will be dispersed according to their wave length. Looking at Fig. 5.2, the hyperbolic tangent function ($\tanh kd$) in the dispersion equation approaches 1 for large depth values (deep water waves) and kd for small values of depth (shallow water waves). Therefore,

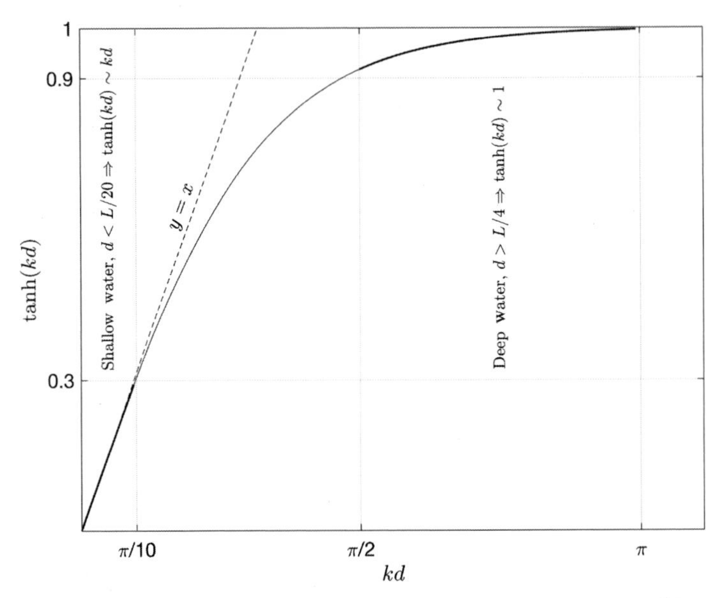

FIG. 5.2 Simplifying the dispersion equation (5.13) in deep and shallow water conditions.

$$C^2 = \frac{g}{k} \Rightarrow \sigma^2 = gk: \quad \text{deep water waves} \tag{5.14}$$

$$C^2 = \frac{g}{k}(kd) \;\Rightarrow\; C^2 = gd: \quad \text{shallow water waves} \tag{5.15}$$

Note that the shallow water wave approximation is nondispersive, because it does not relate the wave celerity to wave length.

5.1.3 Wave Energy and Wave Power

Assume that a regular wave, with a wave period T and wave height H, is propagating in constant water depth d. Power (the rate of energy conversion) is the product of velocity and force. In linear wave theory, the force is generated by pressure (i.e. $F = pA$). Referring to Table 5.1, pressure depends on z. Therefore, wave power can be evaluated by multiplying the pressure and velocity and integrating over depth; the wave power, or more specifically the average energy flux per unit width over a wave period, is given by

$$P = \int_0^T \int_{-d}^{\eta} p_{\mathrm{D}} u \, dz dt = \frac{1}{8}\rho g H^2 C \left\{ \frac{1}{2}\left(1 + \frac{2kh}{\sinh 2kh}\right)\right\} = EC_{\mathrm{g}} = \mathrm{cst} \tag{5.16}$$

$$C_{\mathrm{g}} = C\left\{\frac{1}{2}\left(1 + \frac{2kh}{\sinh 2kh}\right)\right\} \tag{5.17}$$

where p_D is the dynamic pressure and u is the horizontal velocity given in Table 5.1. Note that wave power is conserved in linear wave theory (i.e. the RHS of Eq. (5.16) is constant).

In the above formulation, $E = \frac{1}{8}\rho g H^2$ is the total mechanical energy (i.e. potential and kinetic) of waves per unit surface area averaged over a wave period. In other words, wave power is simply the product of wave energy and wave group velocity. The group velocity C_g (which is dependent on the wave celerity C, wave number k, and water depth) is the speed of wave energy propagation. Referring to the dispersion equation, it is also defined as [1]

$$C_g = \frac{\partial \sigma}{\partial k} = \frac{\partial}{\partial k}\sqrt{gk \tanh(kh)} \tag{5.18}$$

In realistic ocean environments, ocean currents, which are generated by tides, wind, or temperature, affect the propagation of wave energy. Tidal waves and wind waves can be regarded as long and short waves, which interact in various ways. The Doppler shift (which is explained in Chapter 7) is an example where the frequency of waves is influenced by currents:

$$\omega = \sigma + ku \rightarrow \frac{\partial \omega}{\partial k} = \frac{\partial \sigma}{\partial k} + u \tag{5.19}$$

where σ is the relative wave frequency (observed in a coordinate system moving with the same velocity as the ambient current), ω is the absolute wave frequency (observed in a fixed frame), and u is the ambient current velocity. The wave energy flux when we have ambient currents becomes

$$C_g^* = C_g + u \rightarrow EC_g^* = EC_g + uE \neq cst \tag{5.20}$$

where C_g^* is the group velocity in the presence of ambient currents, EC_g is the wave energy transport by the group velocity, and uE is the wave energy transport by tidal currents. When waves propagate in the presence of currents, the wave energy flux is no longer conserved. This happens due to the exchange of energy between wave and current fields. To avoid this issue, wave models apply the conservation of the wave action (E/σ), rather than conservation of wave energy in their formulations (e.g. [2]). Nevertheless, the total energy flux due to waves and currents is conserved. Therefore, the conservation of energy, in a more general way, can be expressed as follows [3]

$$\left[EC_g + Eu\right] + \left\{\frac{1}{2}\rho g h u^3\right\} + \left\{u\left(2\frac{C_g}{C} - \frac{1}{2}\right)E\right\} = \mathrm{cst} \tag{5.21}$$

It is clear that when ambient currents are zero, the earlier equations reduce to $P = EC_g = \mathrm{cst}$.

5.1.4 Irregular Waves

So far, we have considered a wave that is characterized by a single frequency and direction (i.e. a monochromatic wave). Monochromatic waves can be generated in laboratories (wave tanks), but are rarely seen in the ocean. Waves, which have been generated by wind, are irregular and include many wave frequencies. As 'wind waves' travel over long distances, they are dispersed, because the wave celerity/speed is dependent on the frequency. These waves are called 'swell waves' and are more similar to regular waves.

Within the range of validity of linear wave theory (see the next section), we can superimpose several monochromatic/sine waves to construct a more realistic irregular wave. Conversely, we can decompose an irregular wave into a number of sinusoidal waves, which have different frequencies. Mathematically speaking, this is the concept of the Fourier series/integral, which is based on the principal that any function can be constructed by combining a set of simple sine waves. This is a powerful method, because we can also find the wave properties, such as pressure, acceleration, and energy, by combining/superimposing the hydrodynamic field of these individual sine waves.

Two main approaches exist to deal with irregular waves: statistical and spectral. In the statistical approach, statistical variables are calculated and used to represent irregular waves. For instance, parameters such as the average wave period, and the root mean square wave height, can be used to represent the wave period, and the wave height, respectively, for a realistic sea state. In this approach, the sea state is represented by a probability distribution function such as Rayleigh, and statistical parameters are derived from that distribution. The spectral method is based on the concept of the Fourier transform and is the basis of spectral wave models, such as SWAN [2], TOMAWAC [4], or STWAVE [5], which are commonly used for wave energy resource characterization. A detailed discussion about irregular waves is beyond the scope of this book and can be found in many other resources (e.g. [1,6]). Here, basic concepts, with a focus on wave power, are presented.

As mentioned, we can combine several waves with a range of frequencies and phases, to construct an irregular wave, or vice versa. Mathematically speaking, this means

$$\eta(t) = \sum_{i=1}^{N} a_i \sin(\sigma_i t + g_i) \tag{5.22}$$

where a_i and g_i are the amplitude and phase of an individual sine wave, respectively, and η represents the time series of an irregular wave. The total energy of this irregular wave is found by adding up the energy of all sine waves as follows

$$E = \frac{1}{2}\rho g \sum_{i=1}^{N} a_i^2 = \frac{1}{8}\rho g \sum_{i=1}^{N} H_i^2 \tag{5.23}$$

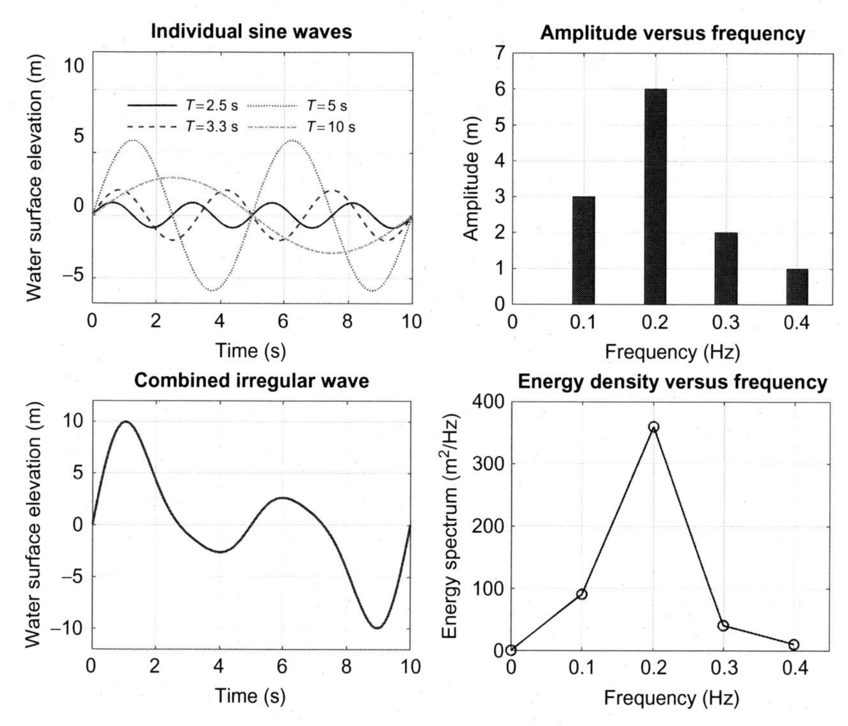

FIG. 5.3 The concept of wave energy spectrum; decomposing an irregular wave into several sine waves.

Each individual sine wave has a different frequency, and different wave energy, which contributes to the total wave energy. Fig. 5.3 shows an example. An irregular wave has been decomposed into four sine waves, with frequencies of 0.1, 0.2, 0.3, and 0.4 Hz and varying amplitudes. The distribution of the amplitudes of these waves versus frequency is shown in this figure. The plot of the energy of these sine waves (or the square of the amplitude) with respect to the frequency represents the energy spectrum. In other words, the energy spectrum shows how the total energy of an irregular wave is distributed amongst various frequencies. If we use many sine waves to decompose the time series of an irregular wave, the difference between the frequencies of these individual sine waves approaches zero; this will lead to a continuous wave spectrum. A continuous energy density spectrum is defined as

$$E_{\text{tot}} = E(\delta\sigma) + E(2\delta\sigma) + \cdots + E(\infty) = \int_0^\infty E(\sigma)d\sigma \tag{5.24}$$

where $E(\sigma)$ is called the 'energy density spectrum', because it is energy per unit angular frequency (J/rad). Alternatively, the 'variance density spectrum' ($S(\sigma)$; m^2/rad) can be used to represent the distribution of energy with respect to frequency. It is defined as,

$$E_{\text{tot}} = \rho g \int_0^{\infty} S(\sigma) d\sigma = \rho g \sum_1^N S(\sigma_i)\delta(\sigma) \rightarrow E(\sigma) = \rho g S(\sigma) \tag{5.25}$$

Considering Eqs (5.23), (5.25), we can see that the variance density spectrum represents the square of the amplitude of each individual sine wave:

$$\frac{1}{2}a_i^2 = S(\sigma_i)\delta\sigma \tag{5.26}$$

The moments of the variance density spectrum are frequently used to compute statistical wave properties. The rth moment (m_r) is defined as

$$m_r = \int_0^{\infty} \sigma^r S(\sigma) d\sigma \tag{5.27}$$

For instance, the significant wave height H_{mo}, which is the mean of the highest one-third of waves in a record, is given by

$$H_{\text{mo}} = 4\sqrt{m_0} \tag{5.28}$$

Wave Power for Irregular Waves

Let us start with the total wave energy of a wave spectrum. By definition, the wave energy (in Joules) can be computed as

$$E = \rho g \int_0^{\infty} S(\sigma) d\sigma = \rho g m_0 \rightarrow E = \frac{1}{16}\rho g H_{\text{mo}}^2 \tag{5.29}$$

in which we used the following relation

$$H_{\text{mo}} = 4\sqrt{m_0} \rightarrow m_0 = \frac{1}{16}H_{\text{mo}}^2 \tag{5.30}$$

Because irregular waves can be represented by a wave energy spectrum, we can assume that the total wave power is calculated by adding the wave power of individual waves, which construct the wave spectrum. The wave power of an individual wave with a frequency of σ is given by

$$\Delta P = \rho g C_{\text{g}}(\sigma)\Delta S(\sigma) = \rho g C_{\text{g}} S(\sigma)\Delta\sigma = \rho g \frac{\sigma}{k}\left\{\frac{1}{2}\left(1 + \frac{2kd}{\sinh 2kd}\right)\right\} S(\sigma)\Delta\sigma \tag{5.31}$$

Consequently, the total power becomes

$$\begin{aligned} P &= \rho g \int_0^{\infty} C_{\text{g}}(\sigma) S(\sigma) d\sigma \\ &= \int_0^{\infty} \frac{\sigma}{k}\left\{\frac{1}{2}\left(1 + \frac{2kd}{\sinh 2kd}\right)\right\} S(\sigma) d\sigma \quad \text{and} \quad \sigma^2 = gk\tanh(kd) \end{aligned} \tag{5.32}$$

The earlier equation is hard to implement, unless we use spectral wave models or measured data that include the full wave spectrum. Let us formulate the wave power equation for deep waters, in which the dispersion equation is much simpler. As mentioned in Section 5.1.2, in deep waters, $\tanh kd$ approaches 1. Therefore, the dispersion equation and the group velocity are given by

$$\tanh kd \approx 1 \rightarrow \sigma = \sqrt{gk} \tag{5.33}$$

$$\rightarrow C_{\mathrm{g}} = \frac{d\sigma}{dk} = \frac{g}{2}\frac{1}{\sqrt{gk}} = \frac{g}{2\sigma} = \frac{gT}{4\pi} \tag{5.34}$$

Replacing the group velocity in Eq. (5.32), results in

$$P = \rho g \int_0^{\infty} C_{\mathrm{g}}(\sigma) S(\sigma) d\sigma = \rho g \int_0^{\infty} \frac{g}{2\sigma} S(\sigma) d\sigma = \frac{1}{2}\rho g^2 \int_0^{\infty} \sigma^{-1} S(\sigma) d\sigma \tag{5.35}$$

or

$$P = \frac{1}{2}\rho g^2 m_{-1} \tag{5.36}$$

For the case of monochromatic waves in deep waters, the wave power becomes

$$P = C_{\mathrm{g}} E = \frac{gT}{4\pi}\frac{1}{8}\rho g H^2 = \frac{\rho g^2}{32\pi} H^2 T \tag{5.37}$$

If we wanted to use simple wave statistical parameters to compute the wave power (something similar to the previous equation, which only depends on the wave height and the wave period), we can also use the deep water approximation for irregular waves.

Assuming the deep water approximation, the group velocity (which is the speed of propagation of the wave energy) is $C_{\mathrm{g}} = \frac{gT_{\mathrm{E}}}{4\pi}$; we called the period T_{E} the energy wave period, as it determines the average speed of wave energy propagation. Further, the total wave energy of a spectrum is given by Eq. (5.29) in terms of m_{o} or the significant wave height. Therefore, we can write

$$P = C_{\mathrm{g}} E = \frac{gT_{\mathrm{E}}}{4\pi}(\rho g m_0) = \frac{gT_{\mathrm{E}}}{4\pi}\left(\frac{1}{16}\rho g H_{\mathrm{mo}}^2\right) \tag{5.38}$$

$$P = \frac{\rho g^2}{64\pi} H_{\mathrm{mo}}^2 T_{\mathrm{E}} = \frac{\rho g^2}{32\pi}\left(H_{\mathrm{mo}}/\sqrt{2}\right)^2 T_{\mathrm{E}} \tag{5.39}$$

which is very similar to Eq. (5.37). In fact, we can use the monochromatic wave equation, if we replace the wave period by the energy wave period, and wave height by $H_{\mathrm{mo}}/\sqrt{2}$, which is called H_{rms} or the root-mean-square wave height.

Finally, we can find out how the energy wave period can be calculated based on a wave spectrum. By comparing Eqs (5.36), (5.38),

$$P = \frac{gT_E}{4\pi}(\rho g m_0) = \frac{1}{2}\rho g^2 m_{-1} \rightarrow T_E = 2\pi \frac{m_{-1}}{m_0} \tag{5.40}$$

which is equivalent to T_{m-10} in spectral models such as SWAN.

In brief, wave power can either be computed using the wave spectrum (i.e. Eq. 5.32) or by using a simplified equation based on the statistical wave parameters (energy/average wave period and significant wave height; Eq. (5.39)). The simplified method is based on the deep water wave approximation, which is generally valid in the vicinity of WECs. The range of water depths where a WEC is installed depends on the device (40 m is a typical depth). Consider a wave with a period of 8 s, propagating in 40 m water depth. The wave length is (from the dispersion equation) $L = 2\pi/k = 99$ m, and therefore $kd = 2.54 < \pi$; for this wave, $\tanh(kd) = 0.98$ which is close to 1 and hence, the deep water approximation is valid.

5.1.5 Nonlinear Waves

Linear wave theory is the basis of many analytical methods and numerical models, which describe the properties and propagation of regular/irregular waves in the open ocean and coastal regions. Nevertheless, we should be aware of the range of validity of this theory, and wave processes which cannot be explained by this theory. Referring to Section 5.1.1, several assumptions were made to develop linear wave theory, which may not be valid in real-word applications. In particular, in the derivation of linear wave theory, it is assumed that the amplitude of the wave is small compared with the water depth and wave length. Fig. 5.4 shows the range of validity of linear wave theory. As you can see, for deep waters $\left(\text{e.g. } \frac{d}{gT^2} > 10^{-3}\right)$ and small amplitude waves $\left(\text{e.g. } \frac{H}{gT^2} < 10^{-3}\right)$, linear wave theory is valid; however, in shallow waters and for large amplitude waves, other wave theories such as Stokes and Cnoidal are more appropriate. Note that in certain depths, or for certain wave steepnesses, the wave is no longer stable and will break. This is discussed further in the following section, and is shown by breaking criteria on this figure (shallow and deep criteria).

Wave Breaking

When the wave height, and consequently wave steepness (i.e. the ratio of wave height to wave length) increase, the water surface profile deviates from a sinusoidal wave shape. The wave eventually breaks if the wave steepness becomes too high.

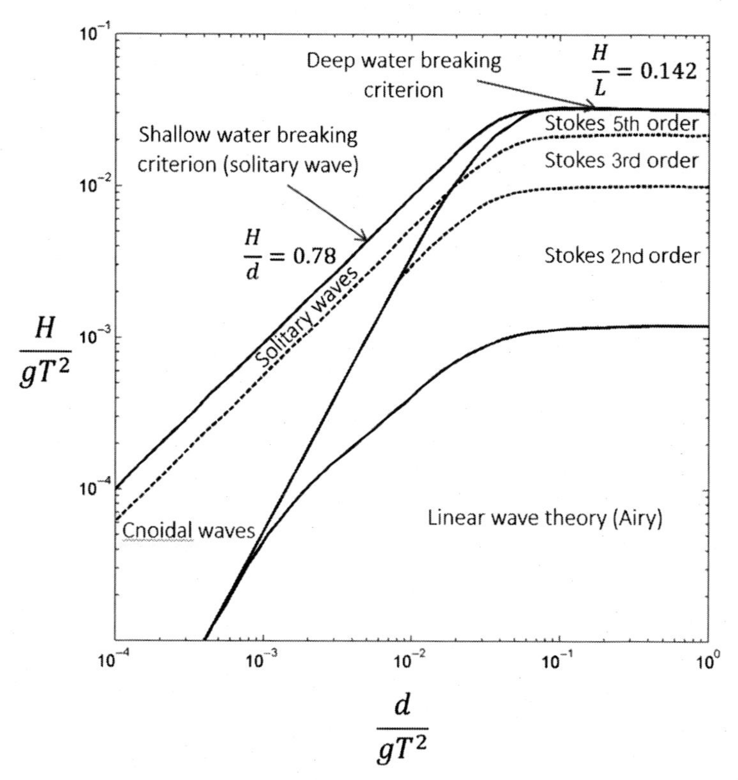

FIG. 5.4 Range of validity of various wave theories. The controlling factors are water depth (x axis), wave height (y axis), and wave breaking. *(Based on a figure first presented by B. Le Méhauté, An Introduction to Hydrodynamics and Water Waves, Springer, New York, NY, 1976.)*

Miche's breaking limit criterion, which was proposed in 1944, can approximately predict wave breaking in deep or shallow waters [7]. It can be written as follows,

$$\frac{kH_b}{\gamma \tanh kd} = 1 \tag{5.41}$$

where H_b is the breaking wave height and γ is a constant parameter. In shallow water, wave breaking occurs due to a decrease in water depth and shoaling (see Section 5.2.1). Looking at Fig. 5.2, in shallow waters, $\tanh kd \approx kd$, which results in the following breaking limit criterion:

$$\frac{kH_b}{\gamma kd} = 1 \rightarrow \frac{H_b}{d} = \gamma \approx 0.7\text{–}0.8 \tag{5.42}$$

In deep waters (see Fig. 5.2), $\tanh kd \approx 1$; therefore, Miche's formula leads to the following relation:

$$kH_b/2 = \gamma/2 \quad \text{and} \quad \gamma/2 \approx 0.44\text{–}0.55 \tag{5.43}$$

$$\rightarrow \frac{H_b}{L} = \frac{\gamma}{2\pi} \approx 0.14\text{–}0.17 \tag{5.44}$$

which is consistent with what we see in Fig. 5.4, and simply indicates that the wave steepness (i.e. $kH/2$ or H/L depending how it is defined) cannot go beyond a threshold; see other resources for more details about this limit [8].

Nonlinear Dispersion Equation

As we discussed, the linear dispersion equation relates the water depth and the wave frequency to the wave celerity (or wave length). If nonlinear effects are considered, the dispersion equation will also be dependent on the wave height. The nonlinear dispersion equation is given by [9]

$$\sigma^2 = gk \tanh kd \left[1 + \left(k\frac{H}{2} \right)^2 \left(\frac{8 + \cosh 4kd - 2\tanh^2 kd}{8 \sinh^4 kd} \right) \right] \tag{5.45}$$

As can be seen, when $kH/2$ or the wave steepness is small, the previous equation approaches what we already saw in the linear wave theory (Eq. 5.13). Fig. 5.5 shows the comparison between the linear and nonlinear dispersion equations for two wave heights (2 and 3 m), assuming a water depth of 10 m. You can try to reproduce these plots by solving the linear and nonlinear dispersion equations.

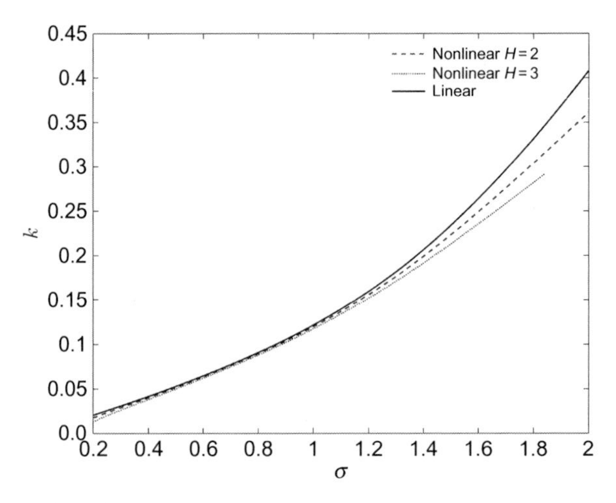

FIG. 5.5 Comparison of linear and nonlinear dispersion equations assuming a water depth of 10 m.

5.2 WAVE TRANSFORMATION DUE TO SHOALING WATER

In general, waves that are generated as a result of a distant storm or localized wind event propagate and 'disperse', relatively unhindered, until they encounter shoaling water. At this stage, two processes influence waves: shoaling and refraction.

5.2.1 Wave Shoaling

As water shoals, remembering that waves transport *energy* in the direction of wave advance, it is important to make use of the principal that wave energy flux per unit surface area, EC_g, is constant. This assumes that no energy is dissipated by the wave train, for example, by wave breaking or bottom friction, as it propagates from deep to shallow water. If we apply a subscript '0' to represent *deep water* values, this implies that

$$EC_g = E_0 C_{g0} = \text{constant} \tag{5.46}$$

Therefore, if we have knowledge of 'deep water' wave properties, we can determine the corresponding wave properties in shoaling water.

Consider a horizontally 1D case, in which the wave crests are everywhere parallel to the depth contours (Fig. 5.6). As the waves propagate through

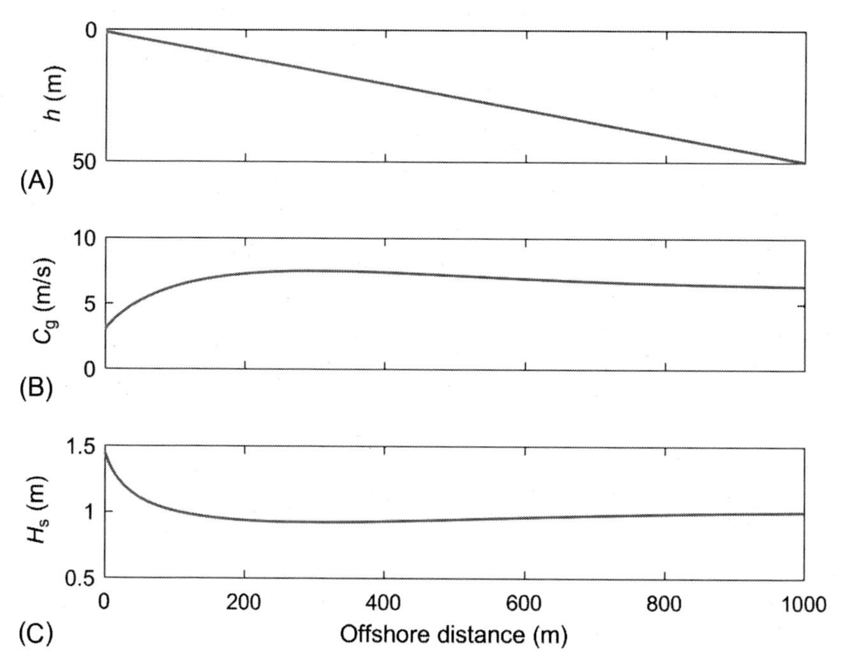

FIG. 5.6 Shoaling wave with an offshore wave height of 1 m (and period 8 s), assuming that waves do not break. (A) Water depth; (B) group velocity; (C) wave height.

TABLE 5.2 Wave Transformation From Deep to Shallow Water, for $T = 8$ s

Environment	L (m)	C (m/s)	C_g (m/s)
Deep	100	12.5	6.2
Shallow	25	3.1	3.1

water of gradually decreasing depth (shoaling water), their phase speed, and consequently their group velocity, reduces (Eq. 5.13), but their frequency tends to remain constant—in other words, waves do not catch one another up in the same way that they would if they were 'dispersing'. By way of illustration, for a wave period $T = 8$ s we can calculate the wave properties shown in Fig. 5.6 and Table 5.2 between 'deep' and 'shallow' water.

Therefore, a fourfold reduction in wavelength and phase speed is accompanied by a halving of the group velocity. Since $E = 1/2\rho g a^2$, when C_g is halved, it follows that E is doubled (Eq. 5.46), and so wave amplitude increases by a factor of $\sqrt{2}$ (Fig. 5.6C). This is known as *wave shoaling*. Note that this illustration is for a 1D case (Fig. 5.6A), but for a more generalized case, another process can complicate wave transformation in shoaling water—*refraction*.

5.2.2 Wave Refraction

In the earlier shoaling example, the wave crests were parallel to the depth contours, and the wave propagated normal to the coastline. In most situations, waves propagate obliquely to the depth contours, and are observed to slowly change direction as they approach the coast. This gradual change in wave direction is known as *refraction*.

Examining Fig. 5.7, which shows wave crests propagating at an oblique angle to the depth contours, there is a variation in water depth experienced along each of the wave crests. Whereas further offshore the wave crest is propagating in relatively deep water, the part of the wave crest closer to the coast is propagating in relatively shallow water. Since wave celerity is related to water depth through the dispersion equation (5.13), this means that the portion of the wave crest in deeper water is travelling faster than the portion of the wave crest that is travelling in shallower water. As a result of this change in phase speed along each wave crest, in a given time interval the crest moves over a larger distance in deeper water than it does in shallower water, and so the waves tend to turn (i.e. refract) towards the depth contours as they propagate.

Variation of ψ (the angle of incidence of the wave orthogonal to the outward normal of the coastline; see Fig. 5.7) is governed by Snell's law. This is the

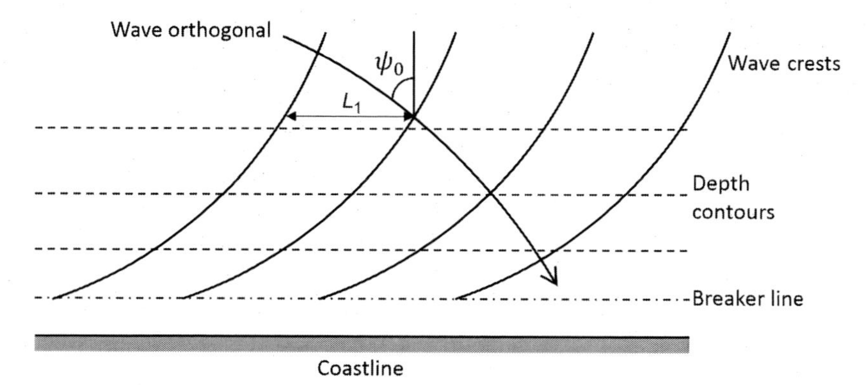

FIG. 5.7 Wave crests that are at an oblique angle to depth contours (*dashed lines*) refract as they propagate towards the coast.

familiar law from studies of optics and, in the present context, expresses the constancy of wavenumber k in the direction parallel to the shore. In other words, the distance between successive crests measured parallel to the shoreline, L_1, remains constant as the waves travel into shallow water (Fig. 5.7). This condition may be written as

$$k \sin \psi = k_0 \sin \psi_0 = \text{constant} \tag{5.47}$$

Based on Eq. (5.47), refraction diagrams can be generated for given offshore wave conditions and maps of bathymetry/topography. However, such analysis is based on the following assumptions [10]:

- The wave energy between wave rays (orthogonals) remains constant.
- The direction of wave advance is perpendicular to the wave crest.
- The speed of the wave of a given period depends only on the water depth at that location.
- Changes in bathymetry are gradual.
- Waves are long-crested, constant-period, small-amplitude, and monochromatic.
- Effects of currents, winds, and reflections from beaches, and underwater topographic features, are considered negligible.

A typical refraction diagram, for a wave of period $T = 10$ s, is shown for an irregular coastline (a headland and two bays) in Fig. 5.8. The wave orthogonals converge at the headland and diverge in the bays. Because wave energy between the wave orthogonals remains constant (see the earlier assumptions), this implies that the wave height will generally increase at the headland and reduce in the two bays.

As we saw in Eq. (5.46), wave energy flux is a constant, provided no energy is dissipated, for example, by wave breaking or friction. Now, in the more

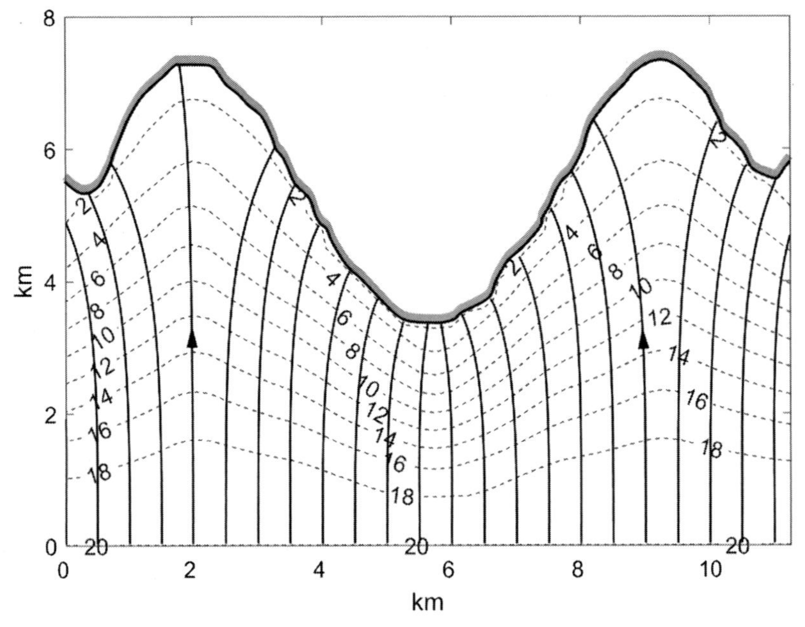

FIG. 5.8 Wave refraction along an irregular coastline. *Dashed lines* represent depth contours, and *solid lines* that propagate towards the coast are the wave orthogonals.

general case, this flux towards the coast is a constant per unit distance parallel to the shoreline

$$\overline{E}C_{\mathrm{g}} \cos \psi = \text{constant} \tag{5.48}$$

Therefore, the variation in wave amplitude may be expressed in terms of deep water reference values as follows

$$a = a_0 \left[\frac{C_{\mathrm{g}0}}{C_{\mathrm{g}}} \right]^{1/2} \left[\frac{\cos \psi_0}{\cos \psi} \right]^{1/2} \tag{5.49}$$

This result shows that, since $\psi_0 > \psi$ in general, waves approaching a coast obliquely are amplified less than those approaching a coast normally (where $\psi = \psi_0 = 0$).

5.3 DIFFRACTION

Diffraction is the process by which wave energy may be transferred laterally along a wave crest, or may be reflected (or scattered). These possibilities are not allowed in refraction. Diffraction occurs in cases in which the assumption of small bottom slope breaks down or in which obstacles or steep-walled boundaries are present.

Consider a wave travelling in water of constant depth, around a headland or breakwater (Fig. 5.9A). In the absence of refraction (since the bed is flat), the

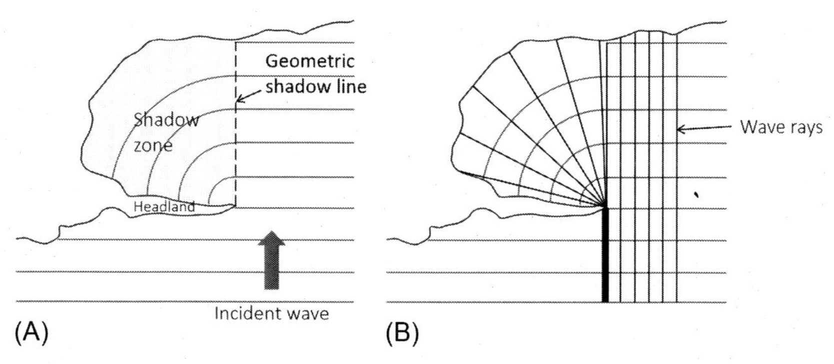

FIG. 5.9 Wave diffraction around a headland (A) with a circular pattern of wave crests in the shadow zone, and (B) represented by wave rays curving around the headland. *(Based on a figure presented by L.H. Holthuijsen, Waves in Oceanic and Coastal Waters, Cambridge University Press, Cambridge, 2010.)*

waves will travel into the shadow of the obstacle in an almost circular pattern of crests with rapidly diminishing amplitudes [6]. Due to the shadowing effect of the headland, large variations in amplitude will occur across the geometric shadow line of the headland. If diffraction were ignored, the wave would propagate along straight wave rays (since depth is constant), no energy would cross the shadow line, and no waves would penetrate the shadow area behind the headland. With diffraction accounted for, the wave rays curve into the shadow area behind the headland (Fig. 5.9B).

5.4 WAVE ENERGY CONVERTERS

5.4.1 Technology Types

There are many WEC technologies (it has been estimated that there are over 1000 device patents), but they can mostly be grouped into one of five technology types:

- Attenuator
- Surface point absorber
- Oscillating wave surge converter
- Oscillating water column
- Overtopping devices

The main features of these WEC types are summarized in Fig. 5.10.

Attenuator

An attenuator is a long-floating device that is aligned with the direction of wave propagation. The device captures the energy of the wave by selectively constraining the movements along its length. The best known attenuator device, and probably the most recognizable of all WECs, is Pelamis.

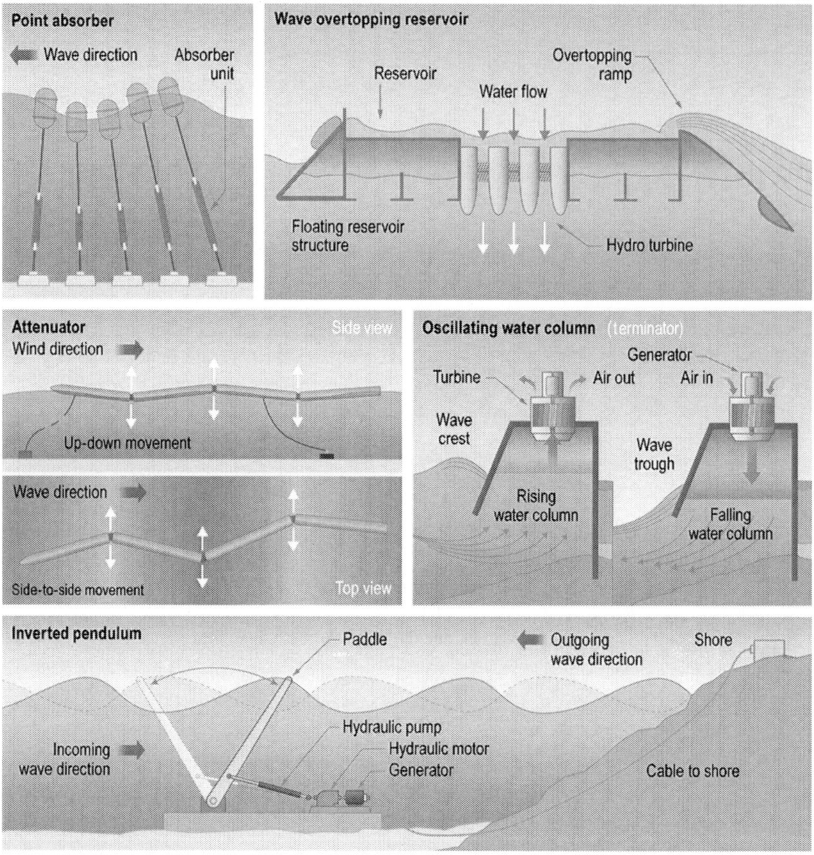

FIG. 5.10 Contrasting WEC technologies [11]. Note that the 'inverted pendulum' is more commonly known as the oscillating wave surge converter.

Surface Point Absorber

Surface point absorbers are floating structures that absorb wave energy from all directions (i.e. in contrast to an attenuator device). Point absorbers have relatively small dimensions relative to typical wavelengths, and so have diameters of a few metres. Examples of point absorbers are WaveBob and PowerBuoy (Fig. 5.11).

Oscillating Wave Surge Converter

Oscillating wave surge converters (OWSCs) are formed of near-surface collectors mounted on an arm pivoted near the seabed. The arm oscillates as an inverted pendulum due to the movement of water particles in the waves. These devices are also known as inverted pendulums. A well-known example of an OWSC is the Oyster.

FIG. 5.11 PowerBuoy: an example of a point absorber. *(Copyright and courtesy of Ocean Power Technologies.)*

Oscillating Water Column

Oscillating water column (OWC) devices are partially submerged, hollow structures, open to the sea below the water surface so that they contain air trapped above a column of water. Waves cause the column to rise and fall, acting like a piston, compressing and decompressing the air. The air is channelled through an air turbine to generate electricity. One example OWC is the Limpet.

Overtopping Devices

Overtopping devices consist of a wall over which the waves wash, collecting the water in a storage reservoir. The incoming waves create a head of water, which is released back to the sea through conventional low-head turbines installed at the bottom of the reservoir. An overtopping device may use collectors to concentrate wave energy. An example of an overtopping device is the Wave Dragon.

5.4.2 Comparison Between WEC Technologies

Issues associated with WECs are survivability (because they will generally be sited in locations where the wave climate is, by its very nature, extreme), and the cost associated with designing the devices to survive such conditions. As an example of the scale of these devices, data extracted from Previsic [12] provides a comparison (Table 5.3).

The scale is immense—every kilowatt of rated power requires around 0.7 tonnes of steel.[2]

5.4.3 Basic Motions of WECs

An object, which is floating on the water surface, or submerged underwater, generally has six degrees of freedom—provided that no constraint prevents its motion. It can translate along or rotate around the three major axes. Assuming that waves are propagating along the x axis (Fig. 5.1), the three translational motions are called, surge (x), sway (y), and heave (z). The rotational motions are called rolling, pitching, and yawing around x, y, and z axes, respectively.

5.4.4 Theory of Heaving Point Absorbers

Mass-Spring-Damper

The motion of a WEC can be simplified and studied using *vibration theory*. In general, most vibrational motions can be simulated/explained by the balance of a restoring force, a damping force, and external forces acting on a body. Depending on the nature of the external forces, a vibration can be periodic or random, similar to what we saw in wave theory (i.e. regular and random waves).

TABLE 5.3 Scale Comparison of Various WEC Technologies

Technology	Length or Height (m)	Width or Diameter (m)	Weight (tonnes)	Rated Power (kW)
Pelamis	180	6	680	750
Power Buoy	10	11	136	150
Wave Dragon (48 kW/m)	220	390	49,000	12,000
Oyster	13	26	408	800

2. Excluding the Wave Dragon, which is a considerably different technology type to the others listed in Table 5.3.

Assuming just one degree of freedom for motion (e.g. a heaving point absorber), the equation of motion can be written as

$$\sum F = ma_z = -k_s z + F_{\text{damp}} + F_{\text{ext}}(t) \tag{5.50}$$

where k_s is the spring constant, m is the mass, and a_z is acceleration, and F_{damp} and F_{ext} are damping and external forces, respectively. The damping force is usually proportional to the velocity and is expressed as $c_d u_z = c_d \frac{dz}{dt}$, where c_d is velocity in the z direction. The spring force is the restoring force, which is a combination of Archimedes/Buoyancy force and gravity, which tends to return the system to its equilibrium position. Therefore, the single degree of freedom (SDOF), mass-spring-damper equation can be written as

$$m\frac{d^2z}{dt^2} + c_d\frac{dz}{dt} + k_s z = F_a \sin(\omega_F t + \phi) \tag{5.51}$$

Note that in the earlier equation, we assumed that the external force is a simple harmonic force with an amplitude of F_a and angular frequency of ω_F. This makes sense if we assume that the force is generated by a harmonic wave. Eq. (5.51) is a linear ordinary differential equation, and so can be solved analytically. The solution can be found by replacing $z = z_o e^{\xi t}$ into the earlier equation, where z_o is a constant.

Analytical Solution of Free Vibration

First, consider a case where the external force is zero. We refer to that case as *free vibration*. This will be equivalent to a WEC in a calm sea, which is displaced from its equilibrium position. It will gradually be brought back to its equilibrium position after a few oscillations. The governing equation for free vibration reduces to

$$m\frac{d^2z}{dt^2} + c_d\frac{dz}{dt} + kz = 0 \tag{5.52}$$

Replacing $z_o e^{\xi t}$ into the previous equation, leads to

$$\left\{m\xi^2 + c_d\xi + k_s\right\} z_o e^{\xi t} = 0 \tag{5.53}$$

Because $z_o e^{\xi t}$ is not zero, $m\xi^2 + c_d\xi + k_s$ should be zero. Using the quadratic formula, the solution becomes

$$\xi = \frac{-c_d \pm \sqrt{c_d^2 - 4mk_s}}{2m} \quad \text{and} \quad \Delta = c_d^2 - 4mk_s \tag{5.54}$$

Consider a special case, where there is no damping (*undamped system*: $c_d = 0$). For an undamped system,

$$\xi = \frac{0 \pm \sqrt{0 - 4mk_s}}{2m} = \sqrt{-\frac{k_s}{m}}$$

$$= i\sqrt{\frac{k_s}{m}} = i\omega_N \rightarrow z = z_o e^{i\omega_N t} \quad \text{or using the Euler's formula} \quad z = z_o \cos(\omega_N t + g_o) \tag{5.55}$$

where $i = \sqrt{-1}$ and $\omega_N = \sqrt{\frac{k}{m}}$ is the natural frequency of the system. This is equivalent to having a WEC in a calm sea with no damping. When it is displaced from the equilibrium position, the system oscillates indefinitely. The solution for an undamped system is a simple harmonic equation (i.e. $z = z_o \cos(\omega_N t + g_o)$), where the amplitude ($z_o$) and phase ($g_o$) of this harmonic motion depend on the initial condition.

When damping is significant, depending on the sign of the Δ in Eq. (5.54), we have three cases as follows (see Fig. 5.12):

- $\Delta < 0$ or $c_d < 2\sqrt{mk}$ or $c_d < 2m\omega_N$: underdamped system; in this case the system will oscillate, but its amplitude gradually decreases until it rests. In the undamped system that we considered first, also $\Delta < 0$, and the system oscillates; however, the amplitude does not decrease, and theoretically never stops oscillating.
- $\Delta > 0$: overdamped system; due to high friction, the system cannot oscillate and returns to equilibrium quickly.
- $\Delta = 0$ or $c_d^* = 2\sqrt{mk}$: critically damped; where c_d^* is defined as the critical damping coefficient. In this case, the system cannot oscillate and quickly returns to equilibrium, similar to the overdamped system.

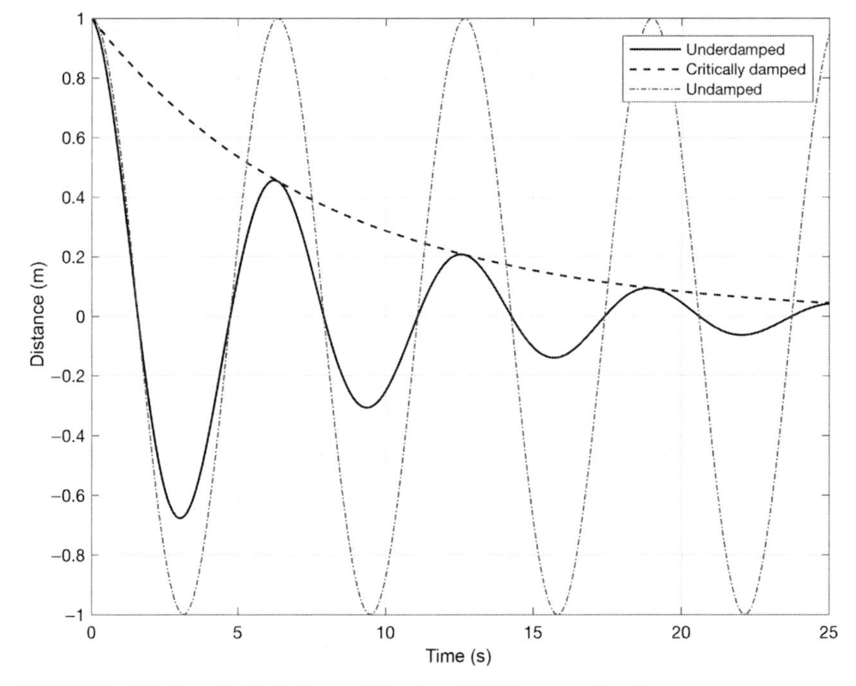

FIG. 5.12 Solution of a single degree of freedom (SDOF) mass-spring damper.

WECs, in general, are underdamped. In the absence of external wave forces, a WEC will oscillate and gradually stop due to frictional damping. The solution of an underdamped system can be found using Eq. (5.54). In order to better formulate the solution, we define the damping ratio (ζ) as

$$\zeta = \frac{c_d}{c_d^*} = \frac{c_d}{2m\omega_N} = \frac{c_d}{2\sqrt{mk}} \rightarrow \zeta < 1 \equiv \Delta < 0 \tag{5.56}$$

Using Eq. (5.54) and the previous definitions for damping ratio and natural frequency, it can easily be shown that

$$\xi = -\zeta\omega_N \pm \left(\sqrt{-1}\sqrt{1-\zeta^2}\right)\omega_N \tag{5.57}$$

and the solution of an underdamped system becomes

$$z = z_o e^{\xi t} = z_o e^{-\zeta\omega_N t} \cos\left(\sqrt{1-\zeta^2}\omega_N t + g_1\right) \tag{5.58}$$

where z_o and g_1 are constants, and depend on the initial conditions. When the damping ratio is zero, the previous solution reduces to an undamped system (Eq. 5.55). When the damping ratio is between 0 and 1 (i.e. underdamped system), the term $e^{-\zeta\omega_N t}$ leads to an exponential decay of the amplitude as can be seen in Fig. 5.12. Finally, when the damping ratio is 1 (i.e. critically damped), the solution is $z_o \cos(g_1)e^{-\zeta\omega_N t}$, which is an exponential decay line.

Analytical Solution of Forced Vibration

In the presence of an external force, the solution of the differential equation is a combination of the solution for free vibration and a particular solution. In the theory of ordinary differential equations, the solution of the free vibration (RHS = 0) is also called *homogeneous* solution, and the forced vibration is called *nonhomogeneous* solution (RHS $\neq$ 0).

In a similar way to free vibration, the solution of the forced vibration case (Eq. 5.51) can also be found by replacing $z = z_1 e^{\xi t} = z_1 e^{i\omega_F t}$ or $z = z_1 \sin(\omega_F t + g_F)$. Note that here the solution frequency is ω_F, which is the frequency of the external force. We do not present the steps for this case, as it is very similar to what we saw in the free vibration case.

$$z = z_1 \sin(\omega_F t + g_F) \quad \text{where } z_1 = \frac{F_a}{\sqrt{m^2(\omega_N^2 - \omega_F^2)^2 + (c_d\omega_F)^2}} \quad \text{and}$$

$$\tan g_F = -\frac{c_d\omega_F}{m(\omega_N^2 - \omega_F^2)} \tag{5.59}$$

In the previous equation, z_1 is the amplitude of the vibration or response of the system to the external forcing. As we can see, this amplitude is dependent on the frequency of the external force compared with the natural frequency of the system, and also the damping coefficient. For instance, we can see that as ω_F

approaches ω_N, the denominator becomes smaller ($\omega_N^2 - \omega_F^2 \rightarrow 0$), and this leads to a larger response amplitude. Also, a larger damping coefficient leads to an increase of the denominator or smaller amplitude.

To perform a more quantitative analysis, let us reformulate the amplitude in Eq. (5.59) as follows (by replacing $c_d = 2\zeta\sqrt{km}$, Eq. (5.56),

$$z_1 = \frac{\mathrm{Fa}/k_s}{\sqrt{\left(1 - \frac{\omega_F}{\omega_N}^2\right)^2 + \left(2\zeta\frac{\omega_F}{\omega_N}\right)^2}} = \mathrm{Fa}/k_s \times \Pi \tag{5.60}$$

The numerator of the previous equation (Fa/k_s) is constant, but the denominator depends on the frequency of the forcing function and damping ratio. Fig. 5.13 shows how the frequency of the external forcing and damping ratio affects the amplitude of the response function. For an energy absorber, the extracted energy would be proportional to the square of the amplitude (i.e. z_1^2) as we will discuss later. Therefore, it is essential that the frequency of the external forces (i.e. waves) are close to the natural frequency of the energy absorber. This will lead to an important topic of phase control and added mass, which is discussed briefly in the next section.

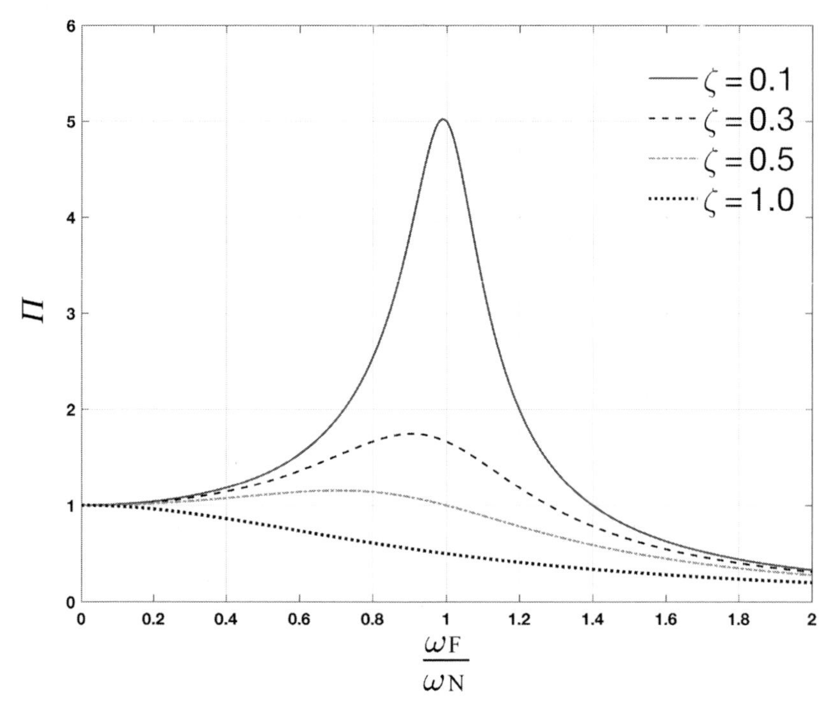

FIG. 5.13 Effect of the forcing function frequency and damping ratio to the amplitude of the vibration. The maximum amplitude corresponds to $\omega_N = \omega_F$.

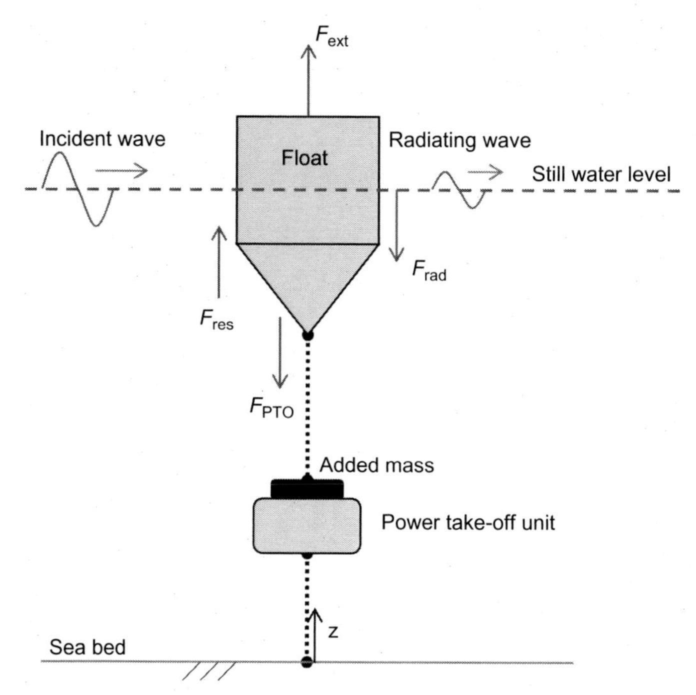

FIG. 5.14 Schematic of a simple point absorber.

Simple Model of a Heaving Point Absorber

Now, consider a simple model of a heaving point absorber as shown in Fig. 5.14. The list of forces acting on the point absorber can be summarized as follows:

$$m\frac{dz^2}{dt^2}_{\text{inertia}} = F_{\text{waves}} + F_{\text{restoring}} + \{F_{\text{hydrodynamic damping}} + F_{\text{power take-off}}\} + F_{\text{tuning/added mass}} \tag{5.61}$$

- Inertia force is proportional to the mass of the floating body.
- Hydrostatic and gravity forces act as a restoring force.
- Damping force: a combination of hydrodynamic damping and power take-off force.
- External force exerted by waves and has a harmonic or random nature.

The previous forces should be formulated using wave theories, or simulated using CFD models.

Considering the simple case of linear wave theory, we start by formulating damping forces. Imagine a heaving point absorber which oscillates in *still water*—the oscillation can begin with an initial displacement of the point absorber from its equilibrium position. If no external force is applied to the point absorber, it will gradually lose its energy and stop after a few oscillation cycles. During oscillation, waves are radiated from the object. These radiated waves are taking energy off the point absorber. According to linear wave theory, the radiating force can be approximated as follows

$$F_{\mathrm{rad}} = -(m_{\mathrm{add}}a_z + b_{\mathrm{hd}}u_z) = -\left(m_{\mathrm{add}}\frac{d^2z}{dt^2} + c_{\mathrm{hd}}\frac{dz}{dt}\right) \tag{5.62}$$

in which the force consists of two parts: added mass inertial force and hydrodynamic damping force. The added mass m_{add} and b_{hd} are dependent on the shape of the point absorber. They are determined by experiment or numerical simulation.

To formulate the restoring force, assume that there is no wave (calm sea) and the point absorber is in equilibrium. The downward force is the weight of the point absorber, and the upward force is the hydrostatic pressure or Archimedes force. When the point absorber is displaced from its equilibrium position, the difference of the Archimedes force and weight tends to bring the system back to equilibrium, and acts as a spring. If the displacement from the equilibrium position is denoted by z, the restoring force would be equal to the mass of the displaced water according to Archimedes law. If the cross-sectional area of the point absorber is A_{a} (waterline area), the weight of the displaced water, or the restoring force, would be

$$F_{\mathrm{s}} = -mg = \rho V_{\mathrm{d}}g = -\rho A_{\mathrm{a}}zg = k_{\mathrm{s}}z \quad \text{and} \quad k_{\mathrm{s}} = \rho A_{\mathrm{a}}g \tag{5.63}$$

The power take-off force, which is used to generate electricity, can be linearly formulated similar to a damping force as follows:

$$F_{\text{power take-off}} = -c_{\mathrm{pto}}\frac{dz}{dt} \tag{5.64}$$

To have more control on the vibration of the point absorber, and to maximize the energy output, we can use tuning or an additional added mass. By using this additional mass, we can change the natural frequency of the system and control the phase lag between the vibration of the point absorber and waves. The inertia force of this additional mass is

$$F_{\mathrm{tuning}} = m_{\mathrm{tun}}\frac{dz^2}{dt^2} \tag{5.65}$$

Replacing all forces into Eq. (5.61), leads to

$$m\frac{dz^2}{dt^2} = F_{\mathrm{w}}\sin(\omega_{\mathrm{F}}t + g_{\mathrm{F}}) - k_{\mathrm{s}}z - \left\{\left[m_{\mathrm{add}}\frac{d^2z}{dt^2} + c_{\mathrm{hd}}\frac{dz}{dt}\right] + c_{\mathrm{pto}}\frac{dz}{dt}\right\} - m_{\mathrm{tun}}\frac{dz^2}{dt^2} \tag{5.66}$$

which after rearranging becomes

$$\{m + m_{\mathrm{add}} + m_{\mathrm{tun}}\}\frac{dz^2}{dt^2} + \{b_{\mathrm{hd}} + b_{\mathrm{pto}}\}\frac{dz}{dt} + k_{\mathrm{s}}z = F_{\mathrm{w}}\sin(\omega_{\mathrm{F}}t + g_{\mathrm{F}}) \tag{5.67}$$

As we can see, the previous equation is exactly equivalent to Eq. (5.51). We can use a similar solution as presented in Eq. (5.59) to find the amplitude and phase of the motion for the point absorber. In order to find the solution, we just need to replace mass (m) with total mass ($m + m_{\text{tun}} + m_{\text{add}}$), and the damping coefficient with $c_{\text{pto}} + c_{\text{hd}}$.

5.5 WAVE RESOURCE ASSESSMENT

For tidal resource assessment, it is generally sufficient to measure and analyse[3] the resource over a relatively short timescale (e.g. 30 days) in order to quantify the resource over a much longer time period (e.g. >1 year). In addition, there is minimal interannual variability in the tidal resource,[4] and so the Annual Energy Yield (AEY) for 1 year will generally provide a good estimate of the AEY for any year. By contrast, waves exhibit variability over a wide range of timescales, from storm (<24 h), to seasonal, and interannual variability. It is therefore essential that any wave resource assessment captures this range of timescales.

Gunn and Stock-Williams [13] analysed 6 years of NOAA WaveWatch III model output to assess the 'theoretical' global wave energy resource (Fig. 5.15). Spatial data of this form, including regional studies at higher resolution [14], are useful for providing an overview of which areas are energetic (e.g. the North Atlantic—Fig. 5.15), and so represent a good opportunity for wave energy conversion, and which regions are less energetic (e.g. the Arabian Sea). Further, many regions, setting aside any practical constraints, can perhaps be identified as being *too energetic*, for example, much of the Southern Ocean, where the wave climate would present significant challenges to device *survivability* (see Section 5.5.2). However, spatial resource assessments of mean wave conditions, even when based on relatively long-time periods (e.g. 6 years used to compile Fig. 5.15), do not give potential developers and investors sufficient information about *temporal variability*.

Temporal wave resource assessment can be performed directly either from observations or from examining time series generated by (validated) numerical models. The problem with relying on instrumentation, for example, directional wave buoys (Section 7.2.1), is, in addition to cost, potential limitations on the length of the deployment available for subsequent analysis. For example, in northwest Europe, it is generally considered that a minimum of around 7 years of data would be required to capture the interannual variability described by the North Atlantic Oscillation (NAO)—a large-scale mode of natural climate variability that has important impacts on the climate of northern Europe [15] (Fig. 5.16). Further, limited measurements during summer months, for example,

3. For example by harmonic analysis.

4. Provided the site is not strongly influenced by wave-tide interaction.

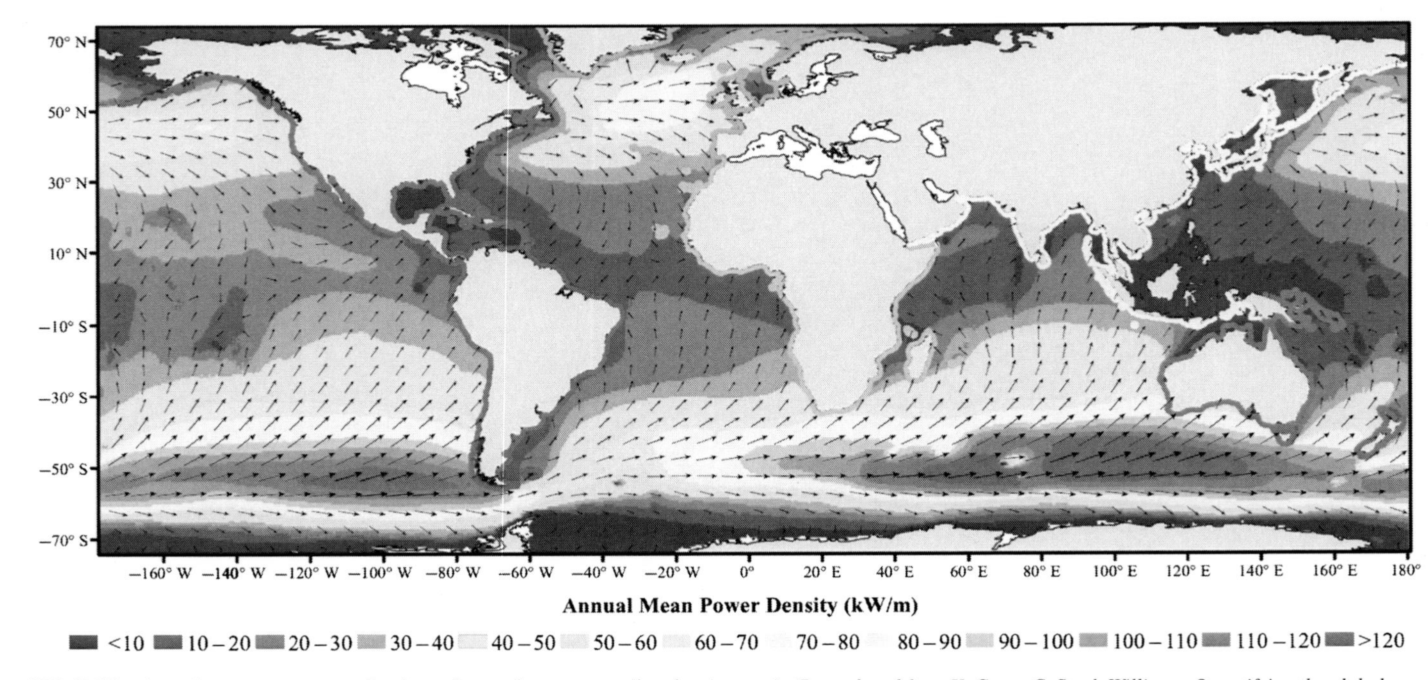

FIG. 5.15 Annual mean wave power density and annual mean wave direction (*vectors*). *(Reproduced from K. Gunn, C. Stock-Williams, Quantifying the global wave power resource, Renew. Energy 44 (2012) 296–304, with permission from Elsevier.)*

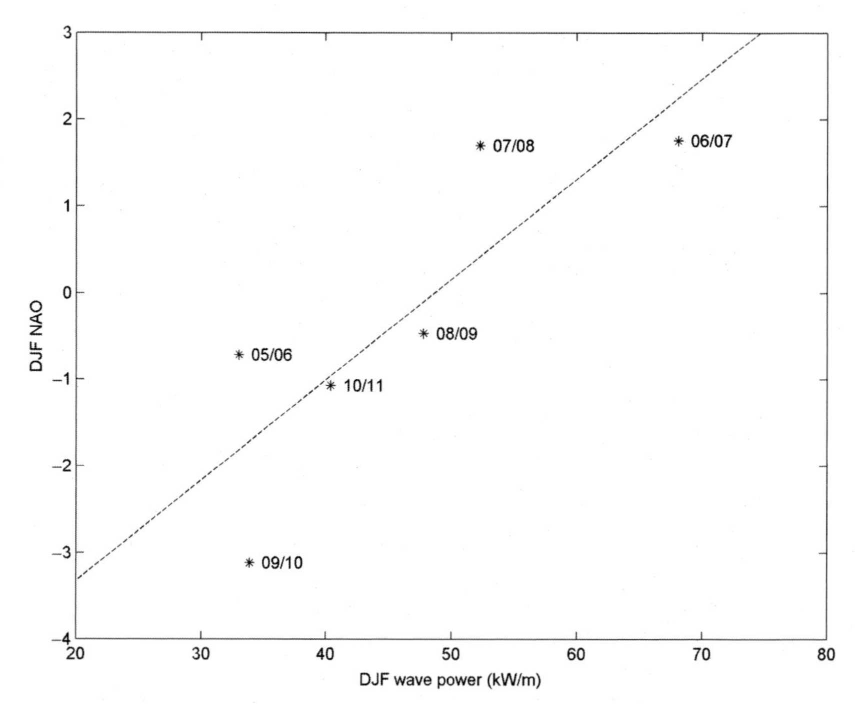

FIG. 5.16 Mean winter (December-January-February) wave power averaged over the NW European shelf seas plotted against the DJF NAO. The *dashed line* is the least squares line of best fit ($r^2 = 0.69$). *(Reproduced from S.P. Neill, M.R. Hashemi, Wave power variability over the northwest European shelf seas, Appl. Energy 106 (2013) 31–46, with permission from Elsevier.)*

cannot be used to infer winter wave conditions due to strong seasonal trends in wave conditions. Although it is important to have local measurements of waves to characterize the resource, in many circumstances it is more usual for limited wave buoy observations to be used to validate a numerical model. The numerical model can then be used to assess the wave resource over much longer timescales.

5.5.1 Theoretical, Technical, and Practical Resources

A wave resource assessment of the type presented in the previous section is what is known as a 'theoretical resource' assessment. A theoretical resource assessment estimates the annual average energy production for a specific wave (or tidal) resource, and is generally based on numerical modelling, with limited validation from historical data records. By contrast, the 'technical' resource is defined as the portion of the theoretical resource that can be captured using a specified technology. It therefore includes device characteristics (e.g. efficiency) and constraints such as interdevice spacings. The calculation of the

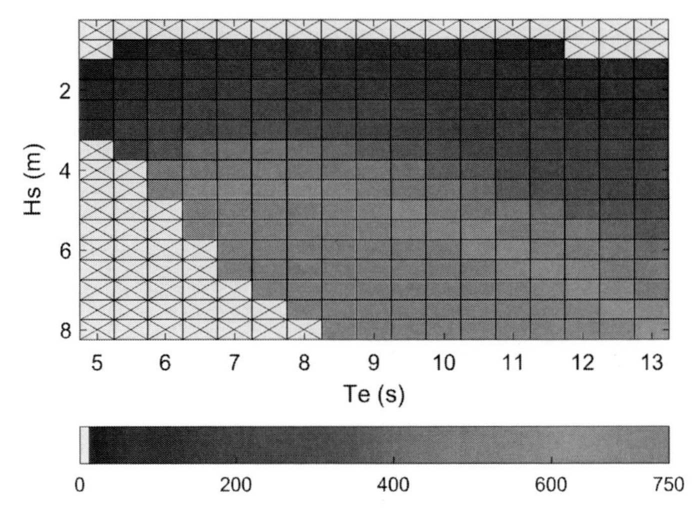

FIG. 5.17 Pelamis power matrix—power output in kW. Hatched regions are either conditions prior to cut-in of the device (low energy) or survival (large or steep waves).

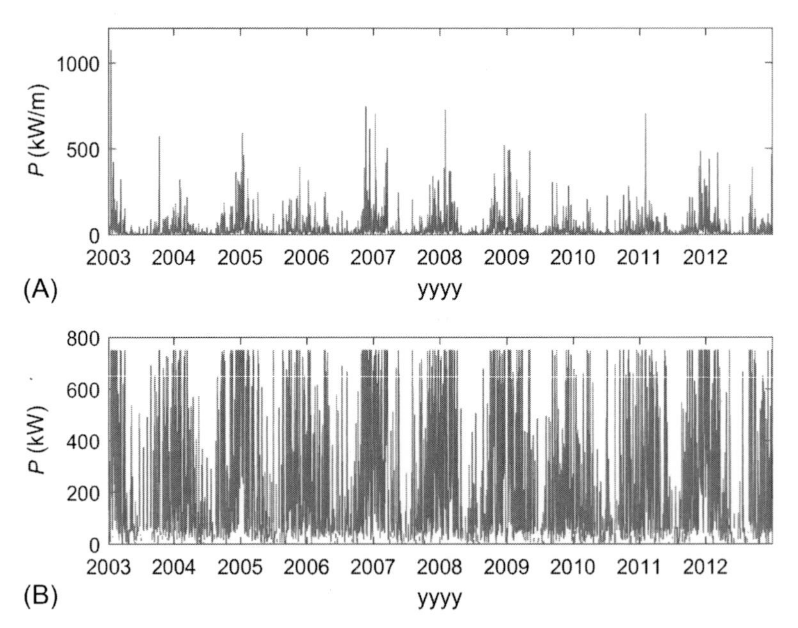

FIG. 5.18 (A) Theoretical and (B) technical wave resources at Bernera, Scotland, based on a single 750 kW Pelamis device applied to a 10-year modelled time series. The wave input data were generated by a modified version of the wave model presented in [15], extended to a longer-time period.

technical wave resource requires access to a power matrix for the selected technology (e.g. Fig. 5.17). A comparison between the theoretical resource and the technical resource (based on a single Pelamis device) at Bernera in the west of Scotland is shown in Fig. 5.18. Although both theoretical and technical resources exhibit strong seasonal and interannual variability, the extremes in the theoretical resource (e.g. regularly over 500 kW/m) are capped by the device characteristics, rated at 750 kW. Out of interest, the capacity factor (see Section 1.5) for the Pelamis device at this location over a decade of simulation is 22%, which is fairly encouraging compared with other renewable technologies (e.g. Table 1.5).

Finally, the practical resource is the portion of the technical resource that is available after considering all constraints such as socioeconomic and environmental factors. The practical resource excludes regions of low power density, regions that are distant from electricity infrastructure, navigation channels, marine protected areas, etc.

5.5.2 Survivability and Maintenance

WECs operating on the water surface are exposed to storms and other extreme events. In particular, high and steep waves, especially breaking waves, are likely to damage WECs. Further, a particular WEC is designed to operate, and convert energy efficiently, over a limited range of conditions (e.g. see the Pelamis power matrix in Fig. 5.17), usually tuned to the most frequent wave conditions that occur at a particular site: extreme waves are not usually within this range. Therefore, it is necessary for most wave devices to have a survival mode; for example, a device could be sunk to a particular depth where wave orbital motion will be reduced (e.g. related to the wave length), or to relax the load on a turbine in an OWC device.

At the other end of the scale, prolonged periods of low wave energy provide opportunities for device maintenance. A site that is consistently energetic may be desirable from theoretical and technical resource perspectives, but is undesirable from a practical resource perspective.

REFERENCES

[1] R.G. Dean, R.A. Dalrymple, Water Wave Mechanics for Engineers and Scientists, vol. 2, World Scientific Publishing, Singapore, 1991.

[2] N. Booij, R.C. Ris, L.H. Holthuijsen, A third-generation wave model for coastal regions: 1. Model description and validation, J. Geophys. Res. Oceans 104 (C4) (1999) 7649–7666.

[3] M.R. Hashemi, S.T. Grilli, S.P. Neill, A simplified method to estimate tidal current effects on the ocean wave power resource, Renew. Energy 96 (2016) 257–269.

[4] M. Benoit, F. Marcos, F. Becq, Development of a third generation shallow-water wave model with unstructured spatial meshing, in: Coastal Engineering 1996, 1997, pp. 465–478.

[5] J.M. Smith, A.R. Sherlock, D.T. Resio, STWAVE: Steady-State Spectral Wave Model User's Manual for STWAVE, Version 3.0, Engineer Research and Development Center, Coastal and Hydraulics Lab, Vicksburg, MS, 2001.

[6] L.H. Holthuijsen, Waves in Oceanic and Coastal Waters, Cambridge University Press, Cambridge, 2010.

[7] J.-F. Filipot, F. Ardhuin, A.V. Babanin, A unified deep-to-shallow water wave-breaking probability parameterization, J. Geophys. Res. Oceans 115 (C4) (2010) C04022.

[8] A. Toffoli, A. Babanin, M. Onorato, T. Waseda, Maximum steepness of oceanic waves: field and laboratory experiments, Geophys. Res. Lett. 37 (5) (2010) L05603.
[9] A. Chawla, J.T. Kirby, Monochromatic and random wave breaking at blocking points, J. Geophys. Res. Oceans 107 (C7) (2002) 3067.
[10] U.S. Army Coastal Engineering Research Center, Shore Protection Manual, vol. 1, US Army Corps of Engineers, Vicksburg, Mississippi, 1973.
[11] C. Augustine, R. Bain, J. Chapman, P. Denholm, E. Drury, D.G. Hall, E. Lantz, R. Margolis, R. Thresher, D. Sandor, N.A. Bishop, S.R. Brown, G.F. Cada, F. Felker, S.J. Fernandez, A.C. Goodrich, G. Hagerman, G. Heath, S. O'Neil, J. Paquette, S. Tegen, K. Young, Renewable electricity generation and storage technologies, in: Volume 2 of Renewable Electricity Futures Study, 2012, NREL/TP-6A20-52409-2.
[12] M. Previsic, The Future Potential of Wave Power in the United States, U.S. Department of Energy, Washington, DC, 2012.
[13] K. Gunn, C. Stock-Williams, Quantifying the global wave power resource, Renew. Energy 44 (2012) 296–304.
[14] S.P. Neill, M.J. Lewis, M.R. Hashemi, E. Slater, J. Lawrence, S.A. Spall, Inter-annual and inter-seasonal variability of the Orkney wave power resource, Appl. Energy 132 (2014) 339–348.
[15] S.P. Neill, M.R. Hashemi, Wave power variability over the northwest European shelf seas, Appl. Energy 106 (2013) 31–46.

FURTHER READING

[1] B. Le Méhauté, An Introduction to Hydrodynamics and Water Waves, Springer, New York, NY, 1976.

Chapter 6

Other Forms of Ocean Energy

Although the focus of this book is tidal, offshore wind, and wave energy, and those topics have been covered in detail in the preceding three chapters, we also recognize that there are other forms of ocean energy. These are primarily ocean currents (specifically western boundary currents), ocean thermal energy conversion (OTEC), and salinity gradient energy. These forms of ocean energy are introduced and discussed in limited detail in this chapter, in which we outline the main principles, technology types, commercial progress, potential environmental impacts, and challenges.

6.1 INTRODUCTION

In addition to the forms of ocean energy that were discussed in detail in the previous three chapters (tidal, offshore wind, and wave), there are three other main types of ocean energy:

- Ocean currents
- Ocean thermal energy conversion (OTEC)
- Salinity gradients

Although these forms of ocean energy conversion are each limited to particular environments (e.g. OTEC is limited to the tropics; see Section 6.3), they all have significant potential for electricity generation.

6.2 OCEAN CURRENTS

Ocean current energy is the kinetic energy that resides in large-scale open-ocean geostrophic surface currents. These currents transfer heat from tropical to polar regions, influence weather and climate, distribute nutrients, and disperse marine organisms [1]. Because these currents are largely driven by the winds, they are actually indirect forms of solar energy.[1] The Ocean Energy Council has estimated that ocean currents have a globally extractable potential of 450 GW,

---

1. Because winds are driven by gradients in air pressure which are a result of variations in solar energy received at the surface of the Earth.

Fundamentals of Ocean Renewable Energy. https://doi.org/10.1016/B978-0-12-810448-4.00006-9

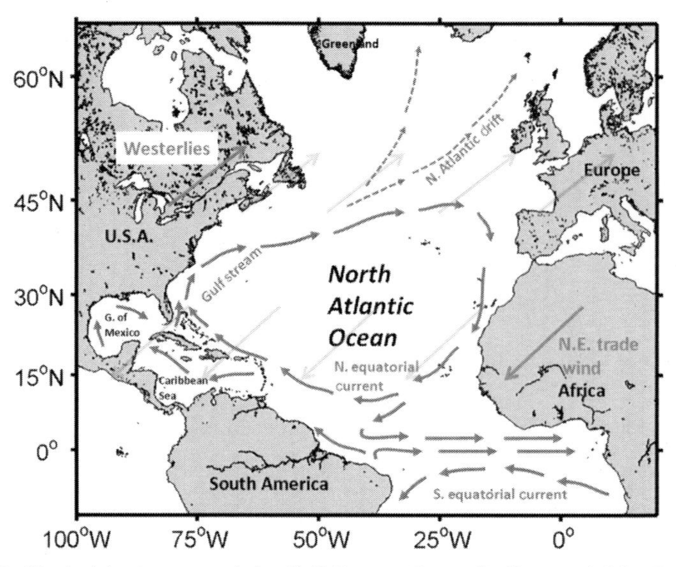

FIG. 6.1 North Atlantic gyre and the Gulf Stream. *(Image kindly provided by Kevin Haas and reproduced from K. Haas, X. Yang, V. Neary, B. Gunawan, Ocean current energy resource assessment for the Gulf stream system: the Florida Current, in: Marine Renewable Energy, Springer, New York, NY, 2017, pp. 217–236, with permission from Springer.)*

and because ocean currents are relatively stable and predictable, they represent a useful form of marine-based electricity generation that could be considered in any future energy mix in countries that border a suitable ocean current resource.

Ocean gyres are driven by trade winds (easterlies) in combination with westerlies (Fig. 6.1). Friction between these relatively persistent winds and the surface of the ocean drives a mass flow of water, leading to basin-scale gyre systems (e.g. the North Atlantic Gyre in Fig. 6.1). The eastward rotation of the Earth offsets the centre of the gyre systems to the west of the ocean basin, and so there is a western intensification of the associated currents. In contrast to eastern boundary currents, which are relatively broad, shallow, and slow-moving, these *western boundary currents* are therefore narrower, deeper, and faster. The five main western boundary currents are

- The Gulf Stream
- Kuroshio Current
- Brazil Current
- Agulhas Current
- East Australia current

It is primarily these five western boundary currents that have been identified as possible candidates for exploiting the ocean current resource. For example, the Gulf Stream (in the North Atlantic) and its counterpart in the North Pacific (the Kuroshio Current) are around 100 km wide, and in some places have surface velocities in excess of 2 m/s (e.g. [2]).

Because it passes relatively close to the state of Florida, much research has been invested in understanding and quantifying the Florida Current—the ocean current that flows from the Straits of Florida, around the Florida Peninsula, and along the southeastern coast of the United States before it merges with the Gulf Stream. The volume of water transported in the Florida Current is around 30 Sv (1 Sv $=$ $10^6\,m^3/s$), in contrast to around 150 Sv in the Gulf Stream itself [3]. By way of comparison, the mighty Amazon, the largest river in the world, has an annual mean flow of 0.2 Sv. The mean energy flux in the Florida Current has been estimated from numerical simulations as around 22 GW, with a standard deviation of around 6 GW [3,4], and over half of the total kinetic energy flux is concentrated in the upper 200 m of the water column. Therefore, from both economic and technical perspectives, any technology that is to exploit this resource would realistically need to be located in the upper part of the water column, such as an array of floating turbines, and in relatively deep water (see Section 6.2.2).

Observations have demonstrated that the northward extension of the Florida Current—the Gulf Stream off the coast of North Carolina—has a fairly persistent current speed of just under 1 m/s at a depth of 75 m, associated with a mean (and maximum) power density of 800 W/m^2 (4500 W/m^2) [5]. At such water depths, subsurface turbines could become feasible as turbine and mooring technologies develop.

The Agulhas Current flows southward along the east coast of South Africa. As with other western boundary currents, it is relatively fast and narrow, with a mean flow of around 70 Sv [6]. As with the Florida Current, its close proximity to the coast and relative stability make it an attractive proposition for electricity generation. However, the trajectory of the Agulhas Current is intermittently interrupted by perturbations known as 'Natal Pulses'—large solitary meanders that form at the Natal Bight between 29 and 30 degrees S [7]. There are, on average, 1.6 Natal Pulses per year, each lasting around 15–20 days. In the region of the Agulhas Current that looks most promising for electricity generation (around 28.8 degrees E and 32.5 degrees S), due to favourable velocity and bathymetry characteristics, the current core is in close proximity to the coast. At this location, any turbine deployed in the Agulhas Current would need to operate in current speeds of 0.6–2 m/s (Fig. 6.2).

6.2.1 Variability

It is often thought that "unlike the wind, [ocean currents] are always flowing" [8]. However, Bane et al. [8] propose an alternative statement: "the [ocean currents are] always flowing, but not always in the same place." Although ocean currents are relatively persistent, they suffer from variability at a range of temporal scales. For example, as mentioned earlier, the Agulhas Current is characterized by 'Natal Pulses' lasting 15–20 days, with 1 or 2 events occurring per year. Most variability in the Florida Current is found at 'high' frequencies,

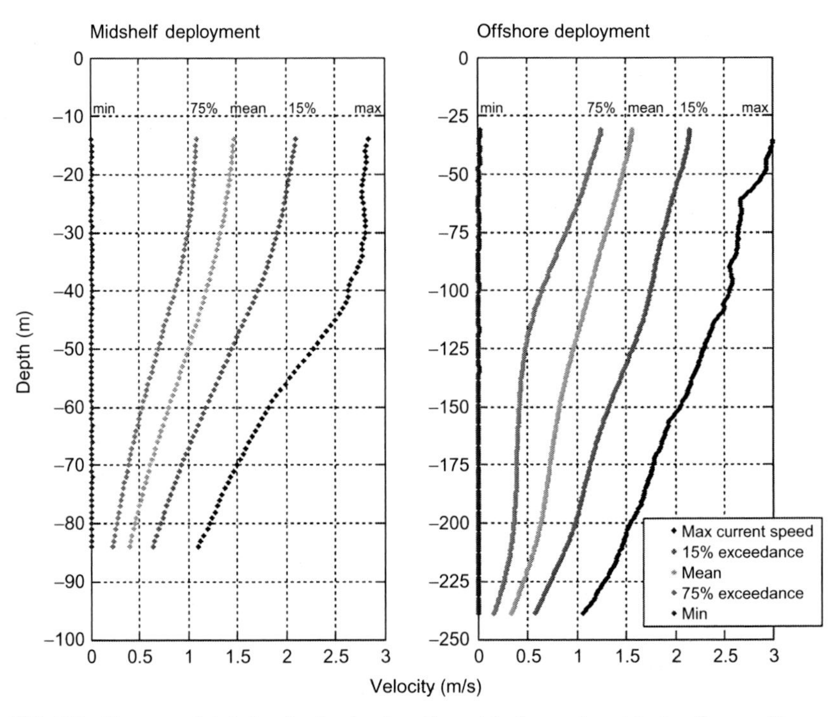

FIG. 6.2 Flow speed statistics for the Agulhas Current in the southwest Indian Ocean. *(Reproduced from I. Meyer, J.L. Van Niekerk, Towards a practical resource assessment of the extractable energy in the Agulhas ocean current, Int. J. Marine Energy 16 (2016) 116–132, with permission from Elsevier.)*

TABLE 6.1 Variance (Square of the Standard Deviation) of the Florida Current Associated With Different Time Periods

	<30 days	1–11 months	Annual	13–42 months	>42 months
Variance (Sv^2)	2.4	4.6	0.9	1.3	0.8
Total variance (%)	24	46	9	13	8

Notes: First row is the variance, second row is the variance as a percentage of the total variance in the time series, which is around 10 Sv^2.
Source: Data from Meinen et al. [9].

with roughly 70% of the total variance occurring at periods less than annual [9]. Annual, interannual, and longer timescales each represent around 10% of the total variance (Table 6.1).

These timescales are in contrast to the intermittency experienced at tidal current sites (Chapter 3). Tidal currents are characterized by intermittency at either diurnal or semidiurnal timescales, in addition to fortnightly (spring-neap) variability (in regions where tides are semidiurnal). Therefore, the relative persistence of ocean currents and their lack of lunar influence (because they are indirectly governed by the Sun, i.e. they are driven by winds) could make them attractive from a daily/weekly and seasonal electricity demand perspective. However, there are clearly significant challenges in attempting to convert the theoretical ocean current resource into practical electricity.

6.2.2 Technology Types

Although there is clearly a significant ocean current resource, including a 'technical resource' where the currents pass relatively close to coastlines and population centres, designing a device (or an array of devices) to economically convert energy within such environments will be challenging. Due to increased water depths (of order several hundred metres) compared with typical tidal energy environments (Chapter 3), it is likely that any device that is to exploit the ocean current resource will be tethered to the sea bed with cables, with the relatively constant current interacting with the turbine(s) used to maintain position and stability [10]. The 'Kuroshio power plant' concept proposed off the coast of Taiwan consists of three clusters of turbines, each individually anchored to a floating platform that can deform with the variable ocean currents, and anchored to the sea bed using a large number of cables [11]. It is recommended, to avoid damage to the power plant from large typhoon-generated waves in this region, that the turbines be placed at least 30 m below the sea surface. Shirasawa et al. [12] propose an ocean current design based on a horizontal axis turbine that is tethered 100 m (hub height) below the sea surface. Although a 2.3 m diameter prototype has been tested using towing experiments at sea, it is claimed that an array configuration of 300×80 m diameter turbines based on this design could produce up to 1 GW of electricity from the Kuroshio Current off Japan; however, clearly the technical challenges associated with such large-scale energy conversion are immense.

From commercial and R&D perspectives, two devices have shown promise for harvesting the ocean current resource. Again, both are tethered designs. The Aquantis 2.5 MW C-Plane device, developed through the US Department of Energy Aquantis project, was specifically designed to harness the energy resource of the Gulf Stream. Although no sea trials have yet been conducted, the device is rated at 1.6 m/s [7], and so operates at considerably lower current speeds than typical tidal stream applications. The Minesto 500 kW Deep Green technology (Fig. 3.26) is also suited to lower current environments and is again a tethered device with a rated speed of 1.6 m/s. Minesto have themselves expressed an interest in adapting the Deep Green turbine (originally intended for low-speed tidal energy environments) for ocean current applications (www.minesto.com).

6.2.3 Environmental Impacts

In the 2004 film 'The Day After Tomorrow', a slowdown in the major ocean currents (caused by melting polar ice) results in a new ice age. Therefore, people tend to be very cautious when looking at a process that could alter the circulation of the major ocean currents. However, compared to other processes like climate change, any exploitation of the ocean currents to generate electricity will be relatively minor, and certainly well within the bounds of natural variability of these systems. Therefore, the environmental impacts of this form of electricity generation are considered to be negligible and localized to the array itself (e.g. [13]).

6.3 OCEAN THERMAL ENERGY CONVERSION

OTEC technologies exploit the temperature difference between warm sea water at the surface of the ocean, and cold sea water at around 1000 m depth, to produce electricity [14]. Between the tropics[2] of the world's oceans, the temperature of the surface waters is significantly higher than the temperature deeper in the water column (Fig. 6.3). By contrast, at higher latitudes, the vertical temperature gradient is considerably less and varies seasonally. It is generally considered that a temperature difference of at least 20°C is required for OTEC plants to operate, and so this form of energy conversion is restricted to the tropics.

The OTEC resource is vast—the International Renewable Energy Agency (IRENA) has estimated that the global resource could be up to 30 TW [14], which would make OTEC by far the ocean energy resource with the largest potential.

The temperature gradient between the sea surface and deep water can be harnessed using different OTEC processes that mainly differ based on the working fluid that is used. The two main types of technology are *closed cycle OTEC*, which mainly uses ammonia as the working fluid, and *open cycle OTEC*, which uses the sea water itself as the working fluid.

6.3.1 Closed Cycle OTEC

Warm tropical sea water, at around 25°C, is clearly at a temperature that is insufficient to drive a steam turbine [15]. However, this warm surface water can be used to vaporize a fluid with a low boiling temperature such as ammonia. In a closed cycle OTEC power plant, this 'working fluid' flows around a closed loop as shown in Fig. 6.4. Warm water from the sea surface flows through a heat exchanger, causing the working fluid to boil and vaporize. This vaporized fluid flows through a turbine, which turns a generator that converts the energy into electricity. Upon leaving the turbine, the working fluid needs to be cooled so that it can be reused, otherwise the efficiency of the system would drop significantly. Cold water pumped from the deep ocean flows through a second heat exchanger that cools the working fluid to its original temperature, ready to enter the cycle again. The cold water from the deep ocean, now at a slightly elevated

---

2. 23.5 degrees S (Tropic of Capricorn) to 23.5 degrees N (Tropic of Cancer).

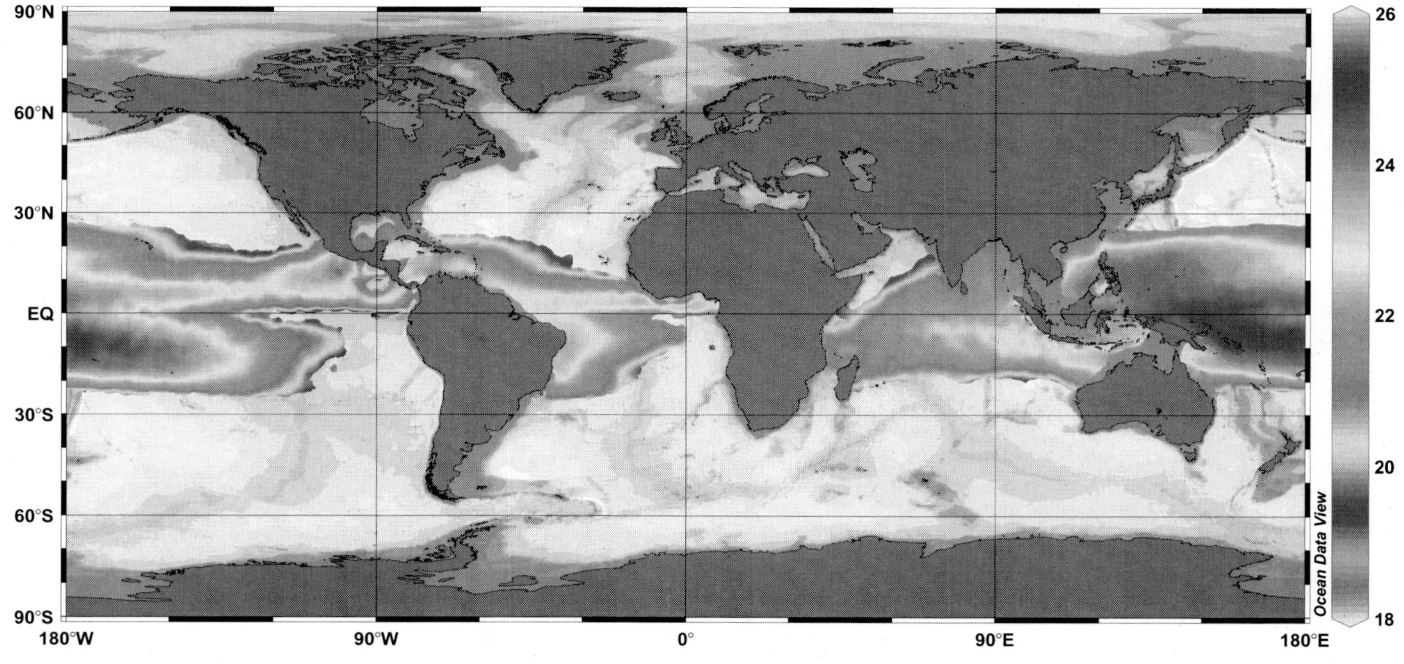

FIG. 6.3 The 'OTEC region', based on a mean annual temperature difference between water depths of 20 and 1000 m. *(Image kindly provided by Gérard Nihous, and reproduced from K. Rajagopalan, G.C. Nihous, Estimates of global Ocean Thermal Energy Conversion (OTEC) resources using an ocean general circulation model, Renew. Energy 50 (2013) 532–540, with permission from Elsevier.)*

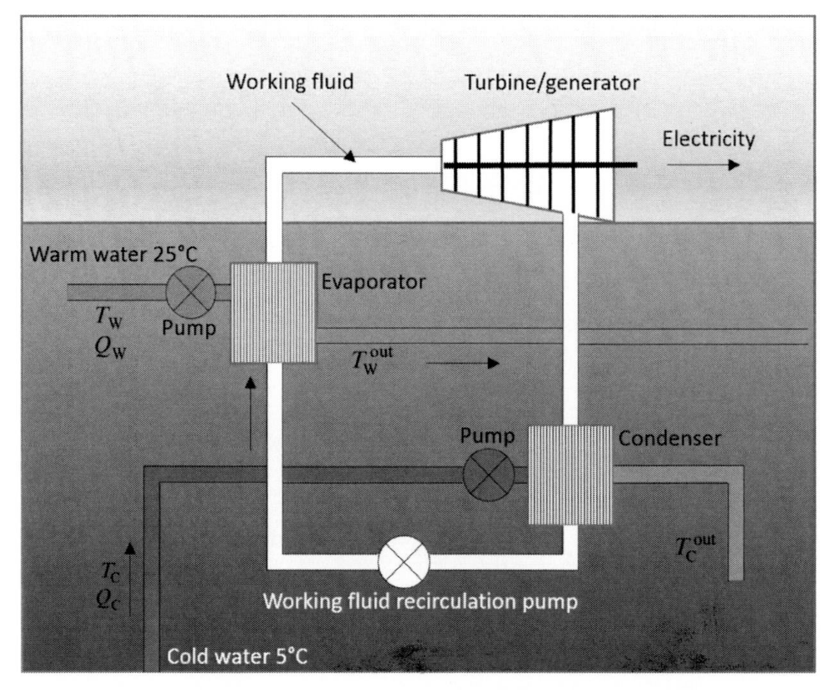

FIG. 6.4 Closed cycle OTEC power plant. *(Based on an image produced by DCNS.)*

temperature, is discharged into the ocean. The warm water from the ocean surface, at a slightly reduced temperature, is discharged into the upper ocean.

The world's oldest OTEC power plant is of the closed cycle design—a 103 kW power plant at the Natural Energy Laboratory on Hawaii, built in 1979. This power plant was land based, which lowers installation and maintenance costs, and facilitates bringing the electricity to the shore, but floating OTEC plants are generally considered to be more efficient and with fewer environmental concerns.

6.3.2 Open Cycle OTEC

In contrast to a closed cycle system, in an open cycle OTEC power plant, the sea water itself is used as the 'working fluid'. Although sea water has a much higher boiling point than a fluid such as ammonia, warm surface water can produce steam by reducing its pressure. This is achieved by pumping the surface water into a low-pressure container (known as a flash evaporator), causing it to boil. The vapour is used to drive a generator. This vapour is then condensed by exposure to cold sea water that is pumped from deep in the ocean.

The largest OTEC project that has been built to date is of the open cycle design—a 1 MW plant in Hawaii, which ran from 1993 to 1998 [14].

6.3.3 OTEC Thermodynamics

To better understand how ocean thermal energy is quantified, let us have a quick review of some basic thermodynamic concepts. The amount of heat energy required to raise the temperature of a unit mass of water by 1°C is called specific heat c_h. The specific heat of water is 4184 J/kg, 1000 calories/kg, or simply 1 calorie/g (i.e. the definition of a calorie). For instance, if we wanted to increase the temperature of a stream flow that has a flow rate of $10\,m^3/s$ (or 10,000 kg/s) by 2.5°C, the amount of power required (in Joules per second or megawatts) would be

$$\begin{aligned} P &= (10\ m^3/s)(1000\ kg/m^3)(2.5°C)(4184\ J/kg°C) \\ &= 104.6 \times 10^6\ J/s = 104.6\ MW \end{aligned} \tag{6.1}$$

which is a very large amount of power (e.g. equivalent to about 100 tidal turbines, each with a capacity of 1 MW!). An ocean thermal energy converter tries to reverse this process. In other words, extracting energy from water by cooling it down or converting heat to work. Unfortunately, the reverse process is much more difficult, and not as efficient. The second law of thermodynamics specifically deals with this concept, and the Carnot heat engine (proposed in 1824) was one of the first attempts to study this process. According to Carnot, the maximum efficiency of this process is given by

$$\eta = \frac{T_w - T_c}{T_w} \tag{6.2}$$

where T_w and T_c are the absolute temperatures of warm and cold water (in Kelvin = Celsius + 273.15), respectively. Consequently, for ocean thermal energy, the upper bound of efficiency is quite low. For instance, if $T_w = 25°C$ and $T_c = 5°C$, the maximum efficiency will be

$$\eta = \frac{20}{25 + 273.15} \approx 7\% \tag{6.3}$$

Nevertheless, the amount of energy is very high, and even a low efficiency power plant can convert a significant amount of energy. Referring to Fig. 6.4, assume that the warm water, with a temperature of T_w, is pumped to the evaporator and returns with a lower temperature T_w^{out} after losing some energy. The amount of heat energy that water loses per unit time (P_h) is given by

$$P_h = \frac{d}{dt}\left[mc_h(T_w - T_w^{out})\right] = c_h\frac{dm}{dt}\Delta T_w \rightarrow P_h = \rho c_h Q_w \Delta T_w \tag{6.4}$$

where ρ is the density of water and Q_w is the volumetric flow rate of warm water (remember that mass flow rate, $\frac{dm}{dt}$, is equal to volumetric flow rate times density). A power plant which works based on ocean thermal energy exploits part of this energy. If we denote the efficiency of the power plant by η, the generated power (in the steady-state case) can be estimated as follows

$$P_{OT} = \eta\rho c_h Q_w \Delta T_w \tag{6.5}$$

Replacing the maximum efficiency from Eq. (6.2), this leads to

$$P_{\mathrm{OT}} = \rho c_{\mathrm{h}} Q_{\mathrm{W}} \frac{(T_{\mathrm{W}} - T_{\mathrm{c}})\Delta T_{\mathrm{W}}}{T_{\mathrm{W}}} \tag{6.6}$$

For an idealized case in which the temperature can be reduced to that of cold water (i.e. $T_{\mathrm{W}}^{\mathrm{out}} = T_{\mathrm{c}}$ or $\Delta T_{\mathrm{W}} = T_{\mathrm{W}} - T_{\mathrm{c}}$), the maximum power is given by

$$P_{\mathrm{OT}} = \rho c_{\mathrm{h}} Q_{\mathrm{W}} \frac{(T_{\mathrm{W}} - T_{\mathrm{c}})^2}{T_{\mathrm{W}}} \tag{6.7}$$

For example, for $T_{\mathrm{W}} = 25°\mathrm{C}$, $T_{\mathrm{c}} = 5°\mathrm{C}$, and $Q_{\mathrm{W}} = 1\ \mathrm{m}^3/\mathrm{s}$, the power that is generated will be

$$P_{\mathrm{OT}} = (1000)(4184)(1)\left(\frac{(25-5)^2}{25+273.15}\right) = 5.6\ \mathrm{MW} \tag{6.8}$$

Moving towards a more realistic case, the efficiency of an OTEC power plant depends on to the ratio of cold water flow rate to warm water flow rate (i.e. $Q_{\mathrm{W}} = rQ_{\mathrm{c}}$), which should be optimized for a project (e.g. see [16]). For instance, $r = 0.5$ means that if the volumetric flow rate of warm water is $10\,\mathrm{m}^3/\mathrm{s}$, the volumetric flow rate of cold water is $5\,\mathrm{m}^3/\mathrm{s}$. For a simplified case, it can be shown that the warm water cools down by $\Delta T_{\mathrm{W}} = [3r/(1+r)](T_{\mathrm{W}} - T_{\mathrm{c}})/8$ and the cold water warms by $\Delta T_{\mathrm{c}} = [3/(1+r)](T_{\mathrm{W}} - T_{\mathrm{c}})/8$. Note that for this simplified case, the heat lost by the warm water is gained by the cold water: $\rho c_{\mathrm{h}} Q_{\mathrm{W}} \Delta T_{\mathrm{W}} = \rho c_{\mathrm{h}} Q_{\mathrm{c}} \Delta T_{\mathrm{c}}$. In other words, the extracted energy is neglected in this simple heat and mass balance equation as the efficiency is small. Referring to Eq. (6.6), and replacing $\Delta T_{\mathrm{W}} = 3r/(1+r)\Delta T/8$, results in,

$$P_{\mathrm{OT}} = 3r\rho c_{\mathrm{h}} Q_{\mathrm{W}} \frac{(T_{\mathrm{W}} - T_{\mathrm{c}})^2}{8(1+r)T_{\mathrm{W}}} \tag{6.9}$$

6.3.4 Commercial Progress

One might question how much electricity can realistically be generated from a power plant that consists of turbines that are driven (for example) by vapour produced from boiling ammonia. The real answer to this is a question of scale. Although many test and research OTEC power plants have been developed around the 30–120 kW scale (Fig. 6.5), it is possible, simply by using larger diameter pipes and a larger volume of working fluid, to generate electricity at around the 100 MW scale. However, part of the journey to such large-scale OTEC power plants is to further develop and refine schemes at around the 1–20 MW scale, and there are many such schemes under development and construction throughout the world (Fig. 6.5), for example, the 1 MW plant in Tarawa Island (Pacific), and 20 MW plants in Qingdao (China) and La Martinique (Carribean).

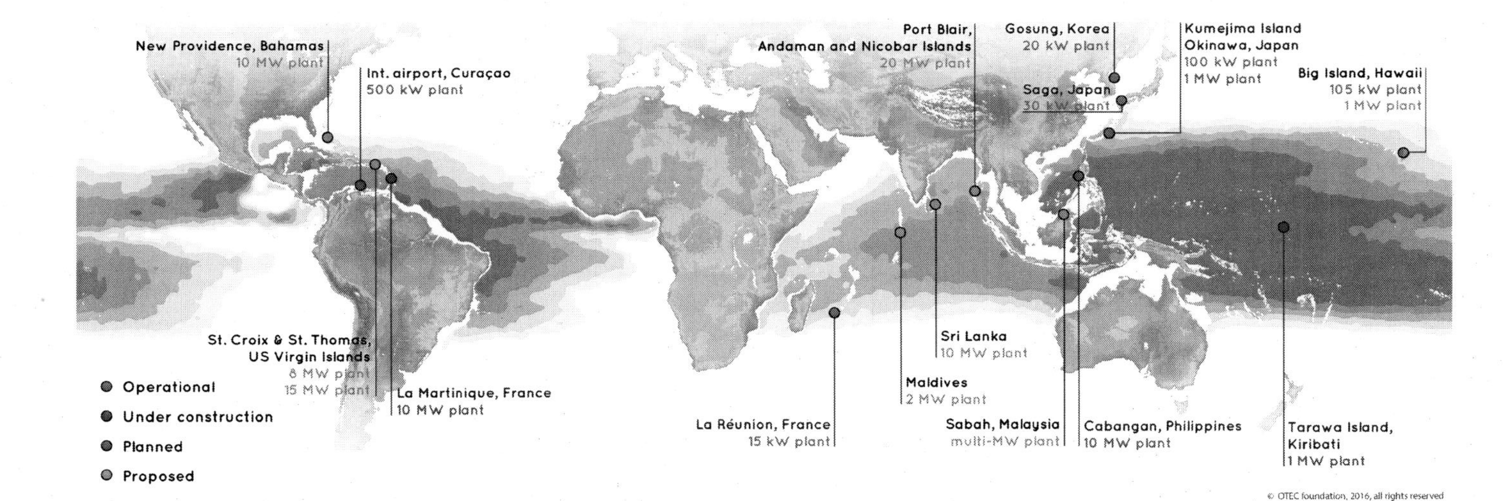

FIG. 6.5 OTEC projects around the world. *(Image kindly provided by the OTEC Foundation.)*

6.3.5 Additional Benefits of OTEC Power Plants

A particular advantage of open cycle OTEC power plants is that they produce desalinated water, which could be particularly beneficial in many island states in the Caribbean and Pacific where there is a good OTEC resource and a paucity of fresh water. In addition, the cold deep water used in both closed cycle and open cycle OTEC plants can be introduced into air conditioning systems, after the water has been used to facilitate condensation. Finally, deep ocean water is rich in nutrients, and after desalination can be a useful product for agriculture and aquaculture.

6.3.6 Environmental Impacts

A large OTEC power plant would require the transport of significant quantities of warm surface water and cold water from around 1000 m depth. In the deep water environment, the cold water tends to be rich in nutrients, and so there could be direct consequences for deep water marine organisms due to the depletion of nutrient-rich water from this environment. The plant operation could alter the stratification of the water column, and this could influence the natural process of upwelling of nutrients, affecting a range of marine species. Further, the discharge of this colder water into warm surface waters could influence marine life in the vicinity of the power plant, for example, it is suggested that 'exhaust' water at 3°C below ambient could trigger algal blooms [14]. The warm and cold water inlets of an OTEC power plant will include screens to prevent debris and larger species from entering the power plant, and this could directly influence marine life that become entrained in the screening system.

6.4 SALINITY GRADIENTS

When there is large difference in salinity between two fluids, for example, where a river flows into the sea, it is possible to harness the energy of the salinity gradient to generate electricity. The global potential for salinity gradient power is surprisingly large—the global technical resource has been estimated by the IRENA as 647 GW, which would represent almost 25% of global electricity consumption [17]. The resource is largely concentrated in South America and Asia, but there is also a considerable resource in North America (Fig. 6.6). The Amazon River alone contains up to 20.8 GW of potential, and the Yangtze up to 2.29 GW (Table 6.2). The resource could be higher still if salinity gradient power were extended to include 'hybrid' applications, for example, by exploiting very large salinity gradients that exist between the brine 'waste' of desalination plants and the ambient water.

6.4.1 Technology Types

There are two salinity gradient technologies: pressure retarded osmosis (PRO) and reversed electro dialysis (RED).

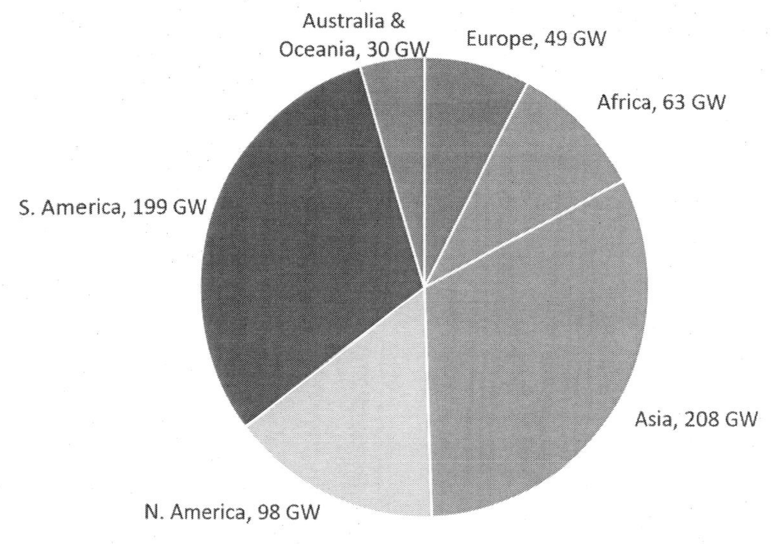

FIG. 6.6 Global distribution of the technical salinity gradient resource. *(Data from P. Stenzel, Potentiale der Osmose zur Erzeugung und Speicherung von Elektrizität (Potential of Osmosis for Production and Storage of Electricity), vol. 4, LIT Verlag, Münster, 2012.)*

TABLE 6.2 Potential for Osmotic Power Production From a Selection of Major World Rivers

Source of Fresh Water	Mean Flow Rate (m^3/s)	Power (GW)
Amazon, Brazil	2×10^5	20.8
La Plata-Paraná River, Argentina	8×10^4	8.32
Congo, Democratic Republic of Congo	5.7×10^4	5.93
Yangtze, China	2.2×10^4	2.29
Ganges, Bangladesh	2×10^4	2.08
Mississippi River, USA	1.8×10^4	1.87
Columbia River, USA	7.5×10^3	0.78

Notes: The power estimate is based on 10% of each river's discharge, and a power output of 1 MW/m^3 per s.
Source: Based on data presented by Helfer and Lemckert [18].

Pressure Retarded Osmosis

In a PRO system, a membrane is used to separate sea water from fresh water. Sea water is pumped into a pressure exchanger, where the osmotic pressure is less than that of fresh water. The freshwater flows through a semipermeable membrane towards the sea water, increasing the pressure inside the sea water chamber. This pressure, which can reach a theoretical pressure equivalent to a water column that is over 100 m in height, is used to spin a turbine and hence turn a generator. A 10 kW PRO power plant in Tofte (Norway) was opened in 2009, consisting of 2000 m^2 of membrane.

Reversed Electro Dialysis

In conventional electrodialysis, salt ions are transported from one solution, through ion-exchange membranes, to another solution under the influence of an applied electric potential difference. As the name suggests, in RED this process is reversed, because the end product is the electric potential difference that can be converted into electricity. Salt water and freshwater flow along alternating cation and anion-exchange membranes. Due to the concentration difference, salt ions permeate through these membranes from salt to fresh water: negatively charged chlorine through one membrane, and positively charged sodium through another. The resulting electric potential can then be converted into electricity. Hundreds or thousands of membranes can be stacked between two electrodes to increase the electricity output. A 50 kW RED pilot plant, opened in 2013, has been constructed on a causeway in the Netherlands (the Afsluitdijk) that separates fresh water from sea water in the Wadden Sea.

6.5 TECHNOLOGICAL CHALLENGES

One of the challenges of salinity gradient power plants is the vast surface area of membrane that is required to generate meaningful levels of electricity. For example, a 2 MW plant using current technologies would require at least 1×10^6 m^2 of membrane surface area,[3] and this membrane would require maintenance and replacement over a time period of around 5 years [17]. The maximum power of membranes is currently less than 3 W/m^2, but recent laboratory experiments have achieved power densities in excess of 14 W/m^2 [17]. Such improvements in membrane efficiencies applied to larger scales are required before the concept of salinity gradient can become commercially viable. Other issues include biofouling, pretreatment of water, and the development of the modules that house the membranes.

REFERENCES

[1] T.S. Garrison, Oceanography: An Invitation to Marine Science, Nelson Education, Toronto, ON, 2015.

[2] S. Imawaki, A.S. Bower, L. Beal, B. Qiu, Western boundary currents, in: Ocean Circulation and Climate: A 21st Century Perspective, 2013.

[3] J.C. Larsen, T.B. Sanford, Florida Current volume transports from voltage measurements, Science 227 (4684) (1985) 302–304.

3. Or less if efficiencies improve.

[4] Z. Defne, K.A. Haas, H.M. Fritz, L. Jiang, S.P. French, X. Shi, B.T. Smith, V.S. Neary, K.M. Stewart, National geodatabase of tidal stream power resource in USA, Renew. Sustain. Energy Rev. 16 (5) (2012) 3326–3338.
[5] C.F. Lowcher, M. Muglia, J.M. Bane, R. He, Y. Gong, S.M. Haines, Marine hydrokinetic energy in the Gulf Stream off North Carolina: an assessment using observations and ocean circulation models, in: Marine Renewable Energy, Springer, New York, NY, 2017, pp. 237–258.
[6] H.L. Bryden, L.M. Beal, L.M. Duncan, Structure and transport of the Agulhas Current and its temporal variability, J. Oceanogr. 61 (3) (2005) 479–492.
[7] I. Meyer, L. Braby, M. Krug, B. Backeberg, Mapping the ocean current strength and persistence in the Agulhas to inform marine energy development, in: Marine Renewable Energy, Springer, New York, NY, 2017, pp. 179–215.
[8] J.M. Bane, R. He, M. Muglia, C.F. Lowcher, Y. Gong, S.M. Haines, Marine hydrokinetic energy from western boundary currents, Ann. Rev. Marine Sci. 9 (2017) 105–123.
[9] C.S. Meinen, M.O. Baringer, R.F. Garcia, Florida Current transport variability: an analysis of annual and longer-period signals, Deep Sea Res. Part I 57 (7) (2010) 835–846.
[10] U.S. Department of the Interior, Technology White Paper on Ocean Current Energy Potential on the U.S. Outer Continental Shelf, 2006.
[11] F. Chen, Kuroshio power plant development plan, Renew. Sustain. Energy Rev. 14 (9) (2010) 2655–2668.
[12] K. Shirasawa, K. Tokunaga, H. Iwashita, T. Shintake, Experimental verification of a floating ocean-current turbine with a single rotor for use in Kuroshio currents, Renew. Energy 91 (2016) 189–195.
[13] K. Haas, Assessment of Energy Production Potential From Ocean Currents Along the United States Coastline, Georgia Tech Research Corporation, 2013.
[14] International Renewable Energy Agency, Ocean Thermal Energy Conversion—Technology Brief 1, 2014.
[15] G.A. Pagnoni, S. Roche, The Renaissance of Renewable Energy, Cambridge University Press, Cambridge, 2015.
[16] G.C. Nihous, An order-of-magnitude estimate of ocean thermal energy conversion resources, J. Energy Resour. Technol. 127 (4) (2005) 328–333.
[17] International Renewable Energy Agency, Salinity Gradient Energy—Technology Brief 2, 2014.
[18] F. Helfer, C. Lemckert, The power of salinity gradients: an Australian example, Renew. Sustain. Energy Rev. 50 (2015) 1–16.

FURTHER READING

[1] K. Haas, X. Yang, V. Neary, B. Gunawan, Ocean current energy resource assessment for the Gulf stream system: the Florida Current, in: Marine Renewable Energy, Springer, New York, NY, 2017, pp. 217–236.
[2] I. Meyer, J.L. Van Niekerk, Towards a practical resource assessment of the extractable energy in the Agulhas ocean current, Int. J. Marine Energy 16 (2016) 116–132.
[3] K. Rajagopalan, G.C. Nihous, Estimates of global Ocean Thermal Energy Conversion (OTEC) resources using an ocean general circulation model, Renew. Energy 50 (2013) 532–540.
[4] P. Stenzel, Potentiale der Osmose zur Erzeugung und Speicherung von Elektrizität (Potential of Osmosis for Production and Storage of Electricity), vol. 4, LIT Verlag, Münster, 2012.

Chapter 7

In Situ and Remote Methods for Resource Characterization

Although wave and tidal energy resources can be either simulated or estimated from various products (e.g. tidal atlases), the resource can only truly be characterized by direct observation. Whilst observational campaigns are costly and logistically challenging, such direct measurements can accurately quantify the resource at high spatial and temporal resolution, and without the assumptions that are necessary when parameterizing numerical simulations. In addition, observations are essential for validating numerical simulations (Chapter 8), which can then be applied to understand processes over longer timescales or hypothetical scenarios such as sea-level rise or assessing impacts (preconstruction) of large engineering projects, in addition to investigating neighbouring regions of interest that have not been directly observed.

In this chapter, we introduce various methods of measuring wave and tidal resources both in situ and remotely. We describe the principal of some of the main instruments used to quantify waves and tides, particularly acoustic Doppler current profilers (ADCPs) and directional wave buoys. We also discuss ship-based sampling techniques (e.g. sea-bed sediment grabs and water column profiling), and remote-sensing technologies, including X-band and high-frequency (HF) radar.

7.1 TIDAL ENERGY RESOURCE CHARACTERIZATION

Desk-based studies can provide useful information about the tidal energy resource of a region. Tidal atlases, such as the ABPmer Atlas of UK Marine Renewable Energy Resources [1], and the NREL Atlas of Marine Renewable Resources in the US, provide information on the spatial distribution of depth-averaged spring tidal currents, albeit at relatively coarse resolution. More detailed regional studies in the peer-reviewed academic literature provide additional information on tidal streams (e.g. [2]), and articles that are *open access* are particularly useful to those working outside academia (e.g. [3]). Examination of Admiralty Charts provides further, detailed, site-specific information, such as the distribution of bathymetric contours, tidal range, and spring and neap current speeds (and directions) in more detail than tidal atlases, at specific points known as *tidal diamonds* (e.g. Table 7.1).

Fundamentals of Ocean Renewable Energy. https://doi.org/10.1016/B978-0-12-810448-4.00007-0

TABLE 7.1 Tidal Streams at Fall of Warness (the EMEC Tidal Energy Test Site), Based on Tidal Diamond Information Extracted From Admiralty Chart 2249 (Orkney Islands—Western Sheet)

Time Relative to HW (h)	Direction of Streams (degrees)	Speed of Tidal Stream (m/s)	
		Spring	Neap
−6	150	3.2	1.2
−5	144	3.7	1.4
−4	141	3.0	1.2
−3	116	1.4	0.6
−2	350	0.2	0.1
−1	308	2.0	0.8
0	329	3.3	1.3
1	329	3.3	1.3
2	320	2.5	1.0
3	325	2.0	0.9
4	324	0.6	0.3
5	160	0.9	0.4
6	153	2.9	1.2

Notes: Times are relative to HW (high water) at Aberdeen.

Such desk-based studies are particularly useful at early *scoping* stages of tidal energy projects, and can help inform site selection, prior to investing in detailed numerical model studies or costly field campaigns. However, the tidal resource can only be truly characterized by conducting in situ measurements.

7.1.1 Water-Level Measurements

The most fundamental and easily quantified property of the tides, and one that has been measured for hundreds of years, is the variation in water elevations over time. The longest accurate time series of tidal elevations in the United Kingdom, and indeed one of the longest in the world, is in Liverpool, extending back to 1768 [4]. Today, many tide gauge networks exist around the world (e.g. Fig. 7.1), and hourly (or shorter timescale) data are widely available, generally over time periods of at least a decade. In addition, tidal analysis of these elevation time series has led to the development of tide tables, which report times and heights of high water (HW) and low water (LW) for a location, generally published for each calendar year (e.g. Table 7.2). However, when specific measurements are required, for example, at locations that are not represented by an existing or historic tide gauge (such as in an estuary or a region that is far from existing tide gauges), several options are available.

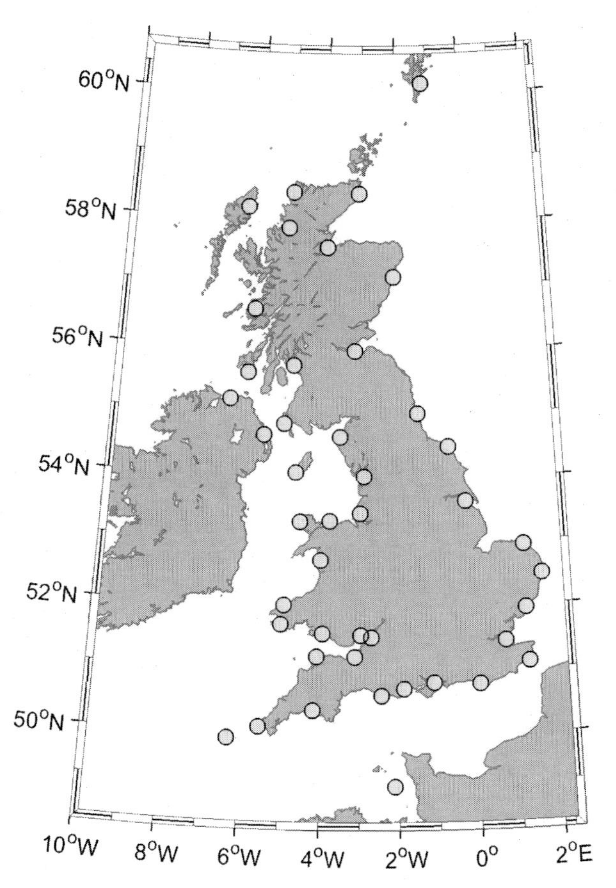

FIG. 7.1 The 44 stations that comprise the UK tide gauge network.

Tidal Poles

The simplest way to measure tidal elevation is to mount a graduated pole on a pier and record the position of water level over time. In calm conditions, it should be possible to read a pole to within 0.02 m [5], but the accuracy will reduce in the presence of waves. Although useful datasets have been obtained using tidal poles for short duration surveys, they are unsuitable for long-term measurements.

Stilling-Well Gauges

A well, called a stilling-well, is connected to the sea via a small hole or narrow pipe. Because the well is partially separated from sea, the limited connection dampens the short-period wave oscillations, and so the water surface inside the well remains relatively calm [5]. A float on the sea surface, inside the

TABLE 7.2 Tide Table for Menai Bridge (UK), January 2008, Based on 17 Tidal Constituents

	High Water				Low Water			
Date	Time	Height	Time	Height	Time	Height	Time	Height
1	04:51	5.51	17:15	5.69	11:16	2.53	23:51	2.44
2	05:55	5.48	18:19	5.56			12:16	2.66
3	06:58	5.57	19:20	5.57	00:49	2.49	13:21	2.66
4	07:54	5.77	20:13	5.69	01:48	2.43	14:24	2.53
5	08:43	6.03	21:00	5.89	02:43	2.27	15:17	2.32
6	09:27	6.30	21:43	6.12	03:31	2.05	16:03	2.07
7	10:07	6.56	22:23	6.36	04:14	1.81	16:44	1.82
8	10:45	6.79	23:01	6.58	04:54	1.60	17:21	1.58
9	11:21	6.99	23:38	6.75	05:32	1.42	17:58	1.37
10	11:57	7.13			06:08	1.31	18:34	1.21
11	00:16	6.86	12:33	7.22	06:45	1.26	19:11	1.13
12	00:54	6.88	13:11	7.22	07:23	1.28	19:50	1.12
13	01:34	6.81	13:52	7.14	08:03	1.37	20:33	1.22
14	02:18	6.64	14:37	6.95	08:46	1.53	21:19	1.40
15	03:06	6.40	15:29	6.67	09:35	1.75	22:12	1.65
16	04:03	6.12	16:31	6.35	10:32	2.00	23:13	1.90
17	05:12	5.90	17:45	6.10	11:40	2.20		
18	06:31	5.86	19:05	6.04	00:26	2.07	13:01	2.24
19	07:47	6.04	20:19	6.17	01:44	2.05	14:23	2.04
20	08:53	6.37	21:22	6.39	02:58	1.85	15:34	1.68
21	09:49	6.74	22:16	6.61	03:59	1.56	16:32	1.30
22	10:38	7.06	23:04	6.76	04:52	1.29	17:23	1.00
23	11:23	7.28	23:47	6.82	05:38	1.11	18:08	0.84
24			12:04	7.35	06:21	1.03	18:50	0.84
25	00:27	6.79	12:44	7.29	07:01	1.08	19:30	0.99
26	01:05	6.67	13:22	7.10	07:39	1.25	20:07	1.24
27	01:42	6.48	14:00	6.80	08:16	1.49	20:44	1.56
28	02:20	6.25	14:39	6.43	08:54	1.81	21:21	1.90
29	03:02	5.98	15:22	6.03	09:34	2.14	22:00	2.22
30	03:49	5.71	16:12	5.64	10:19	2.48	22:46	2.50
31	04:48	5.48	17:15	5.34	11:15	2.76	23:43	2.69

Notes: Predicted tidal height is in metres, relative to chart datum, Menai Bridge. Times are in GMT.

well, tracks the movement of the tide over time, and the level can be recorded on a rotating drum of paper. Stilling wells have probably made the greatest contribution to our records of sea level, and have led to the evidence for sea-level rise over the last two centuries [6].

Pressure Measuring Systems

Rather than tracking the sea surface with a float, the pressure at a fixed point below the sea surface can be measured and converted into a level using the hydrostatic relationship. This can be achieved by a basic pressure sensor installed below lowest astronomical tide (LAT), provided the relationship of the pressure sensor to some local datum can be measured with accuracy. For more precise measurements, two common types of pressure systems suitable for permanent installation are half-tide and full-tide bubblers. In a full-tide bubbler system, a low flow of dry air is fed down an air tube to the top of the pressure point located at some distance below LAT. When the air pressure in the air line is equal to the pressure exerted by the column of water above it, then the excess air is released as bubbles through a nozzle. This means that the pressure in the air line is proportional to the weight of the water column. In a half-tide bubbler system, the measuring point is mounted at the mid-tide height, and so the pressure point is immersed for half of the tidal cycle. During times when the measuring point is exposed, it can be accurately levelled and incorporated into geodetic networks.

Radar Sensor

Finally, because we know the speed of sound in air (based on local measurements of air temperature, pressure, and humidity), a transmitter that is mounted at some height above the water surface can transmit a pulse of sound and measure its double path length from transmitter to reflection at the water surface and back to the instrument. This method is best used in conjunction with a stilling well to eliminate reflections directed away from the receiver (e.g. due to an irregular or sloping sea surface). Alternatively, the transducer can be placed on the seabed looking upwards, using the speed of sound in seawater (again, calculated from local measurements of the sea water properties) as the reference point.

7.1.2 Mechanical and Electromagnetic Current Meters

Traditionally, oceanographers and coastal engineers have used current meters to make measurements of water currents. These are generally of the horizontal axis design, and convert the number of rotations of a calibrated propeller (per unit time) into a current speed (Fig. 7.2). Current meters record measurements at a single point in the water column and, although they have a directional vane to calculate the direction of flow, do not measure the vertical component of velocity, which is generally an order of magnitude lower than the horizontal component.

FIG. 7.2 A mechanical flow meter.

Current meters either can be of the mechanical rotor type described above or can be based on either acoustic or electromagnetic techniques. An acoustic current meter measures the travel time in both directions of an acoustic pulse between a pair of transducers [5]. The travel times will be unequal, because there is a component of the current directed from one sensor to the other. A second pair of sensors allows the complete horizontal current vector to be resolved. One problem with acoustic sensors is the accuracy required to measure very small time differences. For example, for a transducer pair separated by 0.1 m, the timing must be accurate to within 10^{-9} s for a speed resolution of 0.01 m/s. However, acoustic current meters have an advantage, compared with mechanical rotor current meters, in that they are less prone to fouling from seaweed and fishing lines, etc.

Magnetic flow meters are based on Faraday's Law of Electromagnetic Induction. If we consider water particles as a conductor, the motion of water through a magnetic field induces a voltage. This voltage, in general, is proportional to the flow velocity, the magnitude of the magnetic field, and the length of conductor (i.e. the path between electrodes). Magnetic flow meters are widely used for pipe flow measurements as they do not have any moving parts that could interfere with the flow. They measure just one component of velocity.

Tilt current meters are inexpensive and simple mechanical devices, suitable for current measurements near the bed, or at the surface. A subsurface buoy is either anchored to the bed (Fig. 7.3) or hangs from a support that is mounted above the sea surface. The drag force, which is proportional to square of

FIG. 7.3 A tilt current meter in operation made by Lowell Instruments, LLC.

velocity, causes the float to tilt, and the tilt angle can be directly related to the current speed.

As discussed in Chapter 3, tidal currents interact with the sea bed to generate a vertical velocity profile. In addition, the region of the velocity profile near the surface is influenced by wind. Therefore, it is important to consider what depth in the water column is representative of the tidal currents at a location, and this is a factor that must be considered when making measurements in situ with current meters. Alternatively, acoustic *profilers* can be used, as described in Section 7.1.4.

7.1.3 Acoustic Doppler Velocimeter

An acoustic Doppler velocimeter (ADV) is based on the *Doppler effect*, which may be a familiar concept to you from the relative change in frequency of a train whistle or car horn as it travels towards or away from an observer. If an observer is stationary whilst sound waves pass, n waves would pass during a time interval t. If the observer was to walk towards the source of sound, more than n waves would pass during t. Similarly, if the observer was to walk away from the source of sound, less than n waves would pass during t. The Doppler shift is the difference between the frequency heard whilst standing still, and the frequency heard when moving either towards or away from the sound. It is defined as

$$f_{\mathrm{d}} = f_{\mathrm{s}}(V/C) \tag{7.1}$$

where f_d is the Doppler shift, f_s is the frequency of sound when source and receiver are both stationary, V is the relative velocity between source and receiver, and C is the speed of sound, given by

$$C = f\lambda \tag{7.2}$$

where f is frequency and λ is wavelength.

An ADV transmits sound at a fixed frequency, and listens to echoes returning from sound scatterers in a small volume of water: small particles or plankton that reflect the sound back to the instrument. Although the sound is scattered in all directions, a small amount is reflected back to the transducer. The ADV relies on the assumption that these scatterers move at the same velocity as the water. Because the ADV both transmits and receives sound, the Doppler shift is doubled, and so Eq. (7.1) becomes

$$f_d = 2f_s(V/C) \tag{7.3}$$

The Doppler shift only works with radial motion. Although angular motion changes the direction between source and receiver, it does not alter the distance separating them, and so does not cause a Doppler shift. Limiting the Doppler shift to the radial component, Eq. (7.3) becomes

$$f_d = 2f_s(V/C)\cos A \tag{7.4}$$

where A is the angle between the relative velocity vector and the line between the ADV and scatterers.

The reflected acoustic wave is measured by three receiving probes (Fig. 7.4). The changes in the frequency of the reflected signals are the indicator of velocities in three directions, corresponding to the three receiving probes. An

FIG. 7.4 A 10 MHz SonTek ADV.

ADV estimates the three components of velocity (i.e. in the x, y, and z directions) based on these data. ADVs are designed with various signal frequencies, suitable for various ranges of current speeds.

ADVs are popular in laboratory flume studies [7] and are particularly useful for characterizing turbulence due to their HF sampling (e.g. 50 Hz [8]). In the oceanic environment, ADVs are gaining popularity due to their ability to quantify turbulence. For example, the Nortek Vector 3D is capable of measuring velocities up to 7 m/s and, due to the HF sampling associated with ADVs, can measure wave orbital velocities. However, ADVs tend to be deployed in very shallow nearshore environments.

7.1.4 Acoustic Doppler Current Profiler

If a profile of the water column is desired, it would be necessary to deploy a string of current meters (Section 7.1.2) distributed over the water column (e.g. [9]). However, for most practical measurements of marine currents over the water column, particularly when the vertical component of velocity is required, the use of current meters has been more-or-less superseded by the advent, in the 1980s, of ADCPs (Fig. 7.5).

Principal of Operation

The ADV that was introduced in Section 7.1.3 uses a *converging* beam pattern (Fig. 7.4), and so can only sample a small volume of water at the point where the beams converge. By contrast, an ADCP uses a *diverging* beam pattern (Fig. 7.5), usually based on at least three beams, to resolve velocities in three dimensions.

FIG. 7.5 Four-beam (*left*) and five-beam (*right*) acoustic Doppler current profilers (ADCPs). The five-beam ADCP is discussed in Section 7.2.3.

An ADCP works on the assumption that there is horizontal homogeneity; otherwise the trigonometric relations that are used to calculate the velocity components are invalid. Because there are four beams, but only three current directions, one of the beams could be considered to be redundant. However, this fourth beam is used to measure the vertical velocity a second time, and the error between the two measurements gives an indication of the quality of the data (i.e. a measure of how valid the assumption of horizontal homogeneity has been).

The water column is partitioned into vertical *bins*. The ADCP listens to the reflected echos at different time intervals, which correspond to different distances from the transducer (because $v = d/t$), that is different bin depths. This process is known as *range gating* (Fig. 7.6). For example, if the time between the sent and return signal is 8 ms, then the distance to the bin is $0.5 \times 1500\,\text{m/s} \times 0.008\,\text{s} = 6\,\text{m}$ (where 1500 m/s is the speed of sound in sea water, but clearly, in practice, an ADCP will use an accurate local value for the speed of sound, based on a local reading of water temperature and a constant value for salinity).

The echo from a hard surface such as the sea bed or sea surface is much stronger than the signal from scatterers in the water column, and so can dominate the signal. For this reason, data close to the surface (for an upward looking, moored, ADCP) or close to the bed (for a downward looking, hull-mounted, ADCP) is rejected. Also, because the ADCP will be at some height or depth in the water column, there will clearly be a further region of the water column which cannot physically be profiled. Further, the region close to the instrument is affected by *ringing*—an effect where the energy of a transmitted pulse lingers

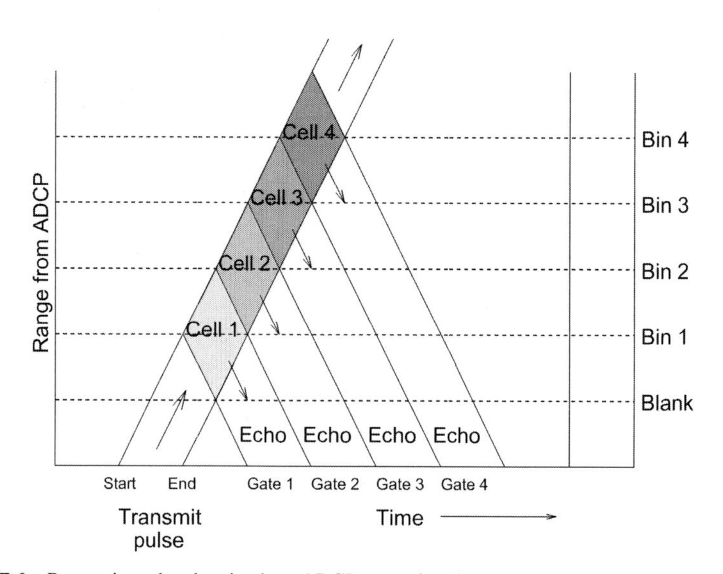

FIG. 7.6 Range-time plot showing how ADCP transmit pulses and echos travel through space.

TABLE 7.3 Nominal Profiling Range and Ringing Distance for a Selection of RDI ADCPs [10]

Frequency (kHz)	Maximum Profile Depth (m)	Ringing Distance (m)
75	700	6
150	400	4
300	120	2
600	60	1
1200	25	0.5

after the transmitted pulse has finished. Therefore, the ADCP must wait until the ringing has decayed before listening to and processing pulses. ADCPs are programmed with a default blanking period to eliminate measurements in this zone (e.g. Table 7.3). Due to the blanking distance, physical height (or depth) of the instrument in the water column, and the echo from the surface (or bed), a significant portion of the water column is not sampled by an ADCP. Therefore, when postprocessing, the velocity profile must be extrapolated into these regions of the water column (e.g. via curve fitting).

To reduce errors from single ping (e.g. 2 Hz) velocity measurements, ADCP data are *ensemble averaged*. The averaging time period requires careful consideration. Larger averaging intervals will reduce uncertainty in the measurements, but at the expense of stationarity—the condition that properties of the flow such as the mean and the variance do not change significantly over the selected time period. For example, a 1 min ensemble average would meet the condition of stationarity for most situations, but at the expense of relatively high instrument noise. A 10-min averaging interval may reduce instrument noise to acceptable levels, but one must consider whether the flow is statistically 'stationary' over this timescale—this may not be the case if the flow is characterized by eddy shedding in the wake of an obstacle, for example.

Because high levels of backscatter in the water column relate to high concentrations of suspended particulate matter (SPM), a secondary application of an ADCP is to *qualitatively* estimate SPM using the backscatter signal. However, only calibrated in situ measurements (e.g. filtered water samples or transmissometer readings) should be used to *quantify* SPM concentrations (Section 7.4.1).

Moored ADCP

ADCPs are most often deployed on a mooring, looking upwards through the water column (e.g. Fig. 7.7). The mooring can either be L-shaped (suitable for

FIG. 7.7 ADCP mooring prior to deployment. This mooring also contains a sediment sensor (in the foreground). This instrument package is deployed as an L-shaped mooring, and the anchor (ballast) and surface marker buoy can be seen in the background.

deploying the instrument package shown in Fig. 7.7) or can be more compact, using an acoustic release system for deployment and recovery (Fig. 7.8). The configuration shown in Fig. 7.8 is particularly suited to deployments where strong current speeds are anticipated, because the low profile 'trawl-resistant' instrument frame sits close to the sea bed, and hence within the near-bed boundary layer where the speed of the current is lower. If a configuration such as that in Fig. 7.7 is deployed in regions of strong tidal flow, the instrument frame should be sufficiently weighted to prevent postdeployment movement. A further consideration is potential for movement of the mooring due to drag on cables, particularly if the deployment is of sufficient duration for seaweed or other marine life to build up on the cables (hence further increasing drag). This is clearly not an issue for the type of mooring shown in Fig. 7.8.

The European Marine Energy Centre (EMEC) standard for the assessment of tidal energy conversion systems [11] suggests that a minimum deployment duration of 30 days is required to adequately resolve the tidal resource of a region. Such guidance can be justified by considering the *Rayleigh criteria* (e.g. [5]), which states that only constituents that are separated by at least a complete

FIG. 7.8 Low profile (trawl-resistant) ADCP mooring. The acoustic release above the instrument is used during the deployment stage; a similar acoustic release, in the base of the instrument frame, is used during recovery. The top *part* of the mooring, which is buoyant, separates from the heavy base when the (recovery) acoustic release is triggered. The *top and bottom parts* of the mooring are connected by a cable, which is coiled inside the base prior to deployment.

period from their neighbouring constituents over the length of the record should be included. For example, to analyse a signal for M2 and S2 requires a time series of length

$$T_{\mathrm{r}} = \left(\frac{1}{T_{\mathrm{S2}}} - \frac{1}{T_{\mathrm{M2}}}\right)^{-1} = \left(\frac{1}{0.500} - \frac{1}{0.517525}\right)^{-1} = 14.77 \text{ days} \qquad (7.5)$$

where T_{r} is the required length of a signal to resolve M2 and S2 constituents.

Table 7.4 lists the record length that would be required to resolve semidiurnal and diurnal constituents. Note that 30 days would be sufficient to resolve at least M2, S2, N2, K1, and O1 constituents, which tend to dominate in most shelf sea regions. However, despite theoretical estimates of the record length required to resolve tidal constituents, deployment length often depends on external factors such as cost, logistics, battery life, internal memory, and weather windows suitable for deployment and recovery.

ADCPs have various modes of operation, one of which is high-resolution pulse-pulse coherent mode; for example, the bin depth of the instrument can be

TABLE 7.4 Record Length Required to Resolve Semidiurnal and Diurnal Constituents, Based on the Rayleigh Criteria

Constituent	Comparison Constituent	Record Length (h)
M2	–	13
S2	M2	355
N2	M2	662
K2	S2	4383
K1	–	24
O1	K1	328
P1	K1	4383
Q1	O1	662

set to a few centimetres to record the flow in detail over a limited range. Such data are suitable for calculating turbulence properties of the flow, for example, using the structure function technique to calculate turbulent dissipation [12].

A typical time series of backscatter and current speed for a 2-week ADCP deployment is shown in Fig. 7.9. Note that ADCPs also contain a pressure sensor, which logs the variation in the water surface. This plot makes both near-bed (constant) and near-surface (variable) blanking regions clear.

Hull-Mounted ADCP

ADCPs are also frequently deployed in the hull of a ship, looking downwards through the water column. When the ADCP data are postprocessed, the ship's movement can be corrected by using either bottom tracking or the ship's GPS. The advantage of using a hull-mounted ADCP is that it introduces spatial variability into the sampling programme. Therefore, a transect, or a series of transects, can be built up over time, provided the timescale for collecting each transect is relatively short compared with the tidal period, and so can be considered as an instantaneous time slice (e.g. Fig. 7.10). Because data collected using a hull-mounted ADCP will be characterized by pitch and roll due to the vessel's movement, this form of data collection is not reliable for either (a) quantifying the vertical component of velocity (other than in regions, such as sills, where this component is very strong), or (b) estimating turbulence properties. In both of these scenarios, a moored ADCP is required.

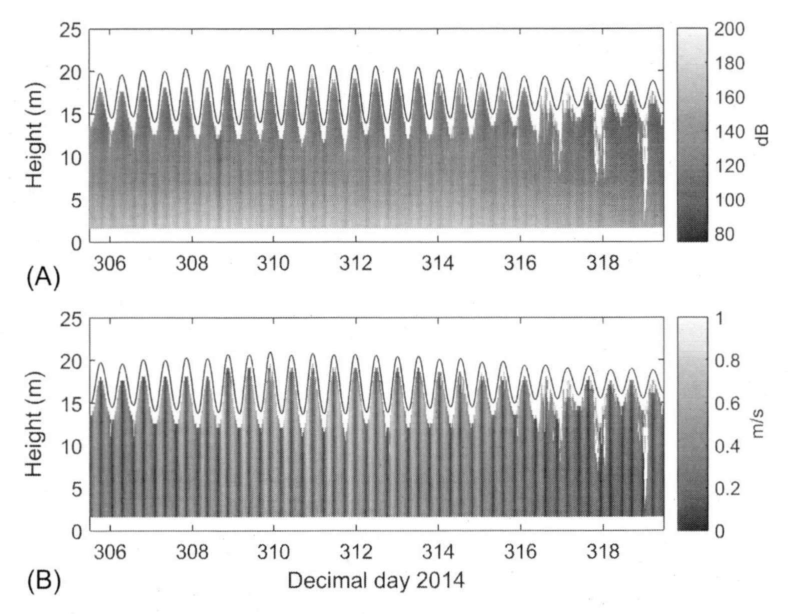

FIG. 7.9 Time series of (A) backscatter and (B) current speed from a 2-week 600 kHz ADCP deployment at the northeast approach to the Menai Strait, UK.

7.1.5 Lagrangian Drifters

Moored current meters and ADCPs measure properties of the sea in an *Eulerian*[1] context. However, it is often desirable to measure properties in a *Lagrangian*[2] framework. This could be useful, for example, when attempting to understand how larvae propagate through a region, how a contaminant such as an oil spill is advected by the current, or for measuring longer-timescale residual flow. Such measurements are conducted using drogues (Fig. 7.11). For short-term deployments, a drogue can be an inexpensive and relatively simple device with two key elements: a surface buoy (so that the drogue can be identified and visually tracked) and a (weighted) parachute with a suitable length of cable appropriate to the representative depth of water that is to be tracked. In the absence of a parachute, the movement of the drogue would simply correspond to the wind-influenced surface currents. Using an inexpensive arrangement and differentiating between surface markers, it is possible to release and track (e.g. with a small boat and a hand-held GPS) many drogues simultaneously over relatively short timescales. There is minimal financial impact if a drogue is lost, and it is more statistically meaningful to track the dispersal of a

1. Observations of fluid motion based on a fixed point in space.
2. Observations of fluid motion where the observer follows an individual fluid parcel as it moves through space and time.

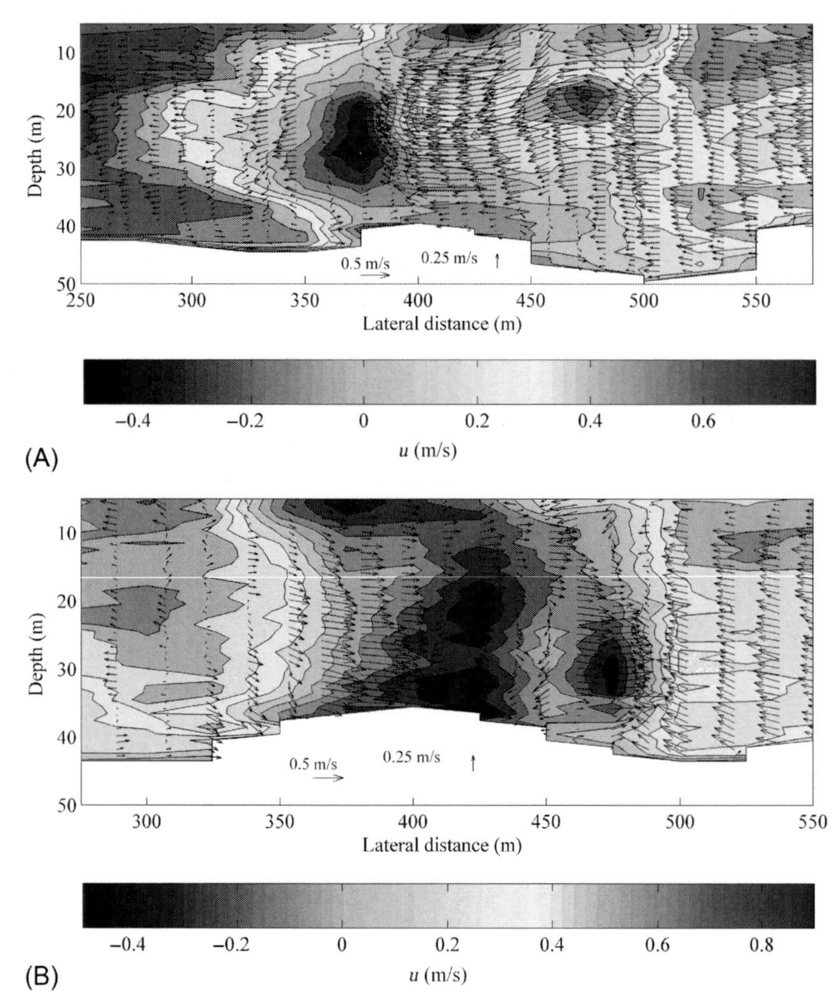

FIG. 7.10 ADCP transect at peak ebb through an island wake in the Firth of Forth, Scotland. Contours are u, and vectors v and w. Each time slice took around 10 min to record using a 300 kHz hull-mounted ADCP, and each slice is located at a different distance downstream from the (50 m wide) island that generates the wake: (A) 125 m downstream, and (b) 75 m downstream. *(Reproduced from S.P. Neill, A.J. Elliott, In situ measurements of spring-neap variations to unsteady island wake development in the Firth of Forth, Scotland, Estuar. Coast. Shelf Sci. 60 (2) (2004) 229–239.)*

large number of inexpensive drogues than to track a few expensive buoys. For longer-term deployments, or better funded projects, it may be desirable to release drogues that have an internal position logging system, or which communicate their position via satellite communication. Additional sensors can also be incorporated with such buoys to record other variables, for example, temperature. A thorough discussion of drogues in the marine environment is provided by Joseph [13].

FIG. 7.11 A Microstar drifter made by Pacific Gyre.

The Global Drifter Programme consists of a global array of drifters, and is funded by the Global Observing System of NOAA. About 1250 satellite-tracked drifters (Fig. 7.12) collect various types of data such as currents, sea surface temperature, atmospheric pressure, winds, and salinity. Data from the drifters can be accessed and downloaded through an online map tool via the project website: www.aoml.noaa.gov/phod/dac/index.php.

7.2 WAVE ENERGY RESOURCE CHARACTERIZATION

At the very early stages of project development, resources such as wave atlases (e.g. the ABPmer Atlas of UK Marine Renewable Energy Resources [1], or the NREL Atlas of US marine renewable energy resources) or the outputs from global wave models (e.g. WAVEWATCH III hindcast reanalysis [14]) may be useful for informing site selection. However, such products have either coarse resolution or do not fully characterize temporal variability in the wave climate (e.g. the ABPmer Atlas provides only annual and seasonal means). Therefore, it is almost certain that at some stage of project development, in situ measurements of waves will be required. The most common way of measuring waves, and

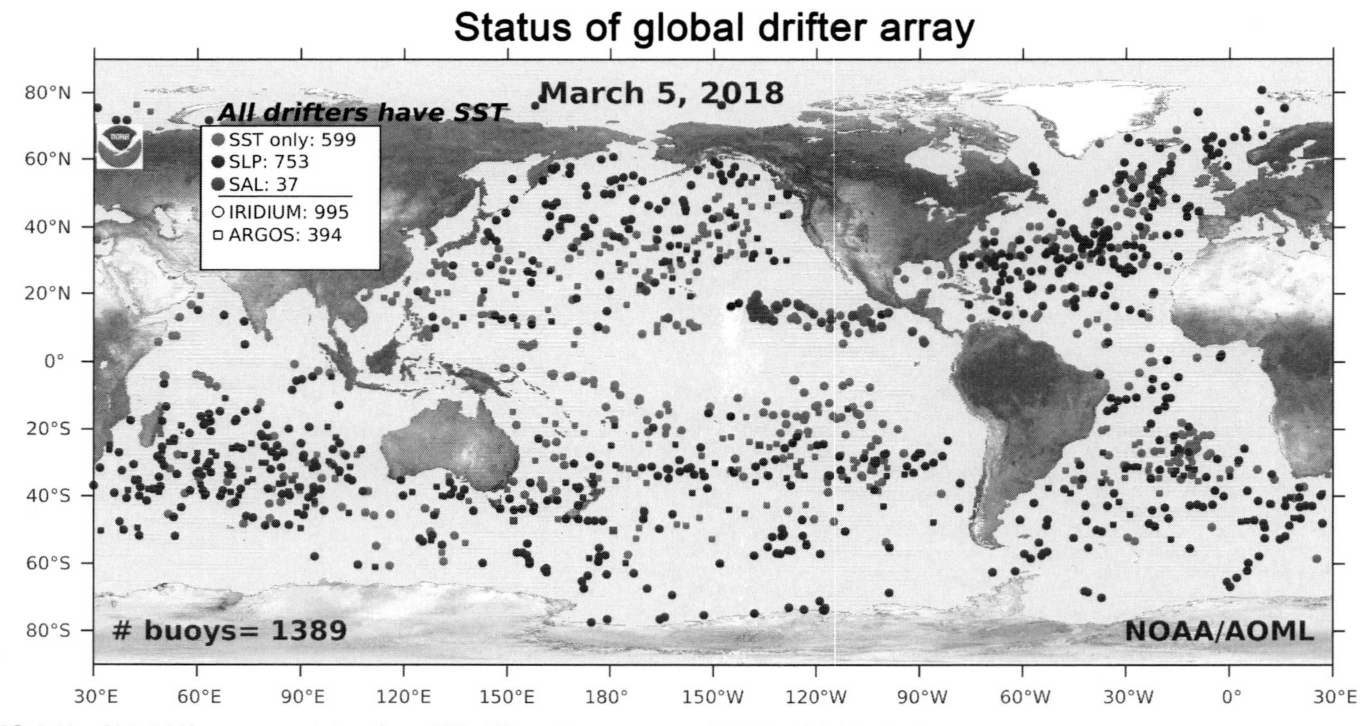

FIG. 7.12 Global drifter array, consisting of over 1000 drifters. *(Image courtesy of NOAA's Global Drifter Programme.)*

hence the main focus of this section, is via surface following wave buoys. However, alternatives such as acoustic devices and pressure transducers are also available, and these are also discussed in this section. In Section 7.3, remote-sensing methods that are suitable for measuring wave properties are also introduced.

7.2.1 Wave Buoys

The most common way of characterizing waves in situ is via a surface following wave buoy. As with measurements of tidal elevations (Section 7.1.1), various networks of such wave buoys exist around the world, such as UK WaveNet (Fig. 7.13), or the Integrated Ocean Observing System (IOOS) in the US. However, wave networks are driven by factors other than quantifying the wave energy resource, such as coastal flood risk, and so there are many energetic regions that are not covered by wave buoys. For example, the west coast of Scotland, despite having the most energetic wave conditions around the United Kingdom, is sparsely populated by wave buoys. Therefore, within many wave resource assessment studies, it is necessary to deploy a project specific wave buoy.

General Principals

Wave buoys measure their vertical acceleration using an on-board accelerometer, supplemented by an artificial horizon that accurately defines the vertical [15]. Modern wave buoys both obtain their position and transmit their data via satellite communications. GPS (i.e. Differential Global Positioning System) is so accurate that wave buoys have now been developed which do not need an on-board accelerometer or other motion sensors [16]. This has several advantages over accelerometer-based buoys, including lower cost and lack of long-term calibration of the instrument. However, GPS-based wave measuring systems have significantly higher energy consumption, and require a constant connection with GPS satellites. This latter requirement can be challenging under rough sea conditions, including whitecapping waves and spray, which affects the reliability of the connection.

In the case of a wave buoy that measures vertical acceleration of the device, the time series of water elevation can be calculated by integrating the vertical acceleration twice, that is, vertical displacement of the sea surface η is

$$\eta = \iint \frac{dw}{dt} \tag{7.6}$$

where w is the vertical velocity.

Wave directional information can be obtained by measuring either the slope of the sea surface or the horizontal motion (surge and sway[3]) of the wave buoy. The sea surface slope can be determined by including additional sensors (inclinometers) to detect the tilt of the buoy in two orthogonal directions, and a

3. Vertical motion is defined as heave.

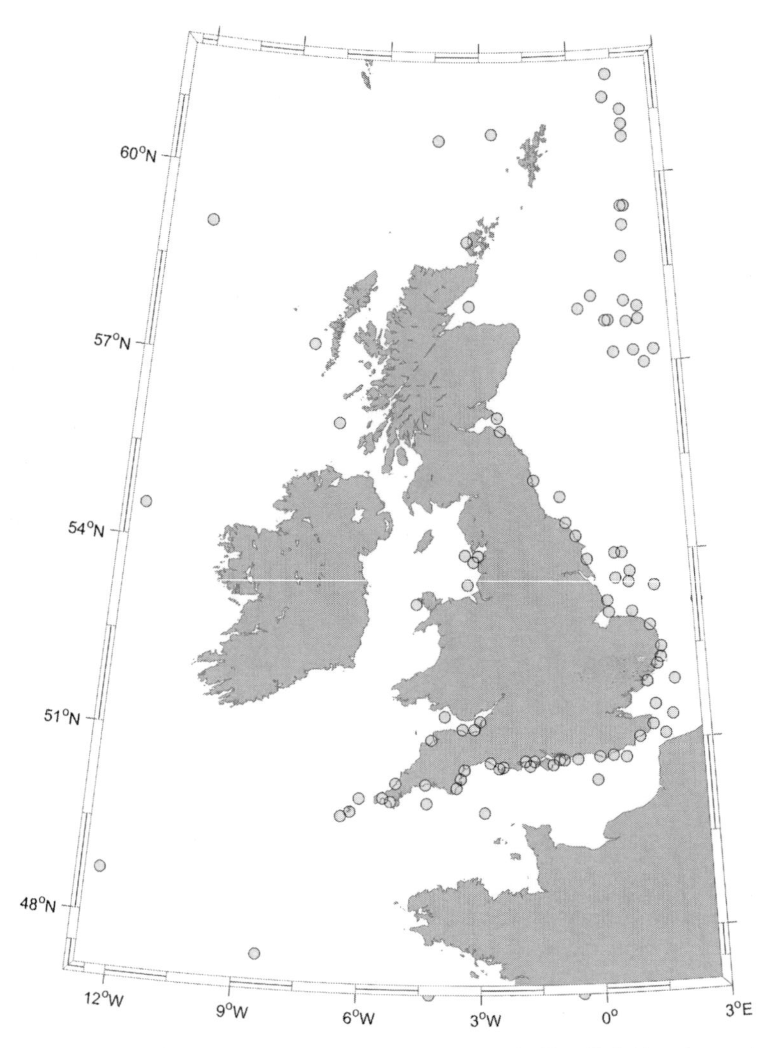

FIG. 7.13 The 96 stations that comprise the UK wave network 'WaveNet'. Note the paucity of wave buoys in the energetic waters to the west of Scotland. The wave buoys that are clustered in the North Sea are part of the network due to collaboration with the oil and gas industry.

sensor to monitor the direction of north [15]—sufficient information to calculate the mean wave direction. Wave buoys which measure their own horizontal motion use the Earth's magnetic field to quantify surge and sway. This surge-and-sway motion indicates the mean wave direction. An example of a directional wave buoy, and probably the most widely used wave buoy around the world, is the Datawell DWR MK III, shown in Fig. 7.14.

FIG. 7.14 Datawell DWR MK III directional waverider predeployment (*left*) and postdeployment (*right*).

Postprocessing and Interpretation

As discussed in Chapter 4, waves exhibit considerable seasonal and interannual variability. To reduce uncertainty in resource assessment, it is therefore valuable to obtain as long a time series as possible from a wave buoy for subsequent analysis and interpretation. However, obtaining a wave record over, for example, a decade, is impractical for the majority of wave energy projects, unless an on-going deployment or historical dataset is available. Therefore, often shorter duration project-specific wave buoy deployments are used to validate numerical wave models (Chapter 8), and the outputs of the validated models (which can be applied to timescales of 10 years or longer, e.g. [17]) can be used for resource assessment. Under such circumstances, a 1-year wave buoy deployment would likely suffice, because such a timescale would capture both seasonal and short-term (e.g. storm) variability. Note that although wave models applied to relatively long-time periods are useful for quantifying the resource, only in situ measurements of wave conditions can truly *characterize* the resource, because observations include, for example, many nonlinear effects that are parameterized within wave models (e.g. wave-wave interactions), and processes that are difficult and computationally expensive to simulate, for example, interaction between wave and tidal resources.

Postdeployment, the three characteristics of the wave data to be considered, in relation to wave power, are the *temporal*, *directional*, and *spectral* characteristics [18].

Temporal characteristics: A wave buoy measures how wave properties such as significant wave height and wave period evolve over time. A wave time

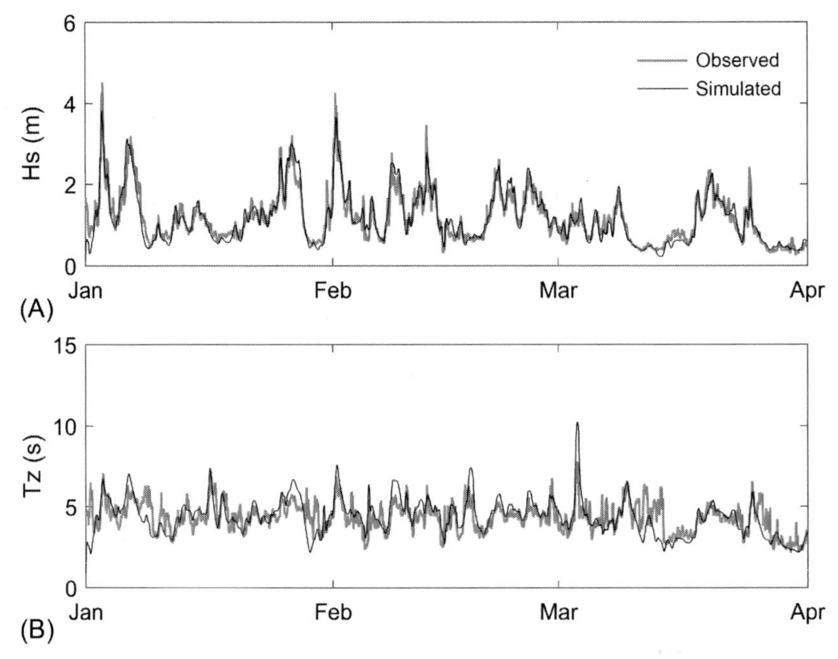

FIG. 7.15 Observed and simulated times series of (A) significant wave height (H_S) and (B) zero upcrossing wave period (T_Z) in Galway Bay, Ireland. *(Data supplied by the Marine Institute, Ireland, and the model is described in Section 8.9.1.)*

series such as that shown in Fig. 7.15 demonstrates, amongst other things, two important temporal properties of waves. First, waves are stochastic, and so time periods of modest wave activity are interspersed with both storms and periods of relative calm. Although they contain vast amounts of energy, and it would be a major breakthrough if we could exploit this energy, storm waves can damage WECs, and hence the majority of WECs would enter *survival model* under such conditions. By contrast, periods of relative calm, when electricity generation would be minimal, represent 'windows of opportunity' for device or array maintenance—particularly prolonged quiescent periods that tend to occur during summer months. Also, note on Fig. 7.15 the qualitative comparison of observations against model simulations. Associated with this qualitative comparison, validation statistics can be derived such as the root-mean-square-error and bias (Chapter 8), giving confidence in the model when applied to longer-time series suitable for resource assessment.

Directional characteristics: The directional information obtained from directional wave buoys discussed in the previous section provide useful information for many wave energy projects. Wave direction is usually presented as a 'waverose,' or by statistical properties such as mean wave direction. Directional information is particularly useful for attenuator-type wave energy convertors (WECs), because such devices should be aligned with the direction of wave propagation for maximum efficiency. Further offshore, although the wave climate will be characterized by a modal wave direction that tends to correspond with the predominant wind direction, waves of different periods and heights propagate from a wide range of directions, in an environment that may be more suited to point absorber devices. However, nearer to the coastline, wave crests that may have originated at oblique angles relative to the coastline in deeper water tend to align themselves parallel to the coastline in the nearshore due to wave refraction (Section 5.2.2). Therefore, a directional wave buoy located in deeper water will generally record a much broader range of wave directions, compared with a wave buoy that is located in shallower coastal waters.

Spectral characteristics: Because wave buoys, in contrast to tide gauges (Section 7.1.1), sample changes in water surface elevation at HF, they can report the spectral properties of waves. Different WECs are tuned to specific parts of the wave energy spectrum. For example, the Pelamis device generates electricity over a relatively wide range of wave parameters, but is optimal within a particular range [19]. To match, or tune, a WEC to the local wave climate, it is therefore important to record spectral properties of waves. An example of a mean 1D wave spectrum is shown in Fig. 7.16. In this example, the spectral peak (over the averaging time period of around 2 weeks) is around 11 s ($f = 0.09$ per s), and the wave energy distribution around this spectral peak varies as shown. Note that if the spectral shape is assumed, for example, by assuming a JONSWAP or P-M spectrum, much of the natural distribution will not be captured. This is important in mean wave spectra, but particularly for instantaneous wave spectra, especially in cases of bi-modal spectra. In addition to 1D spectra, wave buoy postprocessing software also produce 2D

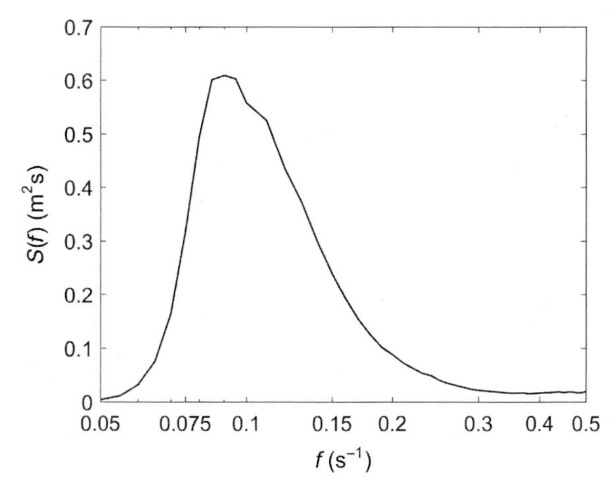

FIG. 7.16 Observed 1D wave spectra from the Pentland Firth, Scotland, averaged over the period January 17–31, 2012. *(Data kindly supplied by Philippe Gleizon, University of the Highlands and Islands.)*

spectra. As mentioned earlier, such directional information is useful for wave energy devices (e.g. attenuators) that must be aligned with the direction of wave propagation.

7.2.2 Pressure Transducers

Wave buoys, as described in the previous section, are relatively expensive, for example, in the region of $60,000 for a directional wave buoy, in addition to operation and maintenance costs. Therefore, under many circumstances, it may be desirable to seek a more economical solution, particularly at the earlier stages of project development. One such alternative is to install a sea-bed pressure sensor. These instruments measure the total pressure, and analysis of instantaneous pressure at high frequencies (e.g. > 1 Hz) provides a measure of wave height and period. However, the accuracy of wave measurements by pressure sensor is strongly influenced by attenuation effects, and so their use is limited to water depths of less than 20 m [20]. Such a pressure sensor can also be used to concurrently determine tidal elevations, but care should be taken to account for changes in the atmospheric pressure when attempting to distinguish the astronomical component [18]. Note that by installing an array of sea-bed pressure sensors, information on the directional characteristics of waves can also be obtained.

7.2.3 Acoustic Waves and Currents

Some ADCPs include an additional vertical-orientated transducer that can measure the vertical velocity (Fig. 7.5), and can measure wave properties or estimate turbulence. Such an instrument is known as an Acoustic Waves and Currents (AWAC). Waves are traditionally measured by wave buoys (Section 7.2.1), but by using an AWAC, currents and waves can be measured concurrently at a location of interest. Because this device is deployed on the sea bed, it does not suffer from constraints such as severe weather impacts or the influence of marine traffic. Such an instrument is particularly beneficial for studies of wave-current interaction. Some researchers have assessed the accuracy of wave measurements by ADCPs compared with wave buoys. Bouferrouk et al. [21] compared the wave data collected by a five-beam ADCP and four wave buoys at the WaveHub (UK)—a site that is dedicated to the testing of wave energy devices. Their results show, in general, good agreement between ADCP and wave buoys. However, unlike wave buoys, the five-beam ADCP could also detect long-period swell waves (0.07–0.08 Hz), for small wave heights, and can only operate in relatively shallow waters.

7.3 REMOTE SENSING

Although, in some ways, an ADCP could be considered a *remote*-sensing platform, remote sensing really refers to observations made at some distance from the medium that is being observed. Remote sensing most often refers to the use of satellite sensors. However, remote-sensing instruments can also be carried by aircraft (airborne remote sensing), or installed on land- or sea-based platforms (e.g. X-band and HF radar).

Remote sensing can be categorized as either *passive* or *active*. Passive sensors gather radiation that has been emitted or reflected by the sea surface, for example, reflected sunlight. By contrast, active sensors emit energy, then detect the radiation that is reflected or backscattered from the target.

7.3.1 X-Band Radar

X-band radars are installed on most large research vessels, and on many offshore installations. Their original use is for ship traffic control and navigation. However, with the addition of some hardware and software components, X-band radars can be used to measure waves and currents (e.g. [22]). X-band is a segment of the microwave radio region of the electromagnetic spectrum, within the frequency range 8–12 GHz.

The images produced by marine radars detect not only hard targets such as ships and coastlines, but also reflections from the sea surface, known as 'sea clutter' [23]. Given a wind speed of more than approximately 3 m/s, the backscatter from the sea surface becomes visible in radar images [24]. Such reflections of waves are mostly due to resonance between the radar waves and the features at the water surface (*Bragg Scatter*) [15]. Because the radar wave length is in the centimetre range, only very short water waves reflect the radar waves. However, the base signal (the very short waves) is modulated by longer waves. For navigational purposes, this sea clutter is treated as a nuisance, and is discarded; but it contains valuable information on the actual sea state. With appropriate software, it is possible to analyse this signal to gain information on wave height, wave length, wave period, and surface currents. X-band radar systems scan the ocean surface in real time at high temporal (1–2 s) and spatial (5–10 m) resolution. An area of sea surface of several square kilometres can therefore be continuously monitored. X-band radar systems can be installed on moving vessels or on fixed platforms offshore or at coastal sites.

The main limitations of X-band radar are the limited range, and the requirement for sufficient sea clutter (and associated significant wave height) to allow imaging of the waves. Due to the limited range of X-band radar, a fixed platform would have to be installed reasonably close to the area of interest, or the system would have to be mounted on a vessel. The need to install a fixed offshore platform would sacrifice some of the inherent advantages of X-band radar as a remote-sensing technique, such as easy installation and onshore maintenance, and would represent considerably increased cost. X-band radar systems are also sensitive to interference from rainfall, and indeed this radar frequency is used by weather radars to monitor rainfall.

7.3.2 HF Radar

The main difference between HF and X-band radar systems is that because the former operates at much lower frequencies (3–30 MHz), the waves are transmitted over much longer distances, and so the region that is sampled is much larger, for example, up to 200 km [25]. However, because the wave length is longer than X-band, the resolution of HF radar is coarser than X-band, with typical resolutions of 300 m to 15 km achieved, depending on the transmission frequency [26].

In contrast to X-band, HF radar systems require considerable infrastructure installation in the form of radar antenna arrays, and so are generally considered as investment in a long-term monitoring campaign. The advantages of HF radars are that they can monitor large areas of the sea surface at high temporal resolution, there is minimal risk of collision (in comparison to sea-based measuring systems), and that they can be installed and maintained from land. However, installation costs are high, and they suffer from interference from other users of the broadcasting band, and a poor signal-to-noise ratio in high winds and stormy seas [25]. Note that HF radar can also be used to measure the vertical distribution of winds via a radar wind profiler.

HF radar stations can also estimate surface currents, including tidal currents. They calculate the Doppler shift that is Bragg scattered at the surface. Because the radial velocity (towards or away) from the radar station is estimated, two or more radial stations are necessary to evaluate the components of water velocity. NOAA has set up a network of HF radar systems around the US coast (ioos.noaa.gov/project/hf-radar/) for real-time monitoring of coastal currents. The depth that an HF radar signal can feel the currents is $\lambda_R/8\pi$, where λ_R is the radio wave length. For a 12-MHz signal (suitable for current measurement of about 2 km resolution), the wave length is 25 m, and the effective depth is about 1 m. An HF radar system with a frequency of 40–45 MHz can measure currents up to 300 m resolution. In terms of tidal energy resource assessment, a combination of HF radar and ADCPs can provide a good assessment of the spatial (by HF radar) and vertical (by ADCP) variability. Some applications of HDF radar in tidal energy resource assessment can be found elsewhere [27].

7.3.3 Satellite and Airborne Remote Sensing

Satellites provide uniform global coverage of the sea surface, with no preference for ports. However, the resolution of satellite data that is useful for resource characterization tends to be coarse, for example, several kilometres resolution global coverage in 1 year of data collection would not be uncommon. Airborne remote-sensing platforms offer the advantage of improved resolution compared with satellite platforms but, because over flights tend to be commissioned, at increased cost to the end user compared with satellite data.

Satellite altimetry measures the time taken for a radar pulse to travel from the satellite to the sea surface and back to the satellite. Radar altimeters map the topography of the ocean surface with unprecedented accuracy. The extra gravitational attraction of sea-bed features such as seamounts produces minor variations in gravity, which in turn produce tiny variations in ocean surface height. Combined with suitable algorithms, radar altimetry can therefore be used to estimate bathymetry, and altimetry-derived data is integrated into popular global bathymetry datasets such as GEBCO.[4] Direct tidal analysis of altimetry data can be used to derive estimates of the ocean tides [28], but altimetry data are generally assimilated within hydrodynamic models to produce accurate global tidal constituent databases such as FES2012 [29] that can be used for direct (coarse spatial resolution) analysis or as model boundary conditions. Further, the strength and shape of the returning radar signal from altimetry satellites provides valuable information on wind speed and the height of ocean waves. Although a single satellite track has high temporal resolution, repeat coverage of a region is widely spaced in time, and so instantaneous altimetry data are best suited to validating the spatial skill of a wave model, rather than directly characterizing temporal properties. Finally, when a satellite altimeter approaches the coast, land entering the radar footprint modifies the shape of the waveforms, making the estimate of range and other derived quantities difficult in such regions. For this reason, satellite altimetry data are usually discarded in the coastal zone, but some recent work is beginning to make data in such regions useable (e.g. [30]).

Synthetic Aperture Radar (SAR) is based on the same principal as X-band radar, but the instrument is carried by a satellite or aircraft. The problem in observing waves from high altitude is that the antenna needs to be very large in order to distinguish individual waves in the modulation pattern. In a SAR system, the satellite or aircraft continuously illuminates the sea surface and receives the scattered signal as it travels along its flight path. Knowledge of this history permits later reconstruction of the reflected signals as if they were received by a single ‘synthetic’ antenna occupying a physical space that is defined by the movement of the satellite or aircraft along its flight path, even though the signals were received by a much shorter antenna [31].

In any remote-sensing technology, it is important to obtain concurrent ground truthing data (e.g. from a vessel or mooring) for either calibration of the remote-sensing algorithms or validation of the remotely sensed data. Because it is logistically difficult to organize such activities, for example, the concurrent mobilization of an aircraft and a ship, it perhaps works best when a relatively long-term mooring (e.g. a wave buoy) is already in place, and the remotely sensed data primarily used to improve spatial coverage of the observed property.

4. See http://www.gebco.net.

7.4 VESSEL MEASUREMENTS

Although vessels are used to deploy instruments, they can also be used as a platform for making direct measurements of the water column and sea bed. Although expensive, they have the advantage, in contrast to moorings, that multiple locations can be sampled during one cruise. For example, using a gridded arrangement of sampling stations, cruises can be organized such that variables can be sampled at several stations over one or multiple tidal cycles at relatively high temporal resolution (e.g. hourly).

The RV Prince Madog[5] at Bangor University (Fig. 7.17) is an example of a good general-purpose research vessel suitable for a range of ocean energy-related activities. With a length of 35 m, draft of 3.5 m and maximum speed of 12 knots, the vessel is sufficiently robust and manoeuvrable to deploy instruments and record data in relatively shallow regions of strong tidal flow. The vessel has nine scientific berths and an endurance of 10 days between ports,

FIG. 7.17 RV Prince Madog.

5. Madog was a Welsh Prince who, according to folklore, sailed to America 300 years earlier than Christopher Columbus.

and so can remain on station for extended periods of time, sampling around the clock.

Various forms of vessel sampling are introduced in the following sections. However, one form of vessel measurement that has already been covered is a hull-mounted ADCP (Section 7.1.4). Therefore, this is not considered further in this section.

7.4.1 Vertical Profiling

In many circumstances, it is useful to make measurements that resolve the vertical structure of the water column. The most widely used instrument for vertical profiling is the conductivity, temperature, and depth (CTD) (Fig. 7.18). A CTD consists of a cluster of sensors that measure properties of the water column and either log internally (for subsequent download) or, more frequently, transmit the data directly to the ship through a data cable attached to the instrument package. The CTD is lowered on a steel cable to the sea bed, and variables continuously sampled throughout the water column. A CTD measures the electrical conductivity and temperature of the water, and from these two properties salinity can be calculated. The depth of each measurement is determined by a pressure sensor. Although regions of strong wave and tidal activity that could be suitable for generating electricity are generally well-mixed, there will be some amount of periodic (e.g. due to regions of freshwater influence) or seasonal stratification. Further, the density of the water is required to make accurate calculations of the marine energy resource and to provide a more thorough characterization of the resource than velocity measurements alone will provide.

It is common for other instruments to be integrated into a CTD cluster, such as a transmissometer or optical backscatter (OBS) (both measure suspended sediment concentrations), or instruments that measure biological properties such as dissolved oxygen or chlorophyll fluorescence (an indication of the concentration of phytoplankton in the water column). In addition, an instrument of the design shown in Fig. 7.18 allows water sampling bottles to be added to the CTD. The operator can *fire* these bottles at predetermined water depths to collect water samples for subsequent processing; for example, the water samples can be filtered for either SPM or chlorophyll concentrations to calibrate the electronic SPM and chlorophyll sensors, respectively.

7.4.2 Point Sampling

Ships are useful platforms for making point measurements of *dynamical* water properties that vary both temporally and spatially (e.g. plankton), and properties of the sea bed that could be considered relatively *static* over longer timescales (e.g. sea-bed sediments).

FIG. 7.18 CTD and carousel water sampler prior to deployment.

Dynamical Properties

As mentioned in Section 7.4.1, a CTD fitted with sampling bottles can be used to collect water samples from a fixed point in the water column (e.g. near the surface or sea bed) for subsequent analysis. This analysis could be relatively routine, and possibly performed on the vessel to reduce the need for transporting large volumes of sea water to a laboratory for postprocessing. These activities include filtering for SPM or chlorophyll. However, sometimes more intensive and laborious postprocessing is required, and under such circumstances, only a limited amount of processing on the ship may be required. A good example of this category is filtering sea water for plankton that has been collected using a plankton net. Three commonly used types of plankton net are a ring net, a bongo net, and a trongo net, the main difference being increasing numbers of nets (and hence concurrent samples) from 1 to 2 to 3, respectively. A vessel tows a plankton net for several minutes (or a shorter-time period when concentrations of plankton are high), after which the sample is concentrated and preserved in formalin for subsequent microscopy work in the laboratory to determine plankton densities, based on the volume of sea water that has passed through the net opening. Clearly, an important consideration is the mesh size, which will be determined by the size of the smallest organisms that are to be sampled. Typical (cruise-averaged) results for a zooplankton survey are given in Table 7.5.

Finally, another simple point sampling instrument is the secchi disk. A secchi disk is simply a 30-cm diameter white disk that is weighted and lowered into the water. The depth at which the secchi disk is no longer visible relates (inversely) to the turbidity of the water, and so is a simple measure of near-surface sediment concentration. However, the secchi disk is subject to operator error, for example, the effects of sunlight and waves on the water surface make it difficult to accurately and consistently record a secchi depth.

TABLE 7.5 Mean Near-Surface Zooplankton Concentrations at the Northeast Approach to the Menai Strait (UK), November 8–13, 2012

Zooplankton	Density (Individuals per m^3)
Diatoms	3082.5
Copepods	369.4
Chaetognaths	44.4
Tunicates	9.2
Echinodrm larvae	7.6
Polycheate larvae	3.8
Caprellids	3.5
Crab larvae	3.2
Barnacle larvae	2.4
Ctenophores	2.4
Fish eggs	1.0
Bivalve larvae	0.5

Notes: These values represent the averages over five stations that were visited each day for 6 days.

Static Properties

To characterize sea-bed sediment, grab samplers are often deployed from vessels. Grab samplers are quick and relatively easy to deploy (provided the operating team is skilled) and provide a sample of the sea bed that is suitable for subsequent grain size analysis in the laboratory. Typical grab samplers are the Shipek, Van Veen, and the day grab (Fig. 7.19). However, the fine-grained sediments are washed out in grab samples, and there is a risk of blockage of the jaws from coarser particles. Further, a grab sampler produces a mixed sediment sample, and so no information about the structure of the sea bed is preserved, and the grab only provides information about the sea-bed surface. Under many of these circumstances, more specialist equipment is required, such as corers (e.g. gravity corer, piston corer, box corer, and vibro corer), or drilling. Drilling, in particular, requires a specialist vessel with dynamic positioning.

7.4.3 Multibeam

All vessels are fitted with a single-beam echo sounder, which measures the double way transit time of an acoustic beam that has been reflected from the sea bed. Because we know the speed of sound in water (and can correct using local measurements of temperature and salinity, etc.), this transit time can be used to calculate the water depth under the vessel. In real-time, this is essential for navigation, for example, to monitor the depth under the vessel in shallow waters, but the data can also be logged to create an xyz (scatter) dataset of water depths, which can later be corrected for temporal changes in tidal elevation, for example, by obtaining measurements of a local tide gauge to reduce the data to mean sea level (MSL) during the postprocessing stage. The resulting corrected xyz dataset can then be gridded and used to generate model bathymetry, etc. However, the

FIG. 7.19 Day grab being deployed from the RV Prince Madog.

main disadvantage of a single beam is that only measurements directly under the vessel are obtained, and that the resolution of the data depends on the speed of the vessel. Therefore, it is rather time-consuming to generate a 2D grid of a region using a single-beam echo sounder, and this instrument tends to be more suited to obtaining multiple 1D transects. This shortcoming is addressed by the use of a multibeam echo sounder.

In a multibeam echo sounder, the acoustic signal is generated in the shape of a fan through a wide angular lateral aperture transducer, and reflections of the sea bed are received along multiple narrow beams. Water depths are extrapolated along a wide band called a swath. This swath varies with water depth, but is typically around two to four times the water depth. For example, in water of depth 10 m, the swath will be relatively narrow (e.g. 20–40 m), and so sequential ship tracks will have to be fairly closely spaced to provide contiguous coverage of the sea bed. However, in deeper water (e.g. 100 m), the swath will be in the range 200–400 m, and so ship tracks can be spaced further apart. Multibeam echo sounder surveys therefore represent a considerable improvement of sea-bed coverage compared with single-beam systems, but at a cost of (a) the need for expensive specialist calibrated equipment, and (b) considerable expertise in deployment and postprocessing. In addition, multibeam surveys generate a considerable amount of data, and allowance must be made for data storage. Note that, as with the single-beam system, resolution in the direction of the ship's transit still depends on the speed of the vessel. Lateral resolution depends on the water depth. In some circumstances, it may be necessary to disable other acoustic instruments on the vessel (e.g. ADCPs) to prevent contamination of the multibeam signal.

7.5 OTHER FORMS OF MEASUREMENT

In addition to the in situ and remote methods of data collection that have been discussed so far in this chapter, complementary data that could be useful for resource characterization can be obtained from a variety of sources. Probably the most important of these is meteorological data, for example, time series of wind speed and direction, and atmospheric pressure. The best source of such data are synoptic weather stations, which record such variables to a consistently high standard. However, often a marine energy site will be far from a synoptic weather station, or will be influenced by local topographic effects (e.g. wind that is channelled through a narrow strait); therefore, an alternative (e.g. local existing personal weather stations that report daily means, or a project-specific weather station) may be preferable, or could be used to complement neighbouring synoptic observations.

Other sources of data that are useful for resource characterization include local authority beach profile surveys [32], useful for quantifying the natural variability of nearshore systems that may be influenced by marine energy extraction. In addition to single-beam and multibeam echo sounder surveys (Section 7.4.3), many bathymetric products exist such as GEBCO and EMODnet—these are discussed in Chapter 8, because they are particularly useful for setting up numerical models, in addition to helping directly characterize sites. Existing sea-bed sediment surveys will be useful for helping to characterize a site, and these can be available as point measurements or via gridded data products, for example, provided by the US Geological Survey or the British Geological Survey.

Finally, discussing plans for an ocean energy project with local users and communities and allaying any fears about the development (plus highlighting opportunities) are invaluable at all stages of project development and can facilitate and inform data collection and sampling strategies. In addition, by engaging with local users of the resource, important 'anecdotal' evidence can be gathered, such as which tidal regions are more energetic during stronger wave conditions (and hence may require consideration of wave/current interaction), which regions are more sheltered, and (especially from surfers), under what wind conditions (and in what regions) are ideal surf conditions experienced (which also tend to be the best conditions for generating electricity from waves!).

REFERENCES

[1] ABPmer, Atlas of UK Marine Renewable Energy Resources: Atlas Pages, A Strategic Environmental Assessment Report, Department for Business Entreprise & Regulatory Reform, 2008.

[2] L.S. Blunden, A.S. Bahaj, Initial evaluation of tidal stream energy resources at Portland Bill, UK, Renew. Energy 31 (2) (2006) 121–132.

[3] P.E. Robins, S.P. Neill, M.J. Lewis, S.L. Ward, Characterising the spatial and temporal variability of the tidal-stream energy resource over the northwest European shelf seas, Appl. Energy 147 (2015) 510–522.

[4] P.L. Woodworth, High waters at Liverpool since 1768: the UK's longest sea level record, Geophys. Res. Lett. 26 (11) (1999) 1589–1592.

[5] D.T. Pugh, Tides, Surges and Mean Sea-Level (Reprinted With Corrections), John Wiley & Sons Ltd, Hoboken, NJ, 1996.

[6] GGOS Portal, 2011, Available from: http://www.ggos-portal.org, Accessed 6 December 2016.

[7] G. Voulgaris, J.H. Trowbridge, Evaluation of the acoustic Doppler velocimeter (ADV) for turbulence measurements, J. Atmos. Ocean. Technol. 15 (1) (1998) 272–289.

[8] L.E. Myers, A.S. Bahaj, Experimental analysis of the flow field around horizontal axis tidal turbines by use of scale mesh disk rotor simulators, Ocean Eng. 37 (2) (2010) 218–227.

[9] J. Sharples, J.H. Simpson, J.M. Brubaker, Observations and modelling of periodic stratification in the Upper York River Estuary, Virginia, Estuar. Coast. Shelf Sci. 38 (3) (1994) 301–312.

[10] R D Instruments, Principles of Operation A Practical Primer, 1996, Available from RDInstruments.com.

[11] R. Swift, Assessment of Performance of Tidal Energy Conversion Systems, EMEC/British Standards Institution, London, 2009.

[12] P.J. Wiles, T.P. Rippeth, J.H. Simpson, P.J. Hendricks, A novel technique for measuring the rate of turbulent dissipation in the marine environment, Geophys. Res. Lett. 33 (21) (2006) L21608.

[13] A. Joseph, Measuring Ocean Currents: Tools, Technologies, and Data, Newnes, 2013.
[14] NOAA WAVEWATCH III—National Weather Center 2016, Available from: http://polar.ncep.noaa.gov/waves/index2.shtml, Accessed 4 January 2017.
[15] L.H. Holthuijsen, Waves in Oceanic and Coastal Waters, Cambridge University Press, Cambridge, 2010.
[16] H.P. Joosten, Datawell 1961–2011; Riding the Waves for 50 Years, Drukkerij Gravé, 2013.
[17] S.P. Neill, M.J. Lewis, M.R. Hashemi, E. Slater, J. Lawrence, S.A. Spall, Inter-annual and inter-seasonal variability of the Orkney wave power resource, Appl. Energy 132 (2014) 339–348.
[18] M. Folley, The wave energy resource, in: Handbook of Ocean Wave Energy, Springer, Berlin, 2017, pp. 43–79.
[19] E. Rusu, C.G. Soares, Coastal impact induced by a Pelamis wave farm operating in the Portuguese nearshore, Renew. Energy 58 (2013) 34–49.
[20] D. Reeve, A. Chadwick, C. Fleming, Coastal Engineering: Processes, Theory and Design Practice, CRC Press, Boca Raton, FL, 2004.
[21] A. Bouferrouk, J.-B. Saulnier, G.H. Smith, L. Johanning, Field measurements of surface waves using a 5-beam ADCP, Ocean Eng. 112 (2016) 173–184.
[22] J.C.N. Borge, C.G. Soares, Analysis of directional wave fields using X-band navigation radar, Coast. Eng. 40 (4) (2000) 375–391.
[23] P.S. Bell, J. Lawrence, J.V. Norris, Determining currents from marine radar data in an extreme current environment at a tidal energy test site, in: 2012 IEEE International Geoscience and Remote Sensing Symposium, IEEE, 2012, pp. 7647–7650.
[24] Coastal Wiki, 2008, Available from: http://www.coastalwiki.org/wiki/Waves_and_currents_by_X-band_radar, Accessed 29 December 2016.
[25] T.J. Cross, A guide to remote sensing: information from the Remote Sensing and Marine Renewable Energy Workshop, All-Energy, 21 May, in: NERC Marine Renewable Energy Knowledge Exchange Programme, 2013.
[26] T. Helzel, M. Kniephoff, L. Petersen, WERA: remote ocean sensing for current, wave and wind direction, in: 2006 IEEE US/EU Baltic International Symposium, IEEE, 2006, pp. 1–8.
[27] M. Thiébaut, A. Sentchev, Asymmetry of tidal currents off the W. Brittany coast and assessment of tidal energy resource around the Ushant Island, Renew. Energy 105 (2016) 735–747.
[28] D.E. Cartwright, R.D. Ray, Oceanic tides from Geosat altimetry, J. Geophys. Res. Oceans 95 (C3) (1990) 3069–3090.
[29] L. Carrère, F. Lyard, M. Cancet, A. Guillot, L. Roblou, FES2012: a new global tidal model taking advantage of nearly 20 years of altimetry, in: Proceedings of Meeting, vol. 20, 2012.
[30] J. Gómez-Enri, S. Vignudelli, G.D. Quartly, C.P. Gommenginger, P. Cipollini, P.G. Challenor, J. Benveniste, Modeling ENVISAT RA-2 waveforms in the coastal zone: case study of calm water contamination, IEEE Geosci. Remote Sens. Lett. 7 (3) (2010) 474–478.
[31] J.B. Campbell, R.H. Wynne, Introduction to Remote Sensing, Guilford Press, New York City, NY, 2011.
[32] S.P. Neill, A.J. Elliott, M.R. Hashemi, A model of inter-annual variability in beach levels, Cont. Shelf Res. 28 (14) (2008) 1769–1781.

FURTHER READING

[1] S.P. Neill, A.J. Elliott, In situ measurements of spring—neap variations to unsteady island wake development in the Firth of Forth, Scotland, Estuar. Coast. Shelf Sci. 60 (2) (2004) 229–239.

Chapter 8

Ocean Modelling for Resource Characterization

In situ measurement campaigns are costly, particularly in the environments, which, by their very nature, are characterized by the strong tidal currents or energetic waves that are suitable for generating electricity. Even the best-funded offshore energy operations suffer from logistical difficulties when it comes to making field measurements in remote locations, and where specialist instruments, vessels, and expertise are required to gather, process, and interpret data. In addition, field campaigns take time; for example, several months of data are often required to characterize tidal conditions at a site, and often several years or longer of wave data to characterize the interannual and intraannual variabilities in the wave climate. Under many of these circumstances, ocean modelling is a commonly used and economic tool in resource characterization, which can be used to generate long-time series or understand how the resource varies under hypothetical scenarios such as climate change, extreme events, or how the resource will be influenced by energy extraction. Ocean models can be used at all stages of project development, but are particularly useful at the early scoping stages, prior to detailed site selection and investment in costly field campaigns. However, it should be stressed that models are only as good as their input data, and it is always important to parameterize and validate such models with in situ measurements.

This chapter will develop your understanding of the terminology and different types of model used for resource characterization, methods of model validation, and limitations to modelling. After studying the chapter, you should be familiar with a wide range of modelling concepts and techniques, and form an appreciation of model preprocessing and postprocessing, and model validation, as well as gaining insights into how ocean models work, rather than treating them as mysterious 'black boxes'.

8.1 GENERIC FEATURES OF OCEAN MODELS

There are many features that are common to the majority of ocean models, for example, grid types, discretization, and boundary conditions; these features are introduced in this section.

Fundamentals of Ocean Renewable Energy. https://doi.org/10.1016/B978-0-12-810448-4.00008-2

8.1.1 Horizontal Mesh Type

There is a vast array of ocean models, and a wide range of ways in which they can be categorized. There are many places we could begin this classification but, because it is of much interest to wave and tidal resource characterization, we begin by differentiating between structured and unstructured meshes.

In a *structured* grid, all grid lines are orientated regularly so that, in a 2D case, the coordinate transformations of curvilinear lines result in a square (Fig. 8.1A). By contrast, this restriction does not apply to *unstructured* grids (or mesh) (Fig. 8.1B); however, this grid is more complex to deal with numerically. The advantage of a structured grid is its simplicity and ease of preprocessing and postprocessing; however, to achieve a desired resolution (e.g. to resolve the curved geometry shown in Fig. 8.1A), it is necessary to use a constant and high-resolution grid throughout the computational domain. Clearly, this resolution is easier to achieve by spatially varying the grid resolution, as in an unstructured grid (Fig. 8.1B), with the added advantage of lower grid storage (e.g. the circle in the centre of Fig. 8.1, which could represent an island, requires no storage using an unstructured grid).

Within the context of ocean energy, it is clear that due to its ability to accommodate a wide range of scales within a single-model domain, for example, a regional model which incorporates a tidal energy array within a narrow strait, unstructured meshes tend to be favoured for resource assessment. However, many ocean models, particularly 3D models, which simulate a wide range of physical processes, are based on a structured mesh. Under such circumstances, a sequence of nested models, with increasing grid resolution from outer to inner nests, will be required. In general, structured meshes tend to be favoured for larger-scale ocean modelling applications.

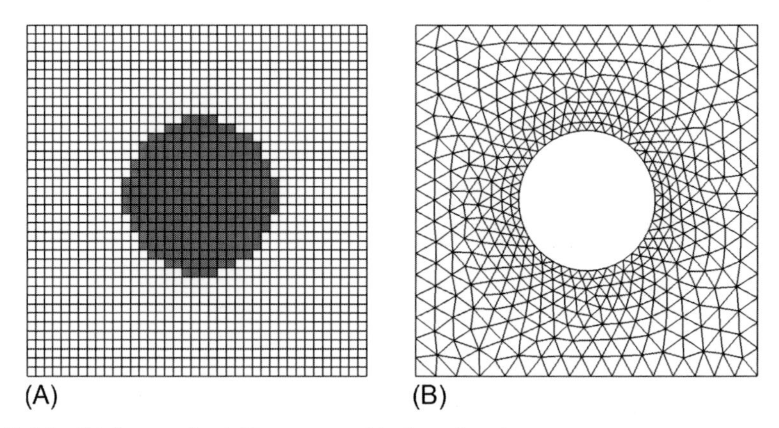

FIG. 8.1 (A) Structured and (B) unstructured horizontal meshes.

8.1.2 Vertical Grid Type

In addition to horizontal resolution, 3D models must also consider the vertical coordinate system. The simplest vertical coordinate system, known as *z*-coordinates, has been used by the ocean modelling community for many decades. The *z*-coordinate scheme divides the water column into a fixed number of depth levels, and these can be distributed to provide higher resolution in any particular region, such as the surface layer (Fig. 8.2A). The disadvantage of the *z*-coordinate system, as demonstrated in Fig. 8.2A, is that it has problems dealing with large changes in bathymetry, which can lead to unrealistic vertical velocities near the bed. Increasing the number of vertical levels will improve the representation of near-bottom flow, but at increased computational cost. This problem is overcome by the sigma coordinate scheme (also known as terrain following coordinate system), where the vertical coordinate follows the bathymetry (Fig. 8.2B). The sigma coordinate system results in the same number of vertical grid points throughout the computational domain, regardless of large changes in bathymetry. The sigma levels do not have to be evenly distributed throughout the water column and could, for example, be more closely spaced near the surface and bed, allowing boundary layers to be better resolved throughout the domain. However, one disadvantage of sigma coordinates is that they can lead to difficulties when dealing with sharp changes in bathymetry from one grid point to another. This can lead to pressure gradient errors, resulting in unrealistic flows [1]. Increased horizontal resolution, or bathymetric smoothing, alleviates the problem.

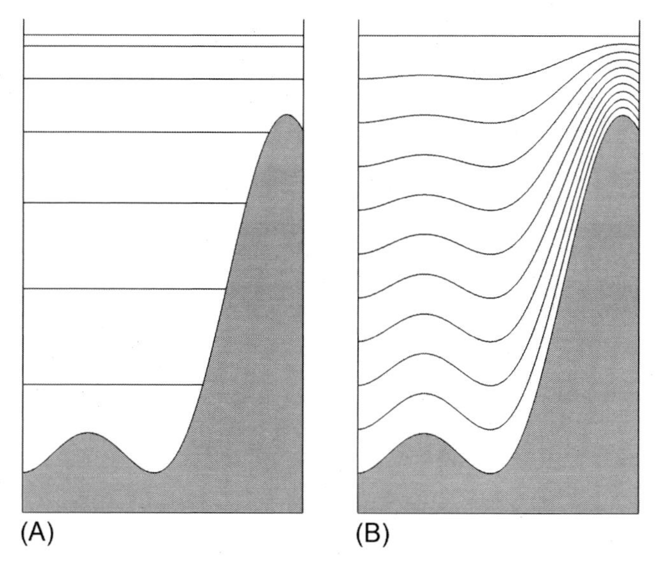

FIG. 8.2 Vertical mesh for (A) *z*-coordinate and (B) sigma coordinate systems.

8.1.3 Sources of Data

As mentioned in the introduction to this chapter, a model is only as good as the data that is used as model input. In addition, data are essential for model validation, which provides confidence in model performance. Model validation is covered in detail in Section 8.5. The main types of data, described in this section, are coastline and bathymetry data, used to set up model bathymetry and mesh, boundary condition data (e.g. astronomical tides), and surface fields such as wind stress and atmospheric pressure.

Coastline Data

The 'closed' model boundary is defined by the coastline, and the source of data will depend on model resolution. For example, for a coarse shelf scale model, the GSHHG (Global Self-consistent, Hierarchical, High-resolution Geography Database) intermediate resolution is sufficient for most applications (Fig. 8.3B), whereas a higher-resolution coastline (e.g. GSHHG full or an alternative dataset) may be desired for smaller-scale coastal applications (Fig. 8.3D). In the example shown, the Pentland Firth (the strait between mainland Scotland and the Orkney archipelago) is adequately resolved using the intermediate resolution, whereas

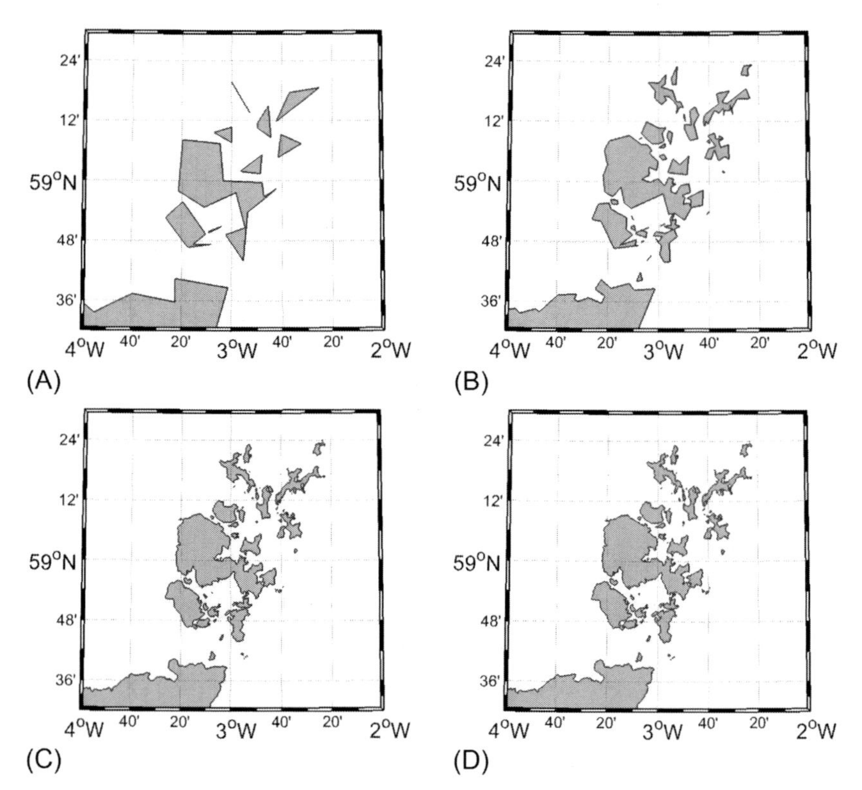

FIG. 8.3 Variable GSHHG coastline resolution applied to the Pentland Firth and Orkney waters. (A) Low, (B) intermediate, (C) high, (D) full.

the interisland channels are resolved using the high-resolution dataset. In more advanced modelling systems, it is not necessary to clearly define a coastline, as it will change by wetting and drying, and this moving boundary is computed by the model. For those cases, a larger domain that includes bathymetry and topography (e.g. 10 m above MSL) will be used as input.

Bathymetry Data

Although it is possible to generate a model grid using even the coarsest of bathymetric data, it is generally desirable for the resolution of the bathymetry data to match or exceed the model grid resolution. For example, there is little point generating a model that has a grid resolution of 25 m if the available bathymetry for the region only has a resolution of 1 km.

There are a wide range of sources of bathymetry data, which vary in scale and resolution from multibeam surveys (resolution <10 m) to global and regional datasets such as GEBCO (global 1/2 arc-min grid), NOAA bathymetry and global relief, and EMODnet (European 1/8 arc-min grid). In general, gridded bathmetry datasets tend to be suitable for shelf-scale or regional-scale models, whereas high-resolution multibeam data may be required for refined, localized, model studies. Finally, it is important to correct available bathymetry data to the desired model datum, usually mean sea level (MSL). In general, bathymetry data are available relative to the lowest astronomical tide, LAT[1] (see Section 3.9), and so accurate information on tidal range is required to correct to MSL.

Boundary Data

Wave and tidal models require boundary conditions. The models may be nested within a (coarser) outer model domain, which directly transfers information from outer to inner nest (e.g. [2]). However, at some stage, and at some level of nesting, a model will likely require boundary information from an external source. For tidal models, many freely available 'tidal atlases' are suitable, such as TPXO. Tidal atlases represent an assimilation between telemetry (satellite) data, in situ data, and global numerical models. One such source, FES2014 [3], is available at a (global) grid resolution of $1/16 \times 1/16$ degrees for 32 tidal constituents, for both surface elevations and tidal currents, as both amplitude and phase components (Fig. 8.4). For wave models, although there are various wave products, which provide statistical wave properties (e.g. NOAA Wavewatch III which provides Hs and Tz), it is generally preferable to seek products, or to run simulations, which transfer 2D wave spectra from outer to inner nest (e.g. [4]).

Surface Forcing

Surface fields are required to force tidal models, which simulate nonastronomical processes (e.g. wind-driven currents or surge), and wind fields are generally

1. LAT is traditionally used on navigational charts, because it represents the shallowest possible water depth.

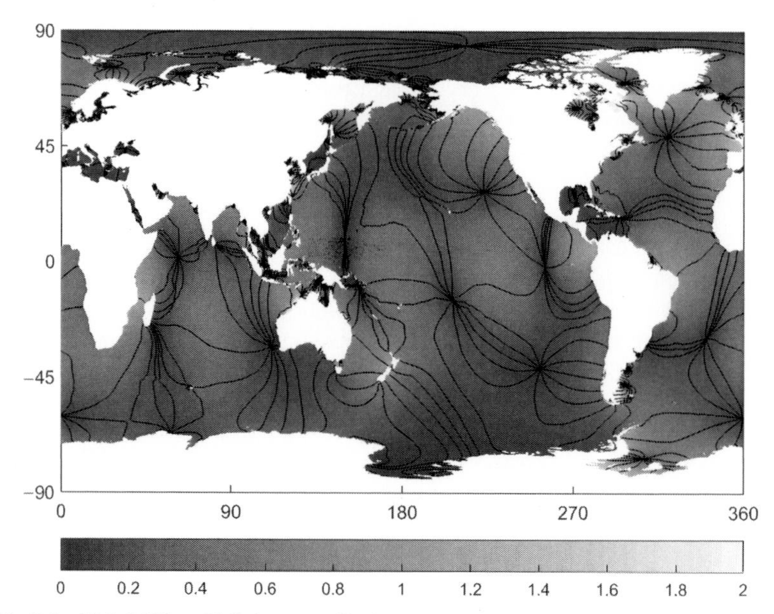

FIG. 8.4 Global M2 cotidal chart (amplitude in m) based on the FES2012 dataset. Contours connect regions that are equal in phase.

essential to force wave models at all but the smallest of scales (where lateral boundary data may suffice). Many sources of gridded meteorological data are available for model forcing, and two of the most popular products are ERA-Interim (European Centre for Medium Range Weather Forecasting [ECMWF]) and CFSR (National Center for Atmospheric Research/University Corporation for Atmospheric Research [NCAR/UCAR]). Both are reanalysis products—an assimilation of historical observational data, using a consistent assimilation (or 'analysis') scheme throughout. By way of example, a snapshot of an ERA-Interim wind field is shown in Fig. 8.5, along with the grid points of the forced wave model for both outer (North Atlantic) and inner (North of Scotland) nests. Unless there are strong topographic effects (e.g. wind that is channelled through a narrow strait), it is generally possible to interpolate relatively coarse wind fields to the ocean model grid points. Another consideration is temporal resolution of the surface fields. In general, 3h wind fields are required to adequately simulate waves. However, for tidal models which include wind-driven currents and surge (in addition to astronomical tides), 6h forcing may suffice.

8.1.4 Time Step

One of the most important considerations in model setup is the time step of the simulation. When selecting a model time step, it is important to consider *accuracy*, *stability*, and computational cost. For example, the model time step

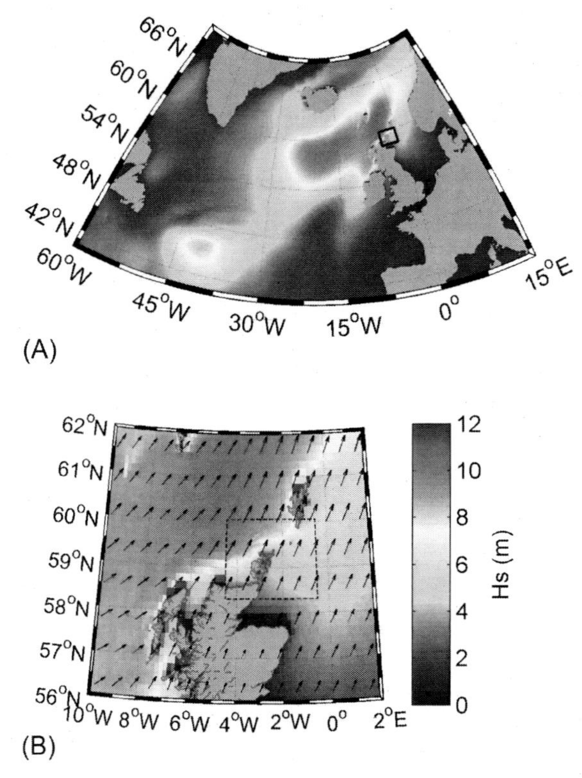

FIG. 8.5 Scales of nesting for wave model simulations, where the variable plotted is a snapshot of significant wave height on January 10, 2009, 12:00. (A) Outer North Atlantic model, and (B) interface between coarser North Atlantic model and (*boxed*) inner nested higher-resolution Pentland Firth and Orkney waters model. The vectors in (B) show the spatial resolution of the corresponding ERA-Interim wind field. *(Reproduced from S.P. Neill, M.J. Lewis, M.R. Hashemi, E. Slater, J. Lawrence, S.A. Spall, Inter-annual and interseasonal variability of the Orkney wave power resource, Appl. Energy 132 (2014) 339–348.)*

must be sufficiently small to capture temporal variability in the physical process. An example of this is shown in Fig. 8.6 where, graphically, a time step of $\Delta t = \pi/8$ is insufficient to capture the physical process, but a time step of $\Delta t = \pi/16$ is sufficient. However, more likely, it is stability which constrains model time step. In most practical ocean models, a wave (e.g. phase speed) is travelling across a discrete spatial grid. To ensure stability, the time step must be less than the time it takes for the wave to travel between adjacent grid points. This condition is known as the Courant-Friedrichs-Lewy (CFL) condition. For example, if the phase speed of a 1D tidal simulation, with grid spacing Δx, is $c = \sqrt{gh}$, then the model time step Δt must satisfy

$$\Delta t \leq \frac{\Delta x}{\sqrt{gh}} \tag{8.1}$$

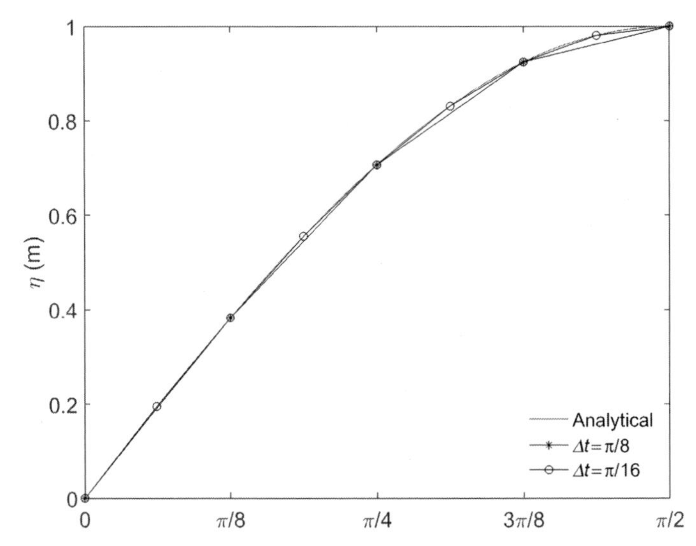

FIG. 8.6 Influence of model time step (Δt) on accuracy.

For a typical shelf sea water depth $h = 50\,\text{m}$, phase speed $c = 22.1\,\text{m/s}$. Therefore, for a typical model grid spacing of $\Delta x = 200\,\text{m}$, time step $\Delta t \leq 9\,\text{s}$, that is, considerably less than any constraint likely to be imposed by accuracy.

Note that halving of the grid spacing requires a halving of the model time step (Eq. 8.1). However, for a 2D modelling problem, halving the grid (in both x- and y-directions) results in a quadrupling of the number of computational grid points. Because the model time step is halved, the computational cost of the problem will have increased by a factor of 8! Both grid spacing and time step are therefore very important criteria when embarking on a model study. Note that the stability of some numerical methods (called implicit schemes) is not controlled by the CFL criteria. For example, this is the case for the SWAN wave model. Implementing implicit numerical schemes in ocean models is usually challenging. For a model that is unconditionally stable, accuracy becomes the limiting factor.

8.1.5 Staggered Grids

Staggered grids are a simple way of avoiding odd-even decoupling between modelled velocity and scalar fields. It also reduces the computational cost, as it is not necessary to compute all variables at all nodes. Odd-even decoupling is a discretization error that can occur on collocated grids, and leads to classical 'checkerboard' patterns in the numerical solution. Ocean models tend to be based on two staggered grid arrangements: the Arakawa B-grid and the Arakawa C-grid (Fig. 8.7). The u and v components in an Arakawa B-grid are computed at the same location (Fig. 8.7A). Because the u (or v) momentum equation contains a v (or u) velocity in the Coriolis term, an Arakawa B-grid is suitable for problems in which the geostrophic balance is important. By contrast, an

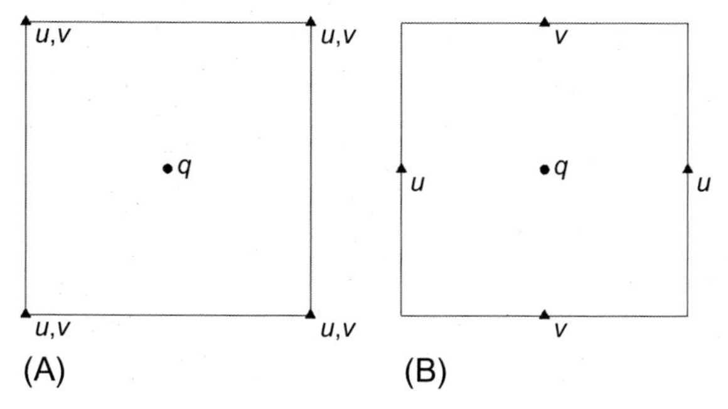

FIG. 8.7 Arakawa (A) B-grid and (B) C-grid. u and v are vectors, and q scalars.

Arakawa C-grid is good for tidal problems, because velocity points are located midway between the elevation points. Because the flow is driven by the surface slope (e.g. $\partial\eta/\partial x$, $\partial\eta/\partial y$), this avoids the need to interpolate elevations.

Most popular finite difference models used for resource assessment use a C-grid arrangement (e.g. ROMS and POM). Incidentally, the simplest grid arrangement, a collocated grid, where velocity and scalar fields are calculated at the same grid points, is known as an Arakawa A-grid.

8.1.6 Discretization

Discretization concerns the process of transferring a continuous function into one that is solved only at discrete points. Therefore, mathematical equations such as the ones included in Chapter 2 are continuous, but we must consider them at discrete points (e.g. points in time and space) before they can be solved numerically, that is, via numerical models.

Discretization: A Simple Finite Differencing Example

We will demonstrate the concept of discretization using a simple finite differencing example. Consider a thin rod of length L (Fig. 8.8). We wish to know the temperature at each point along the rod. We can denote any position along the rod as x. In mathematical notation

$$u(x) = ? \quad 0 \le x \le L \tag{8.2}$$

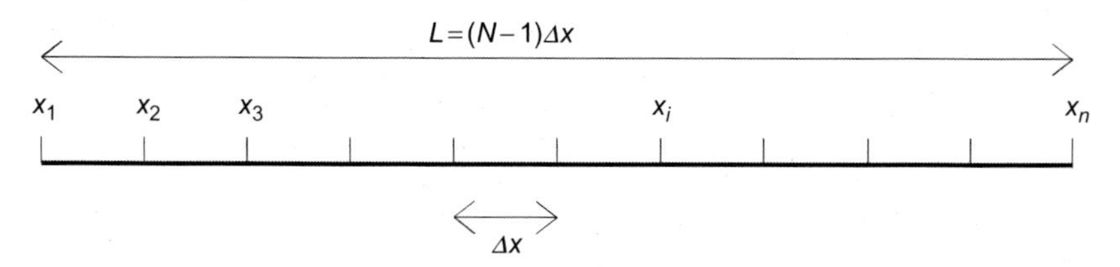

FIG. 8.8 Discretization of a rod of length L, with grid spacing Δx.

where the variable u is the temperature. Because $u(x)$ is a continuous function, there are an infinite number of points along this rod. However, it is much easier if we consider a finite number of discrete points along this rod. In numerical notation

$$u(x_i) = ? \quad x = x_1, x_2, \ldots, x_N \tag{8.3}$$

If we distribute these discrete points uniformly along the rod, the spacing between the grid points will be $\Delta x = L/(N-1)$. Consequently,

$$x_1 = 0,\ x_2 = \Delta x,\ x_3 = 2\Delta x, \ldots,\ x_N = (N-1)\Delta x \tag{8.4}$$

To find the temperature at each point, we need to solve the following differential equation

$$\frac{\partial u}{\partial t} = k\frac{\partial^2 u}{\partial x^2} \tag{8.5}$$

where k is a constant and depends on the material property of the rod.

8.2 NUMERICAL METHODS

8.2.1 Finite Difference Method

To solve a differential equation like Eq. (8.5), we need to evaluate derivatives. Recalling from calculus (Chapter 2), the derivative of a function is defined as:

$$\frac{du}{dx} = \lim_{\Delta x \to 0} \frac{u(x+\Delta x) - u(x)}{\Delta x} \tag{8.6}$$

From a geometrical perspective, the derivative at a point equals the slope of the tangent line at this point.

As an approximation, the derivative can be estimated as

$$\frac{du}{dx} \simeq \frac{u(x+\Delta x) - u(x)}{\Delta x} \qquad \Delta x = \frac{L}{N-1} \tag{8.7}$$

It is clear that as the number of grid points (N) increases, Eq. (8.7) leads to more accurate values. Referring to the simple rod example, the derivative at a general point x_i can be approximated as

$$\left.\frac{du}{dx}\right|_{x=x_i} \simeq \frac{u(x_i+\Delta x) - u(x_i)}{\Delta x} = \frac{u(x_{i+1}) - u(x_i)}{\Delta x} = \frac{u(x_{i+1}) - u(x_i)}{x_{i+1} - x_i} \tag{8.8}$$

which is called a finite difference. Specifically, a finite difference of the form of Eq. (8.8) is called a *forward difference*.

Alternatively, if we use function values at grid points x_{i-1} and x_i, we call it a *backward difference*:

$$\left.\frac{du}{dx}\right|_{x=x_i} \simeq \frac{u(x_i) - u(x_i-\Delta x)}{\Delta x} = \frac{u(x_i) - u(x_{i-1})}{\Delta x} = \frac{u(x_i) - u(x_{i-1})}{x_i - x_{i-1}} \tag{8.9}$$

Another form is the *central difference* and, as shall be shown in the next section on truncation error, it is more accurate than either backward or forward difference. It is defined as

$$\left.\frac{du}{dx}\right|_{x=x_i} \simeq \frac{u(x_{i+1}+\Delta x)-u(x_i-\Delta x)}{2\Delta x}=\frac{u(x_{i+1})-u(x_{i-1})}{2\Delta x}=\frac{u(x_{i+1})-u(x_{i-1})}{x_{i+1}-x_{i-1}} \tag{8.10}$$

Fig. 8.9 shows the geometrical interpretation of central, forward, and backward finite difference schemes.

Truncation Error

Truncation error is defined as the difference between the true (analytical) derivative of a function and its derivative obtained by numerical approximation.

Beginning with the Taylor series expansion

$$f(x+\Delta x)=f(x)+\Delta x\frac{df(x)}{dx}+\frac{(\Delta x)^2}{2!}\frac{d^2f(x)}{dx^2}+\frac{(\Delta x)^3}{3!}\frac{d^3f(x)}{dx^3}+\cdots \tag{8.11}$$

Suppose that we approximate a derivative using a forward difference scheme of the form given in Eq. (8.8). The Taylor expansion gives

$$\frac{f(x+\Delta x)-f(x)}{\Delta x}=\frac{df(x)}{dx}+\left[\frac{\Delta x}{2!}\frac{d^2f(x)}{dx^2}+\frac{(\Delta x)^2}{3!}\frac{d^3f(x)}{dx^3}+\cdots\right] \tag{8.12}$$

Numerical approx. True derivative Error

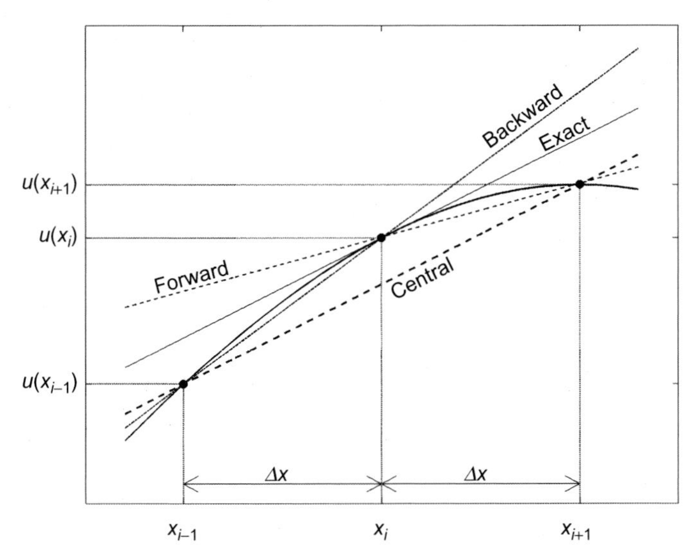

FIG. 8.9 Geometrical interpretation of finite difference method for backward, forward, and central difference schemes.

In this case, the truncation error is

$$\left[\frac{\Delta x}{2!}\frac{d^2f(x)}{dx^2}+\frac{(\Delta x)^2}{3!}\frac{d^3f(x)}{dx^3}+\cdots\right] \tag{8.13}$$

Because it involves terms in Δx and higher powers, we say that it is of order Δx, written as $O(\Delta x)$. Note that as $\Delta x \to 0$ we obtain the true derivative.

The backward difference scheme given by

$$\frac{f(x)-f(x-\Delta x)}{\Delta x} \tag{8.14}$$

also has an error $O(\Delta x)$.

A better approximation can be obtained using the central difference scheme

$$\frac{f(x+\Delta x)-f(x-\Delta x)}{2\Delta x} \tag{8.15}$$

From the Taylor series expansion we obtain

$$f(x-\Delta x)=f(x)-\Delta x\frac{df(x)}{dx}+\frac{(\Delta x)^2}{2!}\frac{d^2f(x)}{dx^2}-\frac{(\Delta x)^3}{3!}\frac{d^3f(x)}{dx^3}+\cdots \tag{8.16}$$

Hence, by subtracting Eq. 8.16 from 8.11

$$\frac{f(x+\Delta x)-f(x-\Delta x)}{2\Delta x}=\frac{df}{dx}+\frac{(\Delta x)^2}{3!}\frac{d^3f(x)}{dx^3}+\cdots \tag{8.17}$$

for the central difference scheme. In this case, the error is $O(\Delta x^2)$. Because Δx is small, $\Delta x^2 < \Delta x$. Therefore, a centred scheme has a smaller truncation error (i.e. is more accurate) than a forward or backward scheme.

8.2.2 Finite Element Method

The majority of ocean modelling problems involve complex geometries such as irregular coastlines, inlets, islands, and headlands. Regular (rectangular or curvilinear) grids cannot conveniently resolve these complex geometries. Finite element method (FEM) is based on an irregular (e.g. triangular) mesh that can easily resolve complex geometries (e.g. Fig. 8.1B). FEM originated in the area of solid mechanics to calculate stress and strain in structures. However, it is today applied to a wide range of multiphysics problems, including fluid mechanics and ocean modelling. FEM can be regarded as a general numerical method to solve partial differential equations (PDEs). The implementation of FEM to fluid/solid mechanics problems involves many steps, and a detailed explanation of those steps is beyond the scope of this book. However, the basic concepts of the method are introduced briefly, because many renewable energy problems use FEM codes for numerical simulations. For instance, ADCIRC is an FEM-based ocean circulation model that can be used to simulate the tidal energy resource of a region (e.g. [5]). Also, FEM is a common technique for the analysis of tidal turbine blades or the structural design of ocean renewable energy devices (e.g. [6]).

In FEM, the domain is first discretized into a number of elements (see Fig. 8.10 as an example). These elements are generated such that higher resolution (i.e. smaller element size) is obtained in places of interest (such as near the coastline in Fig. 8.10).

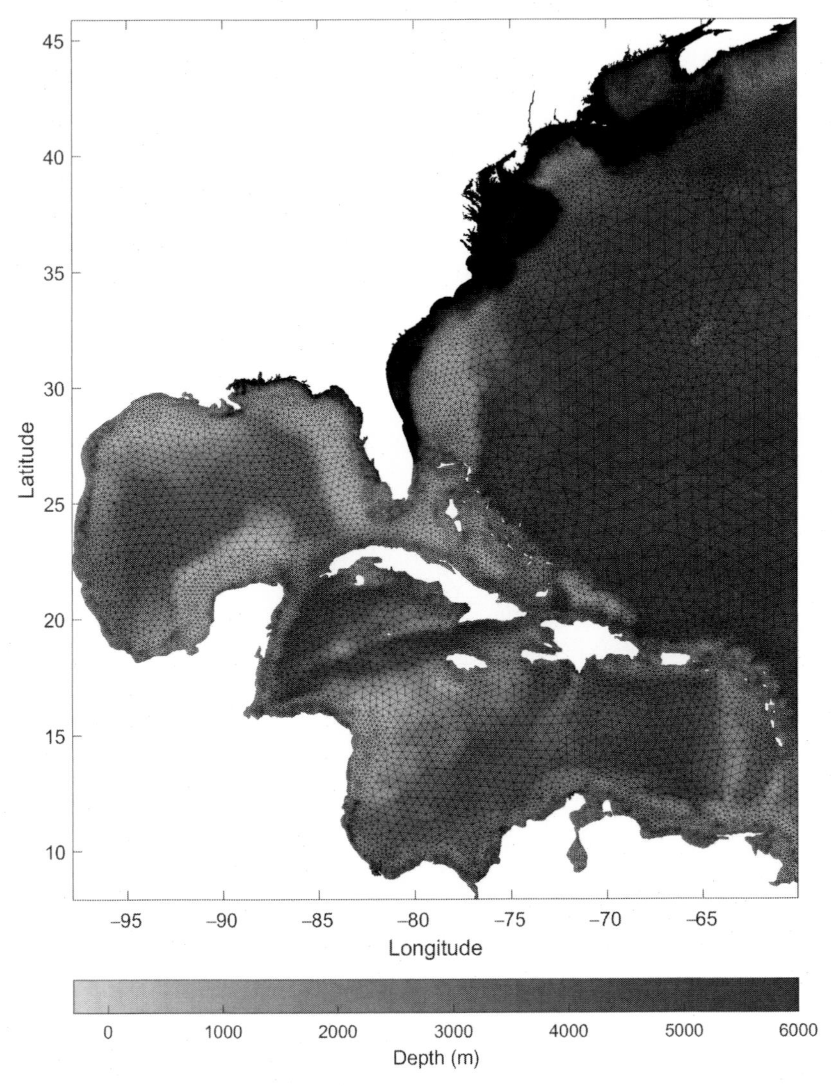

FIG. 8.10 An example of a finite element triangular mesh used to simulate tides and storm surge over Atlantic coast of the United States. The mesh has 3,110,470 nodes and 6,114,065 elements, with a mean size of 333 m (10 m to ~121 km) [7].

Many types of 2D or 3D elements are common in FEM, such as triangular, quadrilateral, and tetrahedron. The triangular mesh is the most common type. After discretization, the state/dependent variable, u (e.g. water depth, velocity, displacement, stress, etc.) is approximated by a function, $\hat{u}$. In other words, u is the exact solution of the PDE of interest, which in most cases does not exist, whilst $\hat{u}$ can approximately satisfy the PDE. $\hat{u}$ is approximated using basis functions

$$\hat{u} = \sum_i u_i \psi_i \tag{8.18}$$

where ψ_i denotes basis functions and u_i is the approximate value of the state variable at node i. For instance, if an element has three nodes, u can be approximated by

$$u \approx \hat{u} = u_1\psi_1(x, y) + u_2\psi_2(x, y) + u_3\psi_3(x, y) \tag{8.19}$$

Note that if (x_1, y_1) is the coordinate of node 1, then basis functions are selected such that $\psi_1(x_1, y_1) = 1$, $\psi_2(x_1, y_1) = 0$, and $\psi_3(x_1, y_1) = 0$. The simplest form of a basis function is the Lagrangian linear function over an element as follows

$$\psi_i(x, y) = a + bx + cy; \quad i = 1, 2, 3 \tag{8.20}$$

The coefficients of this function can be found by imposing the conditions explained previously (i.e. $u(\hat{x_i}, y_i = u_i))$, that is

$$\psi_i(x_k, y_k) = \delta(i, k) \quad i, k = 1, 2, 3 \tag{8.21}$$

$$\delta(i, k) = 1 \quad i = k \tag{8.22}$$

$$\delta(i, k) = 0 \quad i \neq k \tag{8.23}$$

As a next step, after selecting the basis/shape functions, the unknown values u_i should be estimated. FEM forms enough algebraic equations to compute the unknown u_i at all nodes by minimizing the error. Let us consider the exact form of a differential equation

$$L(u) = 0 \tag{8.24}$$

where L is a general differential operator (e.g. $L = \frac{\partial^2}{\partial x^2} + \frac{\partial^2}{\partial y^2} \Rightarrow L(u) = \frac{\partial^2 u}{\partial x^2} + \frac{\partial^2 u}{\partial y^2}$). If we replace $\hat{u}$ into the differential equation, it will not exactly satisfy it (because $u \approx \hat{u}$), and leads to a residual error (R) as follows

$$L(\hat{u}) = R \tag{8.25}$$

In FEM (weighted residual method), u_i is determined such that the residual error is minimized, therefore,

$$\int_\Omega W_j(x, y) L(\hat{u}) = \int_\Omega W_j(x, y) R = 0 \quad j = 1, 2, 3, \ldots, N \tag{8.26}$$

File Edit View Search Tools Documents Help

sample.14

```
! Example of a mesh file (ADCIRC)
80105 42020 ! Number of elements and nodes in the mesh, respectively
!
! Nodes
!
  1      -71.413830   41.472320    ! Node number, x (longitude) and y (latitude) coordinates of the nodes
  2      -71.413415   41.468685
  3      -71.413045   41.464189
  4      -71.413025   41.482322
.....
.....
  42019  -72.001758   41.249907
  42020  -72.001250   41.247105    ! Last node
!
! Elemets
!
  1  3  1  6   7                   ! Element 1 has 3 nodes: 1, 6 and 7
  2  3  2  12  17
  3  3  2  17  6
  4  3  3  15  22
  5  3  3  22  12
  6  3  4  9   19
  7  3  5  18  26
.......
  80102  3  6121   6131  42019
  80103  3  5878   5858  5848      ! Element 8103 has 3 nodes: 5878,  5858, and  5848
  80104  3  42020  6128  6136
  80105  3  6136   6131  42020     |! Last element
```

FIG. 8.11 An example of how a mesh is defined in an FEM model (here ADCIRC).

The previous equation is called the weak form of a differential equation, because the integral (some kind of average) of $L(u)$ is set to zero. Ω is the domain of the problem and W_j are the weight/test functions. Depending on the FEM scheme, several types of weight functions are chosen. For instance, in the Galerkin method, weight functions and basis functions are the same ($W_j = \psi_i$).

After implementing Eq. (8.26), enough algebraic equations are formed to compute the nodal values of the state variable.

FEM models usually use a simple file format to define the mesh. Fig. 8.11 shows a sample mesh file for the ADCIRC model. As this figure shows, first, the total number of elements and nodes is written. Then, the coordinate of each node in a coordinate system is defined. Each element in a triangular mesh is uniquely defined by connecting three nodes. The state variables are computed at each node (or centre of an element) and can be processed using postprocessing software such as Matlab.

8.2.3 Finite Volume Method

Finite volume method (FVM), like FEM, is based on an unstructured (e.g. triangular) mesh. Therefore, it is suitable for irregular and complex geometries. FVM has another advantage over FEM for fluid mechanic problems. So far, the numerical methods that we presented have been based on PDEs. By contrast, FVM is based on the integral form of the conservation laws, rather than their differential form. This leads to more accuracy/stability, especially for sharp gradients (i.e. large derivatives) inside a domain, which is also called *shock-capturing* property. To explain this more clearly, as we mentioned before, the dynamics of flow can be described by conservation of mass, momentum, and energy. These conservation laws can be written as a system of PDEs.

Alternatively, conservation laws can be expressed by integral equations (i.e. Reynolds transport theorem). For instance, the differential form of the continuity equation can be written as

$$\frac{\partial \rho}{\partial t} + \frac{\partial(\rho u)}{\partial x} + \frac{\partial(\rho v)}{\partial y} + \frac{\partial(\rho w)}{\partial z} = 0 \tag{8.27}$$

where u, v, and w are the components of flow velocities in the x, y, and z directions, and ρ is the density. The integral form of continuity for a control volume (or a *finite volume*) can be written as

$$\frac{\partial}{\partial t}\int_V \rho dV + \int_S \rho \mathbf{u} \cdot d\mathbf{S} = 0 \tag{8.28}$$

where V is the control volume, S is the control surface, $\mathbf{u} = (u, v, w)$, and $\mathbf{S}$ is the surface vector.

Therefore, the change of mass inside a control/finite volume plus the net mass fluxes through the control surface should be zero. In FVM, the domain is first discretized into a number of nonoverlapping finite volumes or cells (see Fig. 8.12 as an example). Usually, these finite volumes are triangles (2D) or

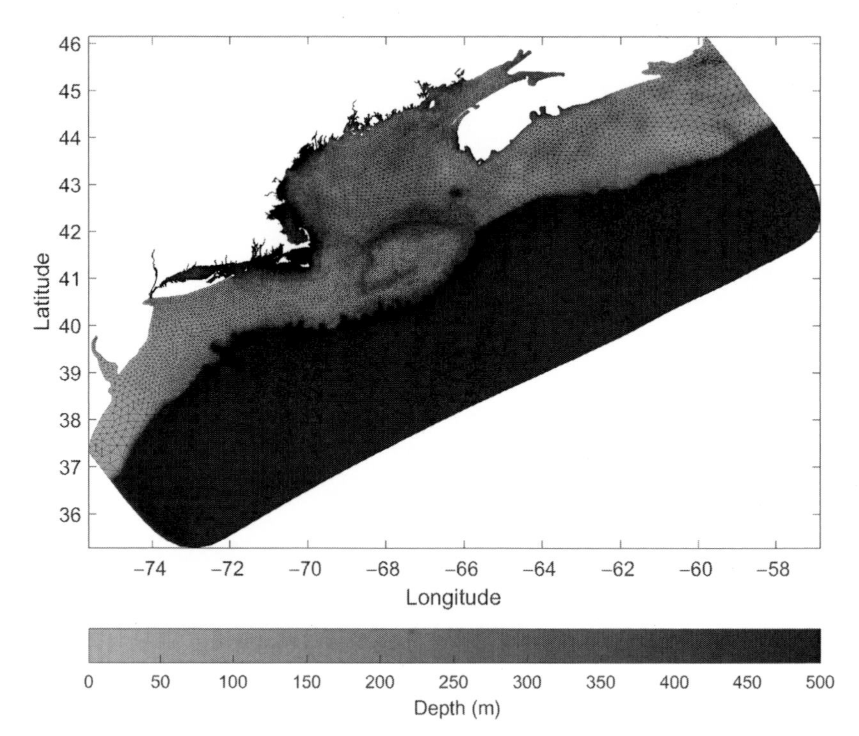

FIG. 8.12 An example of a finite volume triangular mesh used to simulate tides and storm surge over the northeast of the United States. This mesh is used by Finite Volume Coastal Ocean Model (FVCOM [8]), for the Northeast Coastal Ocean Forecast System.

prisms (3D). Next, conservation laws are applied to each individual cell to form enough algebraic equations, which can be solved to compute the state variables.

8.3 TIDAL MODELLING

There are a wide variety of models available for tidal resource assessment, ranging from open-source to commercial codes, and from 2D structured to 3D unstructured. A summary of the main models that are suitable for tidal resource assessment is provided in Fig. 8.13. These models range in complexity of user operation; for example, POM is a relatively simple code, but requires compilation and execution via line commands, whereas commercial codes such as MIKE 3 are controlled by a graphical user interface (GUI) and are suitable for running on desktop PCs. Some of the models (e.g. ROMS and FVCOM) are rather difficult to install and operate, and are generally recommended for more experienced modellers. However, the advantage of such complexity is the flexibility to couple models such as ROMS to many other models (and multiple models simultaneously), for example, the SWAN wave model, the WRF atmospheric model, and the community sediment transport model. Further, because models like ROMS are suited for parallel processing, they can be implemented on supercomputers (see Section 8.7). This is particularly useful for long timescale (e.g. decadal) and high-resolution simulations, but particularly when the tidal model is coupled with other modelling systems, and hence computationally expensive.

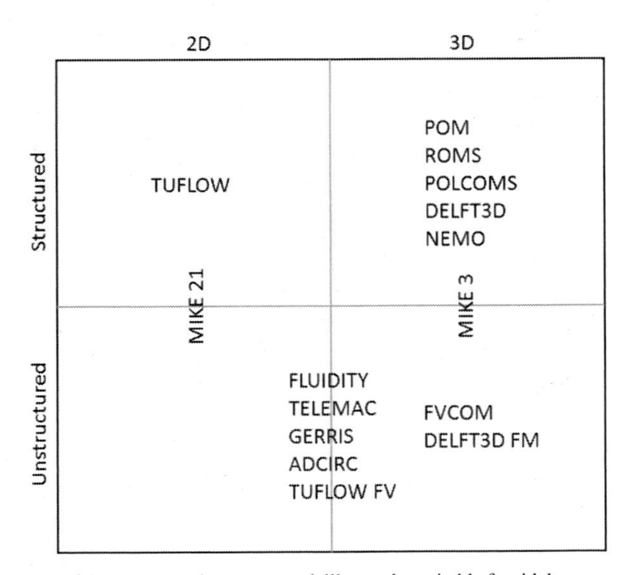

FIG. 8.13 Some of the most popular ocean modelling codes suitable for tidal resource assessment.

8.3.1 Turbulence Closure

For large Reynolds numbers (>2000), the flow velocity experiences fluctuations around the mean velocity. Common numerical models cannot simulate these fluctuations, because they occur over very small timescales (e.g. fractions of seconds). Alternatively, they replace the velocity by mean and fluctuating components as follows

$$u = \overline{u} + \acute{u} \tag{8.29}$$

where $\overline{u}$ is the time averaged and $\acute{u}$ is the fluctuating part. Tidal models simulate the temporal average rather than the actual velocity (u) to avoid this issue. The argument is that these small and rapid fluctuations in velocity are not important, compared with the mean velocity that is of relevance to tidal energy generation. The time average of the fluctuating velocity is zero. Therefore, if the Navier-Stokes equations were linear, we could just replace velocities by time averaged velocities. To make this point clearer, consider the continuity equation, which is a linear equation. Taking a moving time average leads to

$$\int_T \left(\frac{\partial u}{\partial x} + \frac{\partial v}{\partial y} + \frac{\partial w}{\partial z} \right) dt = 0 \tag{8.30}$$

$$\int_T \left(\frac{\partial (\overline{u} + \acute{u})}{\partial x} + \frac{\partial (\overline{v} + \acute{v})}{\partial y} + \frac{\partial (\overline{w} + \acute{w})}{\partial z} \right) dt = 0 \tag{8.31}$$

Also, for the x component of velocity (and similarly for the other components), we have

$$\int_T \frac{\partial u}{\partial x} dt = \frac{\partial \int_T \overline{u} dt}{\partial x} + \frac{\partial \int_T \acute{u} dt}{\partial x} = \frac{\partial \int_T \overline{u} dt}{\partial x} + 0 \tag{8.32}$$

Because the average of the fluctuating velocity is zero, the time averaged continuity equation simply becomes

$$\frac{\partial \overline{u}}{\partial x} + \frac{\partial \overline{v}}{\partial y} + \frac{\partial \overline{w}}{\partial z} = 0 \tag{8.33}$$

Unfortunately, this is not the case for the nonlinear parts of the momentum equation. For instance, if we just consider the convective acceleration term in the momentum equation, (see Chapter 2) we have

$$\int_T u \frac{\partial u}{\partial x} dt = \int_T \left((\overline{u} + \acute{u}) \frac{\partial (\overline{u} + \acute{u})}{\partial x} \right) dt \tag{8.34}$$

$$= \int_T \overline{u} \frac{\partial \overline{u}}{\partial x} dt + \int_T \acute{u} \frac{\partial \acute{u}}{\partial x} dt + 0 + 0 \tag{8.35}$$

$$= \int_T \overline{u} \frac{\partial \overline{u}}{\partial x} dt + \int_T \frac{1}{2} \frac{\partial \acute{u}^2}{\partial x} dt \tag{8.36}$$

Because the time average of $\acute{u}^2$ is not zero (in contrast to $\acute{u}$), additional terms appear in the momentum equation. For instance, the convective acceleration term after time averaging becomes

$$\overline{u}\frac{\partial \overline{u}}{\partial x} + \frac{1}{2}\frac{\partial \overline{\acute{u}^2}}{\partial x} \tag{8.37}$$

Therefore, additional unknowns (e.g. $\overline{\acute{u}^2}$) appear in the governing equations. Because the number of unknowns becomes more than the number of equations after time averaging, the Navier-Stokes equations are no longer closed. This is called the *closure problem*. Additional equations, called turbulence equations, are added to address this issue, and popular choices in ocean modelling are k-ϵ and the Mellor-Yamada 2.5 scheme (e.g. [9]).

8.3.2 Boundary Conditions

Because we generally simulate a particular region of interest (e.g. part of a bay, estuary, or shelf sea environment), we need to specify the conditions at the boundaries of our modelling domain. These boundaries can be forced by what is happening outside a modelling domain, and therefore models need the boundary information as input data. For instance, tidal waves are generated in the deep oceans by gravitational attractions of the Sun and the Moon, which are usually outside the modelling domain for a tidal project. The tidal forcing data (water elevation and velocity) should be specified at the boundaries of a model. The tidal information data are usually extracted from global tidal models or from a coarser outer modelling domain. Sometimes the boundaries of a model are affected by what is happening inside the domain. For instance, a tidal wave that is propagating towards a coastline may be reflected back into the domain, or can be radiated out from another boundary.

From a mathematical point of view, the values of state variables (e.g. time series of velocities) can be specified at a boundary. These are called *Dirichlet* boundary conditions. Alternatively, the derivative of a state variable can be specified at a boundary. For example, at a reflective boundary, the derivative of water elevation can be specified as zero. These are called *Neumann* boundary conditions. In practice, the boundary conditions for a tidal model can be more complex: usually a mix of Dirichlet and Neumann.

There are a variety of types of boundary condition applied to ocean models (e.g. [10]), as follows.

Coastlines

This is a zero gradient condition for surface elevation, and zero flow for the normal component of velocity. For tangential velocities, the coastline can be treated as either no-slip or free-slip, depending on the configuration of the problem; for example, a no-slip condition may have a significant influence on the flow field when simulating strong tidal flows through a narrow strait. Models that simulate a moving boundary (wetting and drying) implement a more complicated and iterative procedure to find the wet part of a domain.

Clamped Elevation

The free surface displacement is set to an externally prescribed value. This is a popular boundary condition for tidal simulations, due in part to the smooth spatial variability of elevation data, and readily available satellite-altimetry-derived data (e.g. Section 8.1.3).

Clamped Normal Velocity

This condition simply sets the normal velocity component in the boundary cell to an externally prescribed value.

Flather

Applied to the normal component of barotropic velocity, the Flather condition is an adjustment to the externally prescribed normal velocity, based on the difference between modelled and externally prescribed surface elevations. This is a form of the classical radiation boundary condition.

Chapman

The corresponding boundary condition for water elevation (assuming an outgoing signal) is Chapman.

8.3.3 Time Splitting

The presence of a free surface in a tidal model introduces waves in the domain that propagate at a speed of $\sqrt{gh}$. These waves impose a more severe constraint on the model time step than any of the internal processes (Section 8.1.4). Therefore, a split time technique is generally used in 3D modelling. The depth-averaged equations are integrated using a 'fast' (barotropic) time step, and the values of $\overline{u}$ and $\overline{v}$ used to replace those found by integrating the full equations on a 'slower' (baroclinic) time step [11]. In general, it is recommended that the ratio between barotropic and baroclinic time steps is in the range 10–20, but this will depend on the scale of the problem. The purpose of time splitting is to reduce the computational effort, because the 3D time step is considerably more costly than the 2D time step.

8.4 WAVE MODELLING

There are two main classes of wave models: phase-resolving models and phase-averaged models. In a phase-resolving model, the domain is discretized onto a grid that is relatively fine compared with the wave length, and the vertical displacement of the sea surface calculated in detail. Because such a process is computationally expensive, phase-resolving models tend to be applied to relatively limited domains, and are most appropriate in cases of strong wave diffraction and reflection due to coastal structures. Examples of wave resolving models are SWASH [12], CGWAVE, and FUNWAVE [13]. Although such models could be useful for wave resource characterization, they are not generally used due to very high computational cost; therefore, the remainder of this section concentrates on the more widely applied phase-averaged models.

8.4.1 Phase-Averaged Wave Models

As discussed in Chapter 5, a wave field can be considered as a wave spectrum, which can be represented by a large number of regular sinusoidal wave components. Wave models work by predicting each of these independent wave components individually, and how they vary in space and time, through the energy balance equation, which has the form (e.g. [14])

$$\frac{dE(\sigma,\theta;x,y,t)}{dt} = S(\sigma,\theta;x,y,t) \tag{8.38}$$

where E is the spectral energy density, σ is the angular wave frequency, θ is wave direction, x and y are the horizontal dimensions, t is time, and S are the source terms, comprising generation, wave-wave interaction, and dissipation.

Energy density E is not preserved in the presence of ambient currents, and so wave models tend to solve the action density (N) balance equation, where $N = E/\sigma$. In spherical coordinates, this can be written as [15]

$$\frac{\partial N}{\partial t} + \frac{\partial c_\lambda N}{\partial \lambda} + \frac{\partial c_\phi N}{\partial \phi} + \frac{\partial c_\sigma N}{\partial \sigma} + \frac{\partial c_\theta N}{\partial \theta} = \frac{S_{\text{tot}}}{\sigma} \tag{8.39}$$

where c_λ and c_ϕ are the propagation velocities in the zonal (λ) and meridional (ϕ) directions, and c_σ and c_θ are the propagation velocities in spectral space.

The numerical solution of Eq. (8.39), without any prior assumption about the spectral shape, is what is known as a *third-generation* wave model [16], and is the most popular type of wave model in use today for resource assessment, including the models SWAN [15], WAM [16], and WAVEWATCH III [17].

8.4.2 Source Terms

Central to third-generation wave models is the calculation of the source terms (RHS of Eq. 8.39), and the key processes are introduced briefly as follows.

Wind Input

There are two mechanisms that describe the transfer of wind energy and momentum into the wave field. Small pressure fluctuations associated with turbulence in the airflow above the water surface are sufficient to induce small perturbations on the sea surface, and to support a subsequent *linear growth* as the wavelets move in resonance with the pressure fluctuations [18]. This mechanism is only significant early in the growth of waves on a relatively calm sea. When the wavelets have grown to a sufficient size to start affecting the flow of air above them, most of the development commences. The wind now pushes and drags the waves with a vigour that depends on the size of the waves themselves, in a phase known as *exponential wave growth*. Wave growth can therefore be described as a sum of linear (A) and exponential (BE) growth terms

$$S_{\text{in}}(\sigma,\theta) = A + BE(\sigma,\theta) \tag{8.40}$$

A drag coefficient is used to transform the wind speed (usually defined at 10 m elevation) into a friction velocity U_*, after which a linear expression can be used to calculate A (e.g. [19]), and an exponential expression to calculate B (e.g. [20]).

Dissipation

In deep water, energy is dissipated from the wave field mainly through wave breaking (whitecapping). In shallow water, it may also be dissipated through interaction with the sea bed (bottom friction) and through depth-induced wave breaking.

Whitecapping is active in wind-driven seas, and is the least understood of all processes affecting waves [14]. A complicating factor is that there is no generally accepted precise definition of breaking and, as you can imagine, quantitative observations of deep water wave breaking are very difficult to achieve. However, the whitecapping source term tends to be based on the theory of Hasselmann [21], in which each white-cap acts as a pressure pulse on the sea surface, just downwind of the wave crest.

In water of finite depth, wave energy is dissipated due to interaction with the sea bed, and this tends to be dominated by bottom friction [22]. This can be expressed as [23]

$$S_{\text{b}} = -C_{\text{b}} \frac{\sigma^2}{g^2 \sinh^2 kd} E(\sigma,\theta) \tag{8.41}$$

where C_{b} is a bottom friction coefficient.

As waves propagate from deep to shallow water, wave shoaling leads to an increase in wave height. Waves tend to steepen at the front and to become more gently sloping at the back, and at some point the waves will break. There are various definitions of wave breaking; for example, breaking occurs when the particle velocities at the crest exceed the phase speed, or when the free-surface becomes vertical. Depth-induced wave breaking is included as a source term in third-generation wave models.

Nonlinear Wave-Wave Interactions

Nonlinear wave-wave interactions redistribute wave energy over the spectrum, due to an exchange of energy resulting from resonant sets of wave components. There are two processes that are important for the inclusion of nonlinear wave-wave interactions in wave models: four-wave interactions in deep and intermediate waters (known as *quadruplets*) and three-wave interactions in shallow water (*triads*). A good explanation of the principal of nonlinear wave-wave interactions is provided by Holthuijsen [14]. Two wave paddles, generating waves of different frequencies and directions, are placed in two corners along one side of a tank of constant water depth. The resulting waves create a diamond pattern of crests and troughs, which has its own wave length, speed, and direction. This diamond pattern would interact with a third-wave component, if this third wave had the same wave length, speed, and direction as the diamond pattern. This is the triad wave-wave interaction, which redistributes wave energy within the spectrum due to resonance. Although each of the individual wave components can gain or lose energy, the sum of the energy at each point in the tank would remain constant. In deep water, it is not possible to meet these resonant conditions (i.e. matching of wave speed, length, and direction), and so triad wave-wave interactions cannot occur in deep water. However, in deep water it is possible for a pair of wave components to interact with another pair of wave components in a quadruplet wave-wave interaction.

Quadruplets transfer wave energy in deep water from the peak frequency to lower frequencies, whereas triads transfer energy from lower to higher frequencies, and transform single-peaked spectra into multiple-peaked spectra as they approach the shore. Both are included as source terms in third-generation wave models, and it is noted that both are computationally expensive. Triads, in particular, are often omitted in wave model simulations, whereas quadruplets are often included. For example, in the SWAN wave model, quadruplets are activated by default in third-generation mode, whereas triads are not included by default.

8.5 VALIDATION

According to the statistician George Box, 'all models are wrong, but some are useful'. Models, however sophisticated, are a representation of reality. To provide confidence in how accurately a model has simulated reality, it is necessary to perform model validation. Of course, validation depends upon the availability of suitable in situ data. Although it is often desirable for such data to be focussed in the region of interest, for example, at the approximate location of a proposed wave or tidal energy array, this is not compulsory. If a model performs well in one region, or under one set of conditions (e.g. during a spring tide), this gives confidence in model performance in another region, or under another set of conditions (e.g. during a neap tide). This is provided that the model parameterizations (e.g. drag coefficient and eddy viscosity) are physically realistic, and the model has not been excessively tuned to fit data in one region or under one unique set of conditions.

8.5.1 Validation Metrics

Various metrics can be used to quantify model validation (e.g. [24]), and some of the most popular metrics are presented in this section.

Correlation Coefficient

Correlation measures the direction and strength of the linear relationship between two variables. If we have a set of n observed (O_i) and simulated (S_i) values, the linear correlation coefficient (r) (also known as Pearson's r) can be calculated as

$$r = \frac{\sum_{i=1}^{n}(O_i - \overline{O})(S_i - \overline{S})}{\sqrt{\sum_{i=1}^{n}(O_i - \overline{O})}\sqrt{\sum_{i=1}^{n}(S_i - \overline{S})}} \tag{8.42}$$

For a perfect model, $r = 1$, whilst for a completely random prediction $r = 0$.

The square of the correlation is often used as a metric, because r^2 indicates the proportion of the variance in the observation that can be predicted by the model.

Root Mean Squared Error

Root mean squared error (RMSE) is the square root of the mean of the square of all of the error. The use of RMSE is very common, and it is considered an excellent general-purpose error metric for numerical predictions.

$$\text{RMSE} = \sqrt{\frac{1}{n}\sum_{i=1}^{n}(S_i - O_i)^2} \tag{8.43}$$

where O_i are the observations, S_i predicted values of a variable, and n the number of observations available for analysis. RMSE is a good measure of accuracy, but only to compare prediction errors of different models or model configurations for a particular variable and not between variables, as it is scale-dependent.

Scatter Index

Scatter index (S.I.) is simply the RMSE (Eq. 8.43) divided by the mean of the observations and multiplied by 100 to convert to a percentage error. As an example of the worth of S.I., if an RMSE for significant wave height is 1 m, this gives us no sense of how well the model is performing, because the mean of the observations could, for example, be either 1 m (S.I. = 100%) or 5 m (S.I. = 20%).

Bias

Bias (also known as mean error) is the mean of the simulated values of the selected variable minus the mean of the observed values, that is

$$BI = \overline{S} - \overline{O} \tag{8.44}$$

It is an index of the average component of the error [25], with a value closer to zero indicating a better simulation This index shows if a model is in general overestimating or underestimating.

Examples of validation metrics applied to a case study of Galway Bay are given in Section 8.9.1.

8.6 TOOLS FOR MODEL PREPROCESSING AND POSTPROCESSING

Preprocessing is one of the most important steps in ocean modelling. Once sources of data have been assembled for model setup (e.g. Section 8.1.3), the model grid and boundary conditions need to be prepared. There are a wide variety of tools which can be used to help set up model grids and boundary conditions, but the two that will be introduced here are Matlab and Blue Kenue.

8.6.1 Matlab

Matlab is a computing environment with a GUI and its own unique, and relatively easy to learn, programming language. An example screenshot of Matlab is shown in Fig. 8.14. Although commands can be entered directly into the Command Window, Matlab is at its most powerful when a Matlab script (i.e. a computer program) is developed and saved in the Editor Window for subsequent execution—an example of which is shown in Fig. 8.14 (this is actually part of the code use to create Fig. 8.7). Matlab is relatively easy to learn, perhaps in contrast to more traditional programming languages such as FORTRAN, as the generated variables instantly appear, and can be visualized, in the Workspace. This aids any necessary debugging, and is particularly useful for the novice user. Matlab also comes with excellent help features linked to extensive documentation.

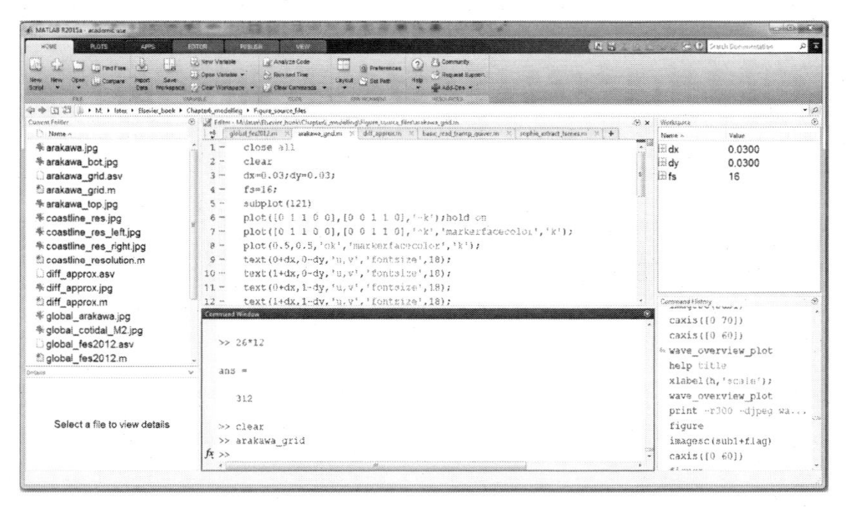

FIG. 8.14 Matlab environment.

One of the key advantages of Matlab is that it is widely used by ocean modellers around the world; hence many existing scripts can be shared and downloaded. For example, Mathworks File Exchange[2] is a forum for finding and sharing custom Matlab applications, classes, code examples, drivers, functions, and scripts, many of which will be useful for model preprocessing and postprocessing. In addition, various researchers have written preprocessing and postprocessing tool boxes in Matlab for some of the most popular ocean models, for example, ROMSTOOLS[3] for setting up and analysing the outputs of ROMS simulations, and OpenEarth tools[4] for setting up and analysing the outputs of the SWAN wave model. Further, many excellent packages have been written and are freely available for Matlab, such as the popular `t_tide` (tidal harmonic analysis) and `m_map` (mapping toolbox, which has been used to generate Fig. 8.3, for example). Matlab is preinstalled with netCDF libraries—a self-describing file format that is popular for storing numerical model inputs and outputs. Because Matlab is based on the principal of matrices, it excels at setting up and analysing structured grids. However, for unstructured meshes, an alternative software package is recommended—Blue Kenue.

8.6.2 Blue Kenue

Blue Kenue is a freely available preprocessing, analysis, and visualization tool that is suitable for a wide range of model applications, and in particular mesh generation for FEM or FVM models. The interface is very intuitive and easy to use (Fig. 8.15), and Blue Kenue is setup to directly import/export data from/

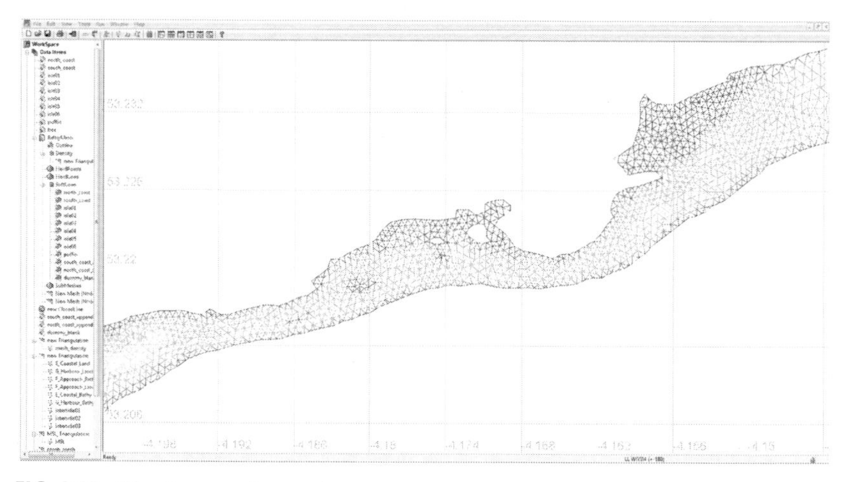

FIG. 8.15 Blue Kenue environment.

2. See https://uk.mathworks.com/matlabcentral/fileexchange/.
3. See http://www.croco-ocean.org/download/roms_agrif-project/.
4. See https://svn.oss.deltares.nl/repos/openearthtools/trunk/matlab/applications/swan/.

to Telemac and ADCIRC—two very popular unstructured models. Blue Kenue is very flexible and can handle a wide range of data types, including GIS files, GRIB, x–y scatter data, and time series. However, one of the disadvantages of Blue Kenue, in comparison to computing environments which incorporate a programming language like Matlab, is that although you can save all the variables in the workspace as you proceed, it is not possible to save a 'line-by-line' sequence of commands. Therefore, if one develops a model grid of a region and wishes to go back several steps in the processing (e.g. to remove an island from the domain), many procedures must be repeated manually, rather than editing and running a script. Note that although primarily used to deal with unstructured (triangular) meshes, Blue Kenue will also generate rectangular meshes. Blue Kenue also has a very easy to implement presentation quality animation facility (including flight paths), helping with model visualization.

8.7 SUPERCOMPUTING

In 1965, Intel cofounder Gordon Moore noticed that the number of transistors per square inch on integrated circuits had doubled every year since their invention. *Moore's Law* predicts that this trend will continue into the foreseeable future. We are now in a golden age of scientific computing, when efficient multicore desktop PCs can be purchased at relatively low cost. However, within the context of ocean modelling, supercomputing refers to running a computational task simultaneously (i.e. in parallel) on hundreds or even thousands of processors. Rather than running a job, representing a single computational domain, in series, the computational domain can be decomposed into several subdomains, which can be run together in parallel (Fig. 8.16). Although such a task is possible on a desktop PC for a limited number of processors, it is when a model that has been optimized for parallel processing is divided into several hundred tasks that supercomputing really excels.

The simplest way of preparing a computational domain for parallel processing is through tiling. However, a 'tile' that contains mostly land cells will represent a considerably lighter computational task than a tile which contains entirely sea points—a situation that is known as uneven load balancing. Therefore, many models use load-balancing techniques for domain decomposition, as demonstrated in Fig. 8.16 for the POLCOMS model. Each of the 10 nodes (0–9) shown in this example has 10,281–10,366 'wet' grid cells, representing a departure of no more than 0.5% from the mean–excellent load balancing.

If we were to run a serial task 1000 times, for example, with slightly differing initial conditions for each simulation, we would complete our simulations in 1/1000 of the time that it would take to complete the same tasks in series—a considerable time saving, in a situation that is known as 'embarrassingly parallel'. This would represent perfect 'linear' speedup where

$$\text{Speedup} = \frac{\text{Time for 1 processor}}{\text{Time for } n \text{ processors}} \tag{8.45}$$

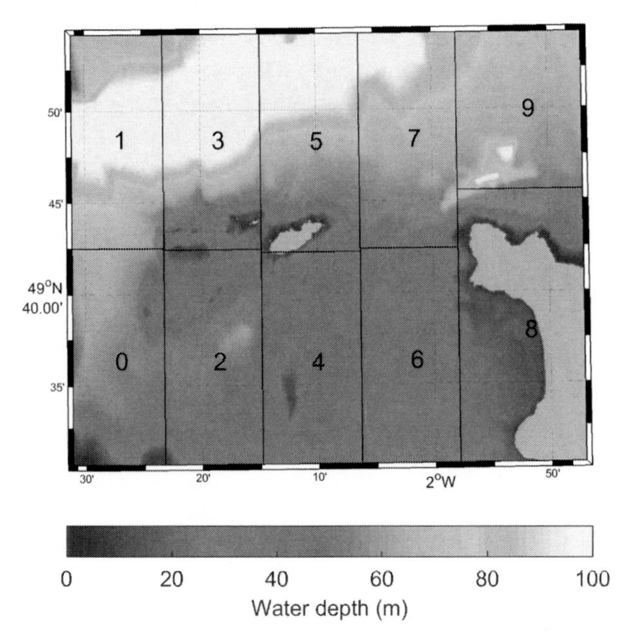

FIG. 8.16 Model domain decomposition for 10 processors. *(Based on a model of the Alderney Race described in S.P. Neill, J.R. Jordan, S.J. Couch, Impact of tidal energy converter (TEC) arrays on the dynamics of headland sand banks, Renew. Energy 37 (1) (2012) 387–397.)*

However, due to load balancing, I/O, and overheads associated with communication between parallel tasks, speedup for ocean models is nonlinear. For example, for the SWAN wave model applied to Galway Bay (Section 8.9.1), there is very little reduction in completion time once around 96 processors are exceeded. However, a job which uses >96 processors may take considerably longer to schedule, compared with a job which requests less resources.

8.8 CFD MODELLING

The dynamics of flow in both large (ocean tides/waves) and small scale (around a device) problems can be simulated by numerical solution of mass, momentum, and energy equations. Computational fluid dynamics (CFD) models (applied to small scales) and ocean models (applied to large-scale oceanic flows) are based on the Navier-Stokes equations. Further, CFD and ocean models use similar numerical techniques such as FDM or FVM. Although, strictly, ocean modelling is a subset of CFD, those working in the field of ocean energy make a clear distinction between what constitutes CFD, and what constitutes ocean modelling. CFD tends to be used for flows that are more isotropic (genuinely 3D) than ocean applications, where the velocities in the horizontal direction are generally greater than the vertical velocities, and CFD will neglect effects of stratification and Earth's rotation (e.g. Coriolis).

CFD is routinely used to deal with fluid/structure interaction and scenarios where there are strong shocks. The temporal and spatial scales are considerably different between CFD and ocean model simulations. In the latter, timescales of lunar, seasonal, and decadal are not uncommon, whereas CFD models may be dealing with simulation timescales of several seconds to few hours. CFD models will resolve fairly limited spatial scales (albeit at high resolution), such as the diameter of a tidal stream turbine, whereas ocean models are applied to regional (e.g. 10 km) and ocean basin (1000 km) length scales. CFD models usually require a special treatment of the free surface as their moving boundary condition. However, ocean models are usually based on a 2D approximation (depth-averaged) and compute water elevation or free surface as a state variable, more conveniently.

8.9 MODELLING CASE STUDIES

To demonstrate some of the principals introduced in this chapter, we examine two modelling case studies—a wave resource model of Galway Bay (Ireland) and a tidal resource model of Orkney (Scotland). We focus on model setup and validation for the Galway Bay case study, and model setup and interpretation of 3D results for the Orkney case study.

8.9.1 Wave Model of Galway Bay (Ireland)

SmartBay[5] is Ireland's national marine test and demonstration facility, suitable for the testing of scaled devices. It is located in Galway Bay, on the west coast of Ireland (Fig. 8.17). The presence of an island chain, the Aran islands, at the entrance to the bay, leads to relatively quiescent wave conditions within the bay, yet a wave climate that is characterized by 'scaled' Atlantic conditions [26].

The purpose of this exercise is to describe the setup of a SWAN wave model of Galway Bay, including sources of data, validation, and some results on wave resource assessment.

Model Setup

The model bathymetry was derived from the EMODnet dataset—available at a grid resolution of $1/480 \times 1/480$ degrees (approximately 200 m) throughout Europe. This bathymetry data was bilinearly interpolated onto a spherical model grid, which covered the inner nested region shown in Fig. 8.17 (which has dimensions approximately 1.0×0.5 degrees). The same bathymetry dataset was used to define the coastline, and was corrected from LAT to MSL.

5. See http://www.smartbay.ie/.

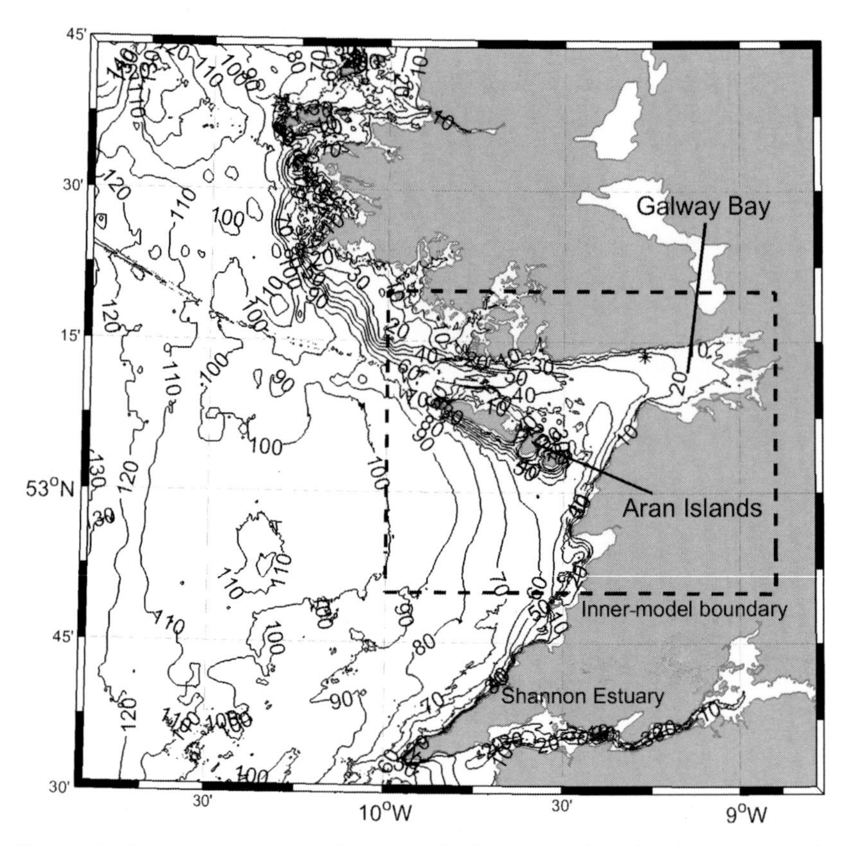

FIG. 8.17 Galway Bay and surrounding region showing the boundary of the inner nested model domain. The *asterisk* shows the location of the wave time series used for model validation. Contours are water depth in metres, relative to MSL.

The model was nested inside a coarser outer SWAN model which covered the North Atlantic at a grid resolution of $1/12 \times 1/12$ degrees. The outer model was used to output 2D wave spectra hourly at the boundary grid locations of the inner nested model domain. Both models were forced with ERA-Interim wind fields, available 3-h at a grid resolution of 0.75×0.75 degrees. There was no feedback from the inner to outer nest, that is, the nesting process was one-way—this is common practice in wave modelling studies.

The spatial resolution of the inner nested model was $1/400 \times 1/400$ degrees (approximately 270 m), and the directional resolution was set to 12 degrees. Forty discrete frequency intervals were simulated over the range 0.04–4.0 Hz (i.e. 0.25–25 s), and these were distributed logarithmically, leading to increased resolution at lower frequencies (Fig. 8.18).

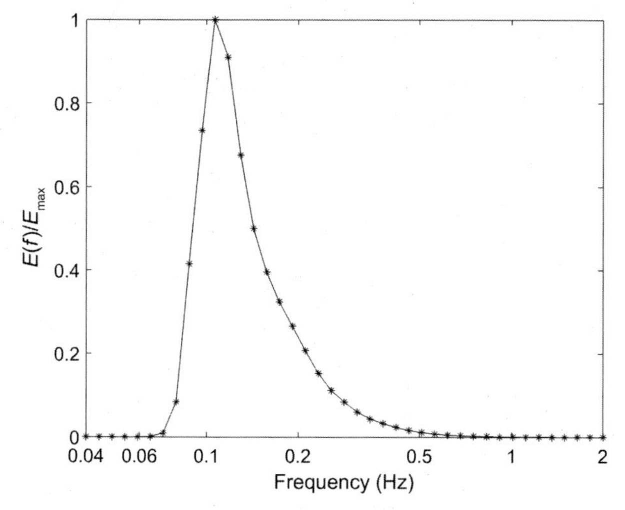

FIG. 8.18 Discrete frequency distribution for typical normalized wave energy density distribution output from the SWAN wave model.

Model Validation

The model was validated against a half-hourly wave buoy time series archived by the Marine Institute in Ireland. The validation was performed from January 1, 2014 to March 31, 2014 (3 months) for both significant wave height (*Hs*) and zero upcrossing wave period (*Tz*).

Time series of observed and simulated variables are shown in Fig. 8.19. The model has clearly captured the temporal variability of the wave climate, but there are a few instances (e.g. beginning of January and February) when the model has under-estimated *Hs* by almost a metre. The scatter plots (Fig. 8.20) show qualitatively that there is very good agreement with *Hs*, and more scatter in *Tz*—this is a fairly typical trend in wave model simulations. The r^2 value for *Hs* is 0.893, and the corresponding value for *Tz* is 0.417. The RMSE for *Hs* is 0.213 m (0.883 s for *Tz*), and S.I. for *Hs* is 18.2% (20.3% for *Tz*). Finally, bias is excellent, almost suspiciously so, for both *Hs* (0.0016 m) and *Tz* (0.035 s).

A final way of assessing model performance is to compare observed and modelled joint probability distributions between *Tz* and *Hs* (Fig. 8.21). Although this comparison is qualitative, it shows the success of the model in capturing both local and swell components of the wave climate.

Results

The 2014 annual mean simulated wave power distribution demonstrates clearly how Galway Bay is sheltered from the North Atlantic (Fig. 8.22). The wave direction further offshore is predominantly westerly, but waves are refracted as they propagate into Galway Bay. One important consideration for a model,

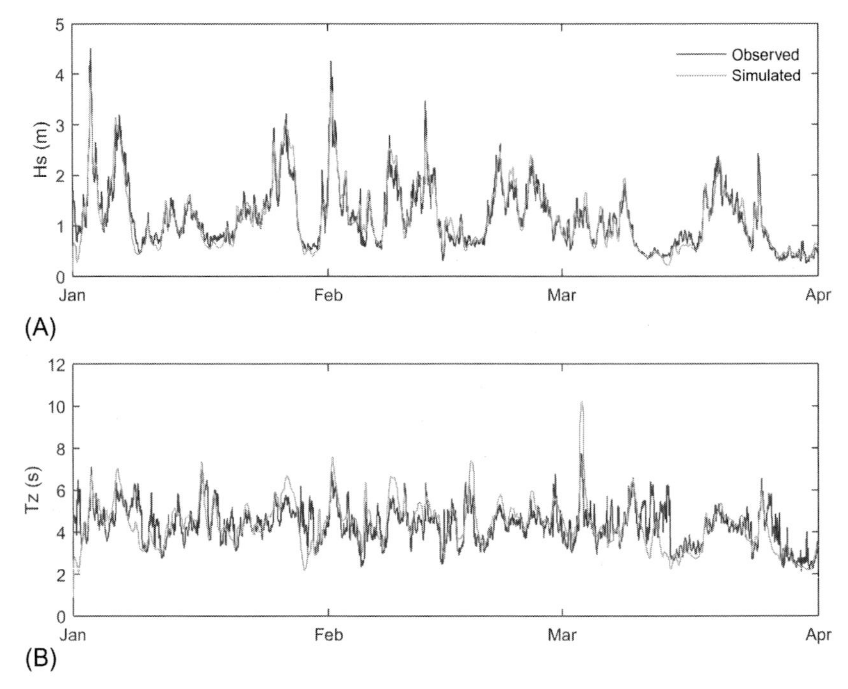

FIG. 8.19 Time series of observed and simulated (A) significant wave height (*Hs*) and (B) zero upcrossing wave period (*Tz*) in Galway Bay, January to March 2014.

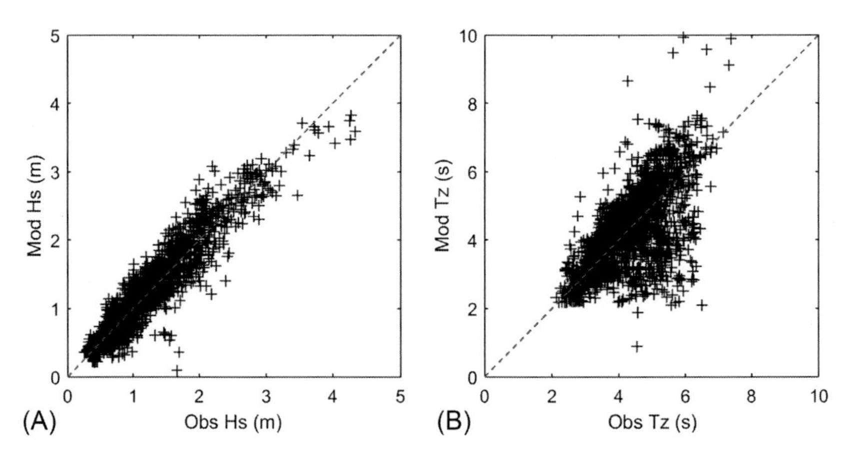

FIG. 8.20 Scatter plots of observed versus modelled (A) *Hs* and (B) *Tz* in Galway Bay over a 3-month period.

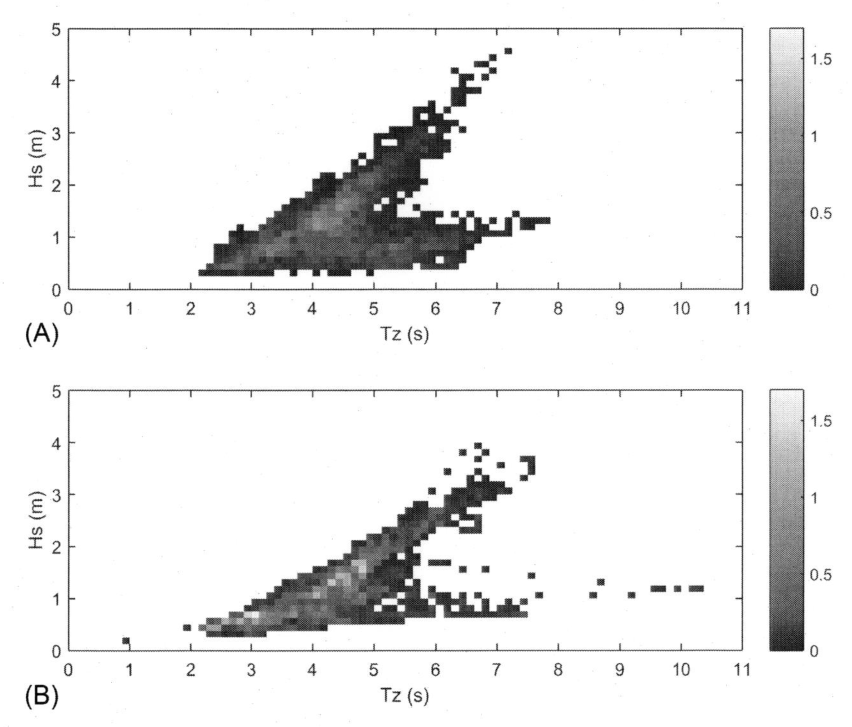

FIG. 8.21 Percentage joint probability distribution between *Tz* and *Hs* for Galway Bay. (A) Galway Bay wave buoy January to March 2014. (B) Galway Bay wave model January to March 2014.

which includes complex topography at this scale (e.g. the Aran Islands), is choice of grid resolution. The resolution selected for this simulation ($1/400 \times 1/400$ degrees) was appropriate for running an annual simulation in ~48 h using 96 processors of a supercomputer. This resolution seems suitable for this duration of simulation, for example, the validation was successful, and the wave energy propagating between the Aran Islands appears to have been captured well (Fig. 8.22). However, for more detailed studies of shorter duration (e.g. storm) events, a higher-resolution simulation may be more appropriate; but at increased computational cost, since a higher grid resolution would require a smaller model time step (see Section 8.1.4).

8.9.2 Tidal Model of Orkney (Scotland)

Orkney is an archipelago in the north of Scotland, separated from the Scottish mainland by the 12 km width of the Pentland Firth. Orkney is comprised of around 70 islands, separated by a series of bays and energetic tidal channels (Fig. 8.23). Orkney is mesotidal; however, tidal waves in the region, dominated

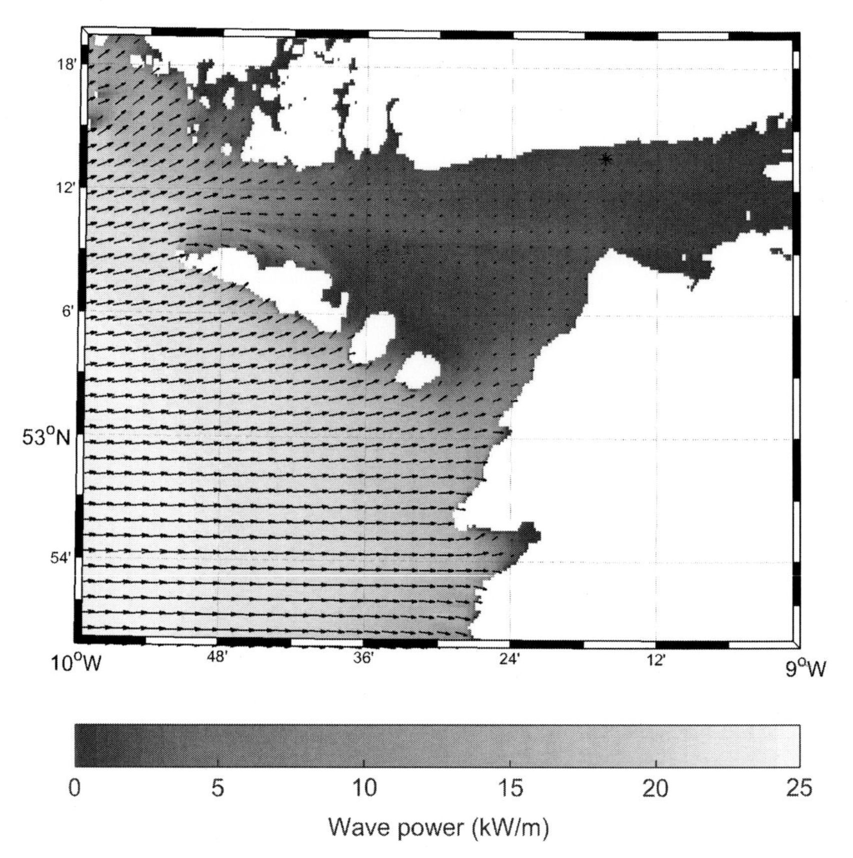

FIG. 8.22 Simulated annual mean (2014) wave power distribution (in kW/m) for Galway Bay. *Arrows* show mean wave direction at selected grid points.

by the principal semidiurnal lunar (M_2) and solar (S_2) constituents, take around two-and-a-half hours to propagate around Orkney from the western to the eastern approaches of the Pentland Firth (Fig. 8.24), leading to a considerable phase lag across Orkney. This phase lag results in a strong pressure gradient across Orkney, driving very strong tidal flows through the Pentland Firth and along the Firths of Orkney. For example, the M_2 phase lag between the Atlantic approach of Westray Firth and the North Sea approach of the connecting Stronsay Firth (Fig. 8.23) is around 65 degrees, that is 2.25 h,[6] and the resulting flow is channelled through constrictions, which narrow to around 5 km, with water depths in the range 25–50 m.

6. Recall from Section 3.3 that a 30-degrees phase lag in M_2 represents around 1 h.

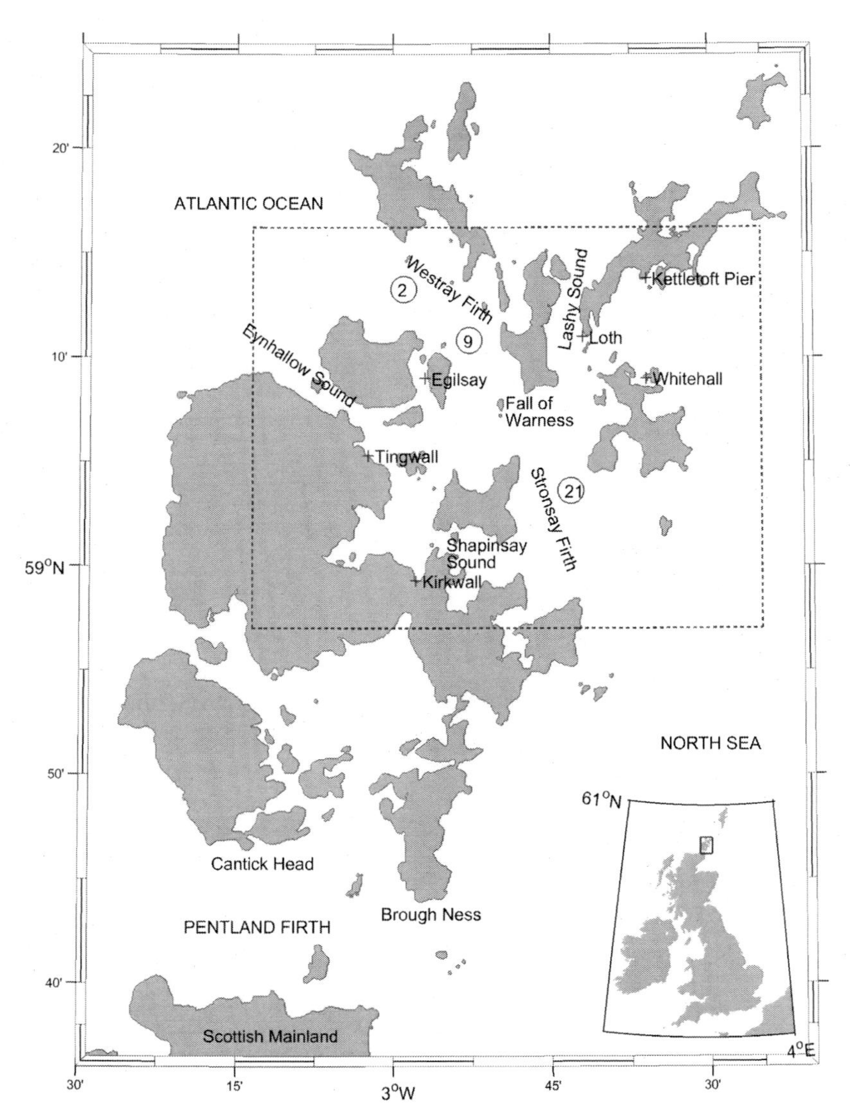

FIG. 8.23 Principal locations in Orkney and surrounding waters. The *dashed box* shows the boundaries of the high-resolution nested model, and tide gauge stations used for model validation (labelled) are shown as *crosses*. *Numbered circles* are locations where time series are presented in the 'Results' section. *Inset* shows the location of Orkney in relation to the British Isles. *(Adapted from S.P. Neill, M.R. Hashemi, M.J. Lewis, The role of tidal asymmetry in characterizing the tidal energy resource of Orkney, Renew. Energy 68 (2014) 337–350, which also includes details of model validation.)*

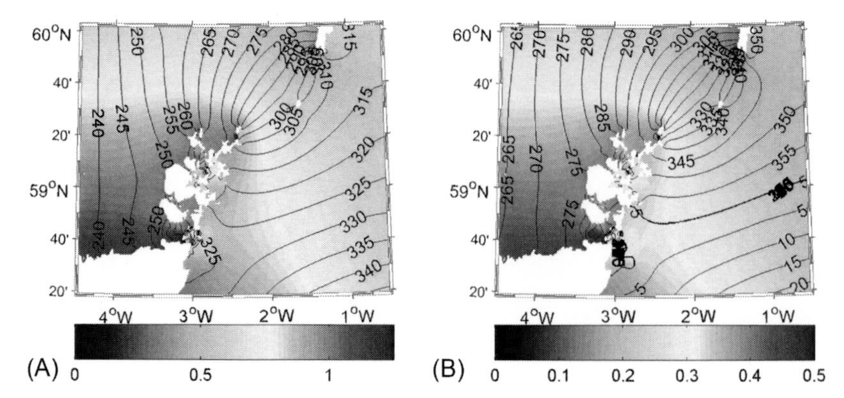

FIG. 8.24 (A) M_2 and (B) S_2 cotidal charts (amplitude in m) calculated from the north of Scotland regional ROMS model. Contours are phase (degrees relative to Greenwich).

The tidal currents flowing through the interisland channels of Orkney exceed 3 m/s in many regions [27], in conjunction with water depths in the range 25–50 m, suitable for the deployment of TEC devices. The marine renewable energy potential of Orkney has been recognized by the formation in 2003 of the European Marine Energy Centre (EMEC), which provides a wave test site to the west of Orkney, and a tidal test site in the Fall of Warness, situated where Westray Firth joins Stronsay Firth (Fig. 8.23).

The original aim of this model study was to investigate the role of tidal asymmetry on net power output (Section 3.10). However, the case study is presented here to demonstrate the setup of a 3D tidal model, and typical model outputs.

Model Setup

To provide boundary conditions to the high-resolution Orkney model, it was necessary to first run a north of Scotland regional model at coarser resolution. The regional model extended from 4 degrees 30′W to 0 degrees 30′W, and from 58 degrees 18′N to 60 degrees 03′N, encompassing the Pentland Firth, Orkney, and part of Shetland (Fig. 8.24). The regional model had a horizontal grid spacing of $1/120 \times 1/228$ degrees (approximately 500×500 m), and was forced at the boundaries by FES2012 ($1/16$ degrees resolution) currents and elevations for the M_2 and S_2 constituents [28]. Bathymetry for the regional model was interpolated from $1/120$ degrees GEBCO data. The regional model was run with 10 equally distributed vertical (sigma) levels for a period of 15 days, and tidal analysis of the elevations and depth-averaged velocities used to generate astronomical boundary forcing for the inner nested high-resolution Orkney model.

The high-resolution Orkney model extended from 3 degrees 13.5′W to 2 degrees 25′W and from 58 degrees 57′N to 59 degrees 16′N at a grid resolution of $1/750 \times 1/1451$ degrees (approximately 75×75 m) (the dashed box shown in Fig. 8.23). Bathymetry was interpolated from relatively high-resolution (approximately 200 m) gridded multibeam data provided by St. Andrew's University. The model domain encompasses the principal high tidal flow regions of Orkney, including Westray Firth and Stronsay Firth, and the EMEC tidal test site at the Fall of Warness. The model configuration used the GLS turbulence model, tuned to represent the k-ε model, and included horizontal harmonic mixing to provide subgrid scale dissipation of momentum [29], and quadratic bottom friction, with a drag coefficient $C_D = 0.003$. This value for the drag coefficient is consistent with previous ROMS studies which simulate the flow through energetic tidal channels, and these studies have demonstrated that the ROMS model is not particularly sensitive to the value of C_D [30,31]. The model was again run with 10 vertical levels for a period of 15 days.

Results

The peak depth-averaged tidal currents and the corresponding peak velocity vectors are shown in Fig. 8.25. Clearly, tidal flow is strongest at the constrictions of narrow channels (e.g. Lashy Sound, Eynhallow Sound, and, in particular, the Fall of Warness—see Fig. 8.23 for locations). The peak current speed reaches 3.7 m/s in Lashy Sound and the Fall of Warness, and because the model was

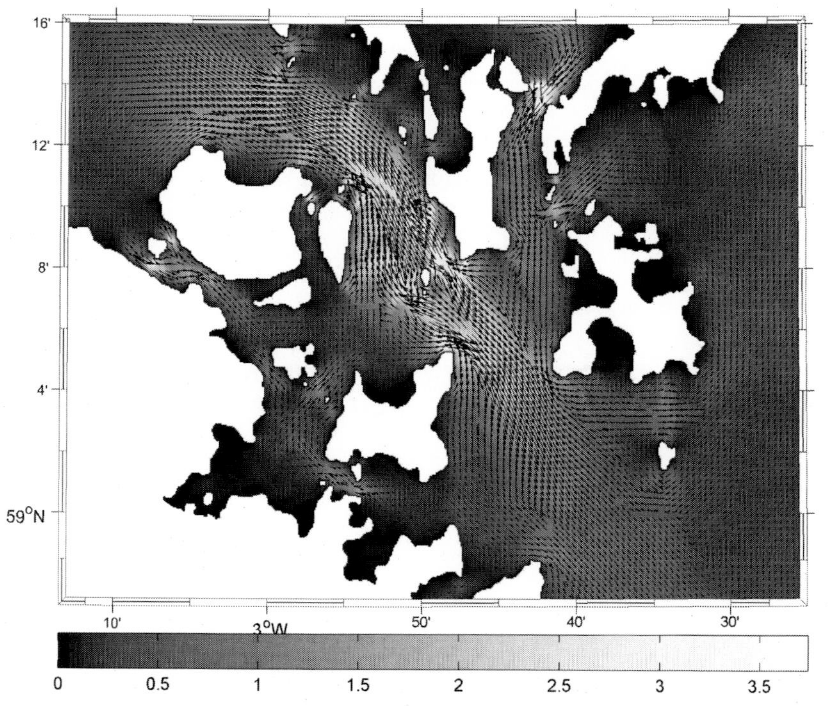

FIG. 8.25 Peak current speed (*colour scale*) and the associated peak spring velocity vectors. For clarity, only every fifth vector in both zonal and meridional directions has been plotted.

forced with the two principal semidiurnal tidal constituents, M_2 and S_2, these represent peak spring tidal currents. The peak velocity vectors also provide a qualitative overview of the tidal asymmetry [32], and the tidal flow appears to be largely ebb-dominant (peak currents directed northwest) in Westray Firth, and flood-dominant (peak currents directed southeast) in Stronsay Firth.[7] Where these two Firths join (i.e. in the vicinity of the Fall of Warness), the circulation is further complicated by the presence of strong residual eddies with length scales of around 4–5 km [33]. However, the tidal currents in this region appear to be more symmetrical, and divergence of the peak velocity vectors indicates that this is a bed-load parting zone [32]. The bathymetric/topographic restriction of the Fall of Warness impedes the flow along the channel, which is driven by the tidal pressure gradient. This impedance results in an acceleration of the flow downstream of the restriction. This leads to a divergence of the residual flow centred on the Fall of Warness, which is further complicated by variability in bathymetry and geometry along and across the channel (Fig. 8.25).

The advantage of a 3D tidal model, compared with a depth-averaged 2D model, is that the vertical distribution of variables in the water column is calculated directly, rather than parameterized. Here, we show detailed results for velocity and turbulence properties for all depths in the water column at three locations which exhibit varying degrees of asymmetry: site 2 (ebb-dominant), site 9 (symmetrical), and site 21 (flood-dominant) (Fig. 8.23).

Velocity time series for the three selected locations are plotted in Fig. 8.26. Above the boundary layer, the asymmetry for sites 2 and 21 is evident at all depths in the water column, and is more pronounced during spring tides. Therefore, for all practical scenarios of energy extraction, that is device hub placed at some height in the water column that is above the near-bed boundary layer, strong velocity asymmetry will translate into an even stronger asymmetry in power output (because power is related to velocity cubed). Such asymmetry in the flow field is clearly undesirable from an electricity generation perspective, and so symmetrical sites (such as site 9) are more attractive for commercial development.

Turbulent kinetic energy (k) per unit mass is defined as

$$k = \frac{1}{2}(u'^2 + v'^2 + w'^2) \tag{8.46}$$

k is plotted at three contrasting locations (Fig. 8.27). The corresponding plot for the rate of dissipation (ϵ) of k is shown in Fig. 8.28, defined as

$$\frac{dk}{dt} \cong P - \epsilon \tag{8.47}$$

7. Because the tidal wave propagates eastwards across the north of Scotland (Fig. 8.24), the tidal currents are directed approximately southeastward along these Firths during the flood phase of the tide.

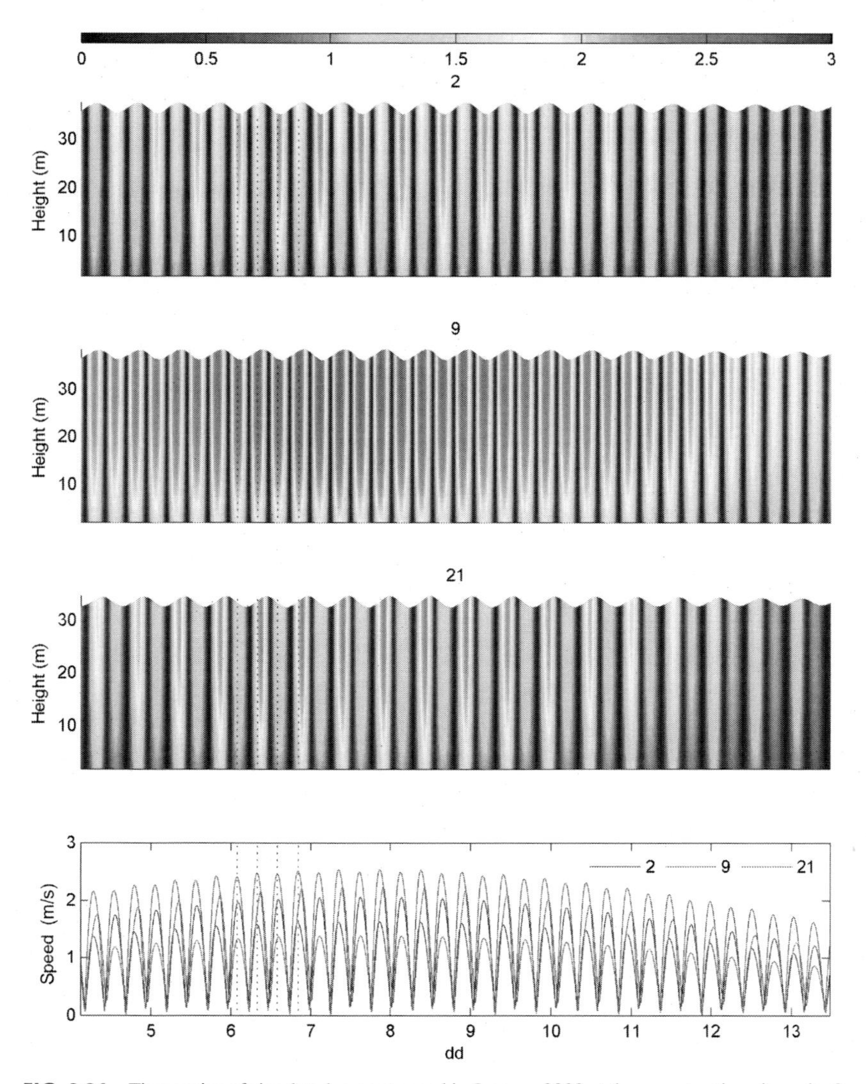

FIG. 8.26 Time series of simulated current speed in January 2000 at three contrasting sites: site 2 (ebb-dominant), site 9 (almost symmetrical), and site 21 (flood-dominant). The *lower panel* shows a time series of depth-averaged current speed. For reference, *vertical dashed lines* are located at peak flood and ebb conditions for site 9.

where P is the rate of production. By comparing these turbulence metrics with the velocity time series (Fig. 8.26), it is clear that even a relatively modest asymmetry in velocity can translate into a large asymmetry in turbulence properties. Although sites 2 and 21 have the strongest asymmetry, this point is made clearer by considering the velocity time series at site 9 (Fig. 8.26) (an almost symmetrical site). This almost indiscernible asymmetry in the velocity

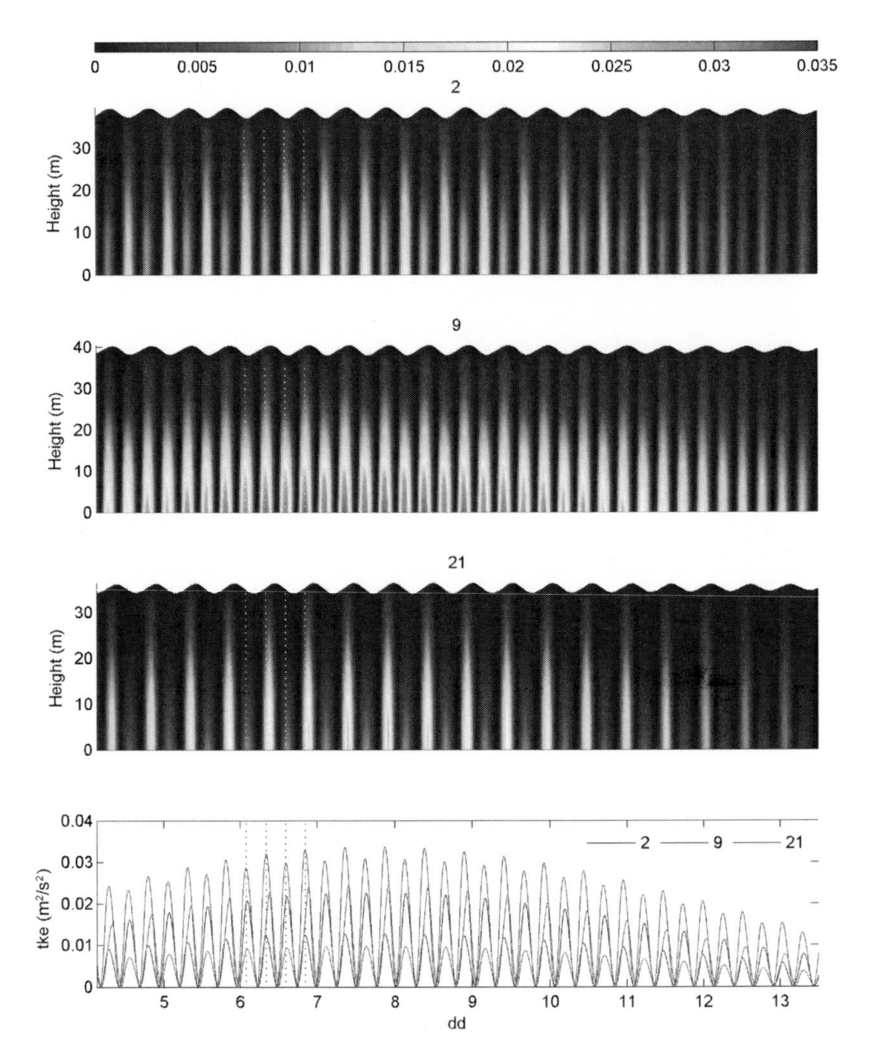

FIG. 8.27 Time series of simulated turbulent kinetic energy in January 2000 at three contrasting sites: site 2 (ebb-dominant), site 9 (almost symmetrical), and site 21 (flood-dominant). The *lower panel* shows a time series of TKE at the model cell closest to the bed. For reference, *vertical dashed lines* are located at peak flood and ebb conditions for site 9.

time series manifests itself as a strong asymmetry in TKE (Fig. 8.27). The production of turbulent kinetic energy is generally proportional to the magnitude of the velocity gradient $\left(P \propto \left[\left(\frac{\partial U}{\partial z}\right)^2 + \left(\frac{\partial V}{\partial z}\right)\right]^2\right)$. For a specific velocity distribution (e.g. a logarithmic distribution), this gradient will be proportional to the current strength. Therefore, we expect magnified turbulence asymmetry due to asymmetry in the current strength.

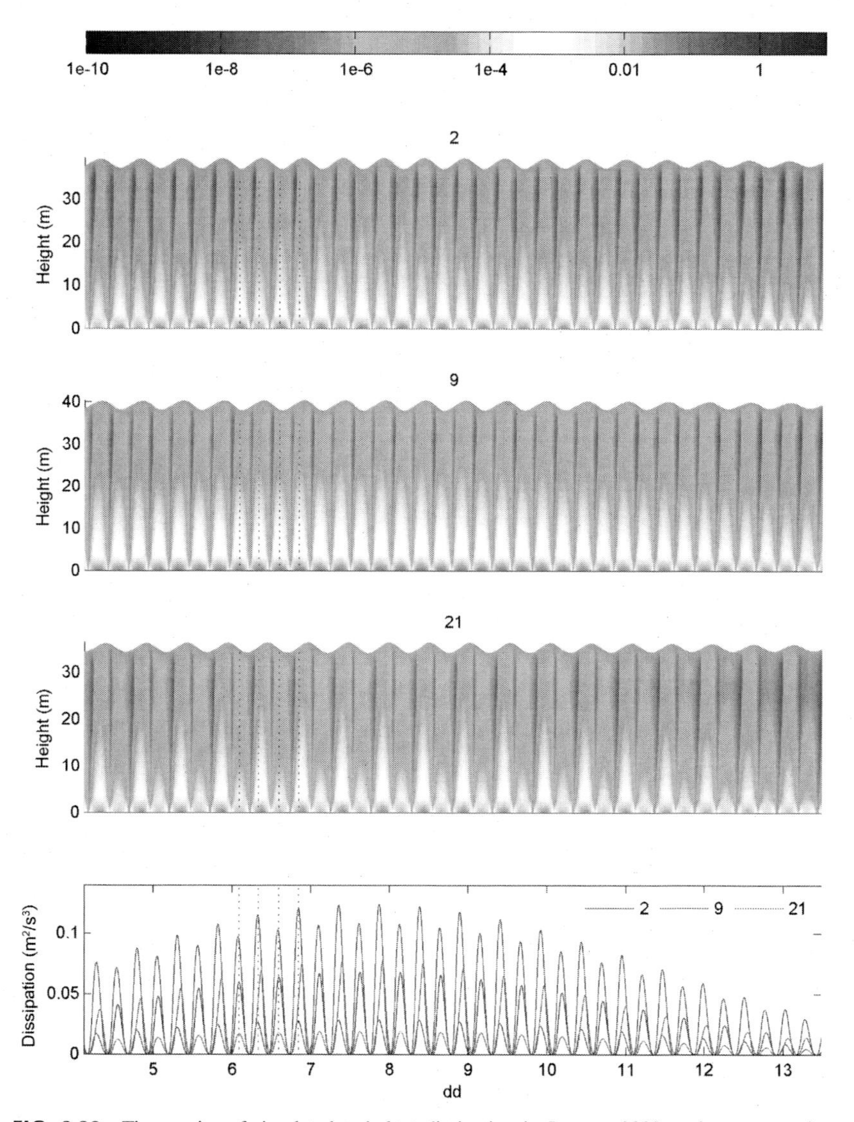

FIG. 8.28 Time series of simulated turbulent dissipation in January 2000 at three contrasting sites: site 2 (ebb-dominant), site 9 (almost symmetrical), and site 21 (flood-dominant). The *lower panel* shows a time series of turbulent dissipation at the model cell closest to the bed. For reference, *vertical dashed lines* are located at peak flood and ebb conditions for site 9.

REFERENCES

[1] G.L. Mellor, T. Ezer, L.-Y. Oey, The pressure gradient conundrum of sigma coordinate ocean models, J. Atmos. Ocean. Technol. 11 (4) (1994) 1126–1134.

[2] S.P. Neill, M.J. Lewis, M.R. Hashemi, E. Slater, J. Lawrence, S.A. Spall, Inter-annual and inter-seasonal variability of the Orkney wave power resource, Appl. Energy 132 (2014) 339–348.

[3] L. Carrere, F. Lyard, M. Cancet, A. Guillot, FES 2014, a new tidal model on the global ocean with enhanced accuracy in shallow seas and in the Arctic region, in: EGU General Assembly Conference Abstracts, vol. 17, 2015, p. 5481.

[4] S.P. Neill, M.R. Hashemi, Wave power variability over the northwest European shelf seas, Appl. Energy 106 (2013) 31–46.

[5] N. Yates, I. Walkington, R. Burrows, J. Wolf, Appraising the extractable tidal energy resource of the UK's western coastal waters, Philos. Trans. R Soc. Lond. 371 (1985) (2013) 20120181.

[6] D.M. Grogan, S.B. Leen, C.R. Kennedy, C.M.Ó. Brádaigh, Design of composite tidal turbine blades, Renew. Energy 57 (2013) 151–162.

[7] M.A. Cialone, T.C. Massey, M.E. Anderson, A.S. Grzegorzewski, R.E. Jensen, A. Cialone, D.J. Mark, K.C. Pevey, B.L. Gunkel, T.O. McAlpin, N.N. Nadal-Caraballo, J.A. Melby, J.J. Ratcliff, North Atlantic Coast Comprehensive Study (NACCS) Coastal Storm Model Simulations: Waves and Water Levels, Coastal and Hydraulics Laboratory, U.S. Army Engineer Research and Development Center, Vicksburg, 2015.

[8] C. Chen, H. Liu, R.C. Beardsley, An unstructured grid, finite-volume, three-dimensional, primitive equations ocean model: application to coastal ocean and estuaries, J. Atmos. Ocean. Technol. 20 (1) (2003) 159–186.

[9] M.R. Hashemi, S.P. Neill, A.G. Davies, A coupled tide-wave model for the NW European shelf seas, Geophys. Astrophys. Fluid Dyn. 109 (3) (2015) 234–253.

[10] G.S. Carter, M.A. Merrifield, Open boundary conditions for regional tidal simulations, Ocean Model. 18 (3) (2007) 194–209.

[11] Regional Ocean Modeling System (ROMS), 2016, Available from: https://www.myroms.org, Accessed 11 November 2016.

[12] M. Zijlema, G. Stelling, P. Smit, SWASH: an operational public domain code for simulating wave fields and rapidly varied flows in coastal waters, Coast. Eng. 58 (10) (2011) 992–1012.

[13] F. Shi, R.A. Dalrymple, J.T. Kirby, Q. Chen, A. Kennedy, A fully nonlinear Boussinesq model in generalized curvilinear coordinates, Coast. Eng. 42 (4) (2001) 337–358.

[14] L.H. Holthuijsen, Waves in Oceanic and Coastal Waters, Cambridge University Press, Cambridge, 2010.

[15] N. Booij, R.C. Ris, L.H. Holthuijsen, A third-generation wave model for coastal regions: 1. Model description and validation, J. Geophys. Res. Oceans 104 (C4) (1999) 7649–7666.

[16] The Wamdi Group, The WAM model—a third generation ocean wave prediction model, J. Phys. Oceanogr. 18 (12) (1988) 1775–1810.

[17] H.L. Tolman, B. Balasubramaniyan, L.D. Burroughs, D.V. Chalikov, Y.Y. Chao, H.S. Chen, V.M. Gerald, Development and implementation of wind-generated ocean surface wave modelsat NCEP, Weather Forecast. 17 (2) (2002) 311–333.

[18] E. Bouws, L. Draper, E.D.R. Shearman, A.K. Laing, D. Feit, W. Mass, L.I. Eide, P. Francis, D.J.T. Carter, J.A. Battjes, Guide to Wave Analysis and Forecasting. WMO-No. 702, World Meteorological Organization, Geneva, Switzerland, 1998.

[19] L. Cavaleri, P.M. Rizzoli, Wind wave prediction in shallow water: theory and applications, J. Geophys. Res. Oceans 86 (C11) (1981) 10961–10973.

[20] G.J. Komen, K. Hasselmann, K. Hasselmann, On the existence of a fully developed wind-sea spectrum, J. Phys. Oceanogr. 14 (8) (1984) 1271–1285.
[21] K. Hasselmann, On the spectral dissipation of ocean waves due to white capping, Bound.-Layer Meteorol. 6 (1–2) (1974) 107–127.
[22] R.C. Ris, Spectral Modelling of Wind Waves in Coastal Areas, TU Delft, Delft University of Technology, 1997.
[23] SWAN Team, et al., SWAN Scientific and Technical Documentation, SWAN Cycle III Version 40.81, Delft University of Technology, 2010.
[24] A. Zhang, K.W. Hess, F. Aikman III, User-based skill assessment techniques for operational hydrodynamic forecast systems, J. Oper. Oceanogr. 3 (2) (2010) 11–24.
[25] L. Mentaschi, G. Besio, F. Cassola, A. Mazzino, Problems in RMSE-based wave model validations, Ocean Model. 72 (2013) 53–58.
[26] A.R. Bento, P. Martinho, C.G. Soares, Numerical modelling of the wave energy in Galway Bay, Renew. Energy 78 (2015) 457–466.
[27] I.G. Bryden, S.J. Couch, ME1—marine energy extraction: tidal resource analysis, Renew. Energy 31 (2) (2006) 133–139.
[28] L. Carrère, F. Lyard, M. Cancet, A. Guillot, L. Roblou, FES2012: a new global tidal model taking advantage of nearly 20 years of altimetry, in: Proceedings of Meeting, vol. 20, 2012.
[29] R.C. Wajsowicz, A consistent formulation of the anisotropic stress tensor for use in models of the large-scale ocean circulation, J. Comput. Phys. 105 (2) (1993) 333–338.
[30] K.M. Thyng, J.J. Riley, J. Thomson, Inference of turbulence parameters from a ROMS simulation using the k-ε closure scheme, Ocean Model. 72 (2013) 104–118.
[31] D.A. Sutherland, P. MacCready, N.S. Banas, L.F. Smedstad, A model study of the Salish Sea estuarine circulation, J. Phys. Oceanogr. 41 (6) (2011) 1125–1143.
[32] R.D. Pingree, D.K. Griffiths, Sand transport paths around the British Isles resulting from M2 and M4 tidal interactions, J. Mar. Biol. Assoc. UK 59 (02) (1979) 497–513.
[33] J. Lawrence, H. Kofoed-Hansen, C. Chevalier, High-resolution metocean modelling at EMEC's (UK) marine energy test sites, in: Proc. of the 8th European Wave and Tidal Energy Conference, vol. 7, 2009.

FURTHER READING

[1] S.P. Neill, J.R. Jordan, S.J. Couch, Impact of tidal energy converter (TEC) arrays on the dynamics of headland sand banks, Renew. Energy 37 (1) (2012) 387–397.
[2] S.P. Neill, M.R. Hashemi, M.J. Lewis, The role of tidal asymmetry in characterizing the tidal energy resource of Orkney, Renew. Energy 68 (2014) 337–350.

Chapter 9

Optimization

To make the best use of an ocean renewable energy resource, for example, to maximize electricity generation whilst minimizing environmental impacts, it is necessary to perform optimization. Optimization theory has wide-ranging applications, but here we consider applications to offshore wind, tidal energy, and wave energy. Many intraarray optimization topics are common across the majority of ocean renewable energy resources, for example, optimizing device spacing and minimizing variability. We also consider interarray optimization within the context of tidal energy plants—a predictable resource with strong potential for phasing of power stations along a tidal wave, hence minimizing variability in aggregated power output. Finally, we introduce tools available for optimizing wind farms (HOMER) and tidal energy arrays (OpenTidalFarm).

9.1 INTRODUCTION TO OPTIMIZATION THEORY

Optimization is closely related to decision making; hence, we face many optimization problems in our daily life as an individual, company, organization, or government. When faced with several alternatives, we are always interested in the *best* alternative or decision. For instance, if you are planning a trip, you can choose amongst several alternative forms of transport, including flights, trains, rental cars, or bus. The best mode of transport would depend on your criteria, which is technically called the *objective function*. If minimizing the cost is the criteria, the best option may be transport by bus. If the objective is to minimize travel time, flights might be the best option. Any optimization problem needs an objective function. Usually, the objective function will be either minimized or maximized. Depending on the application, other names such as cost function (minimize in economy), profit function (maximize in marketing), fitness (maximize in genetic programming through natural selection), energy function (minimize in physics), or error (minimize in numerical modelling) are used. As an ocean science example, model parameter estimation (in calibration, inverse modelling, or data assimilation) is an optimization problem. Given a set of observed data, the best model parameters (e.g. bottom drag coefficient, wind drag coefficient, boundary conditions, etc.) are selected such that the error between model simulations and observations is minimized.

Fundamentals of Ocean Renewable Energy. https://doi.org/10.1016/B978-0-12-810448-4.00009-4

The objective function of an optimization problem depends on the *independent variables*, which can also be called *decision* or *control variables*. Independent variables in the ocean model parameterization example are physical properties of the ocean, such as bottom and surface drag coefficients. The objective function should be sensitive to independent variables. In some complex problems, a sensitivity analysis is performed to eliminate parameters or variables that do not affect the objective function. Independent variables can be discrete or continuous. For example, finding the optimum number of blades for a wind turbine is a discrete optimization problem; whereas finding the optimum distance between two wind turbines is a continuous optimization problem. Optimization problems are mostly formulated as deterministic problems, in which independent variables and objective function(s) are treated deterministically. In contrast, in stochastic optimization problems, due to uncertainties, random variables are used to formulate the problem. The expected value of the objective function is either maximized or minimized in such problems. For instance, optimizing the operational cost of a renewable energy microgrid can be treated stochastically due to uncertainties in supply of energy (i.e. intermittency of the resource [1]).

An optimization problem may be subjected to constraints. Referring to our earlier example of finding the best mode of transport for a journey, assume that minimizing the time is the objective function, but you may not exceed a certain budget for the transportation cost. In that case, the problem becomes more complicated. For instance, direct flights may exceed your budget. Technically, any alternative that does not satisfy constraints is not *feasible*, or out of the *feasible region* or *search space*. The best solution to the transportation problem may be found by comparing indirect flights and trains that do not exceed your budget. In the model parameterization example, an acceptable/feasible range for each parameter is applied as a constraint. For instance, bottom friction cannot be a negative number.

An optimization problem may have one, two, or multiple objective functions that are conflicting. For instance, assume that you are trying to determine the best type of energy system for an off-grid island community. Your first objective function is to minimize the cost; your second objective function can be minimizing pollution (i.e. emission of pollutants such as CO_2). These objective functions are conflicting; the best solution may not be the least expensive or the cleanest type of energy for the island. The majority of real life decision-making problems, including renewable energy systems, have multiple objective functions, and are subjected to various constraints.

Optimization problems are either static or dynamic. In dynamic optimization problems, the parameters of the problem change with time. Finding the best trajectory of an aeroplane over a fixed range (minimum fuel cost and minimum travel time) is a dynamic optimization problem, because the environmental conditions that affect the fuel consumption and travel time of the aircraft (e.g. wind) change with time. In energy-related projects, several variables, such as the price of energy, demand, and interest rates, vary with time; therefore, many renewable energy optimization problems are dynamic in nature.

Simple optimization problems were first introduced in calculus: finding the maximum or minimum of a function. More realistic applications of optimization theory were introduced in World War II in order to reduce the cost of an army and maximize loss to the enemy. Since then, and mainly due to advances in computing power, numerous methods/algorithms have emerged to solve optimization problems, and have been applied across various disciplines. These methods can generally be categorized as linear programming, nonlinear programming, iterative methods, and metaheuristic techniques. In linear programming problems, the objective function and constraints are linear—as opposed to nonlinear programming. Metaheuristic approaches are the most recent suite of optimization techniques, and are mainly based on artificial intelligence and machine learning. Evolutionary algorithms, genetic algorithm, and particle swarm optimization are popular metaheuristic approaches.

Solution techniques for optimization problems can be classified as trajectory- and population-based. In the former method, a single solution is used to search for the optimum. The optimum solution is also a single-optimized solution at the end of the optimization process. Classical iterative schemes such as Newton's method fall into this category. In population-based techniques, a population of solutions are used that evolve through each iteration of the optimization. In general, a set of optimum solutions are provided at the end. An example is genetic algorithms that provide a set of solutions that have the best fitness (e.g. above an acceptable value). For decision-making applications, population-based techniques are more attractive, because they give decision makers a chance to select amongst a list of optimum solutions.

There are several excellent books describing various classical and modern optimization techniques (e.g. [2,3]). Here, an overview of the basic concepts, some examples, and tools to optimize ocean renewable energy projects are introduced.

9.1.1 Mathematical Formulation of an Optimization Problem

An optimization problem can be mathematically formulated as follows:

$$\begin{aligned} &\text{minimize} \quad f(x_1, x_2, \ldots, x_n) \\ &\text{subject to} \quad (x_1, x_2, \ldots, x_n) \in \Omega \end{aligned} \tag{9.1}$$

in which f is the objective (cost, fitness, etc.) function, x_i are independent or decision variables, and Ω represents the feasible region or search space. As mentioned before, we could also *maximize* the objective function (e.g. fitness). Any maximization problem can simply be converted into a minimization problem by multiplying the objective function by -1. The feasible, or search space, is controlled by the constraints of the problem.

For instance, assume that the location of four wind turbines in an offshore wind farm needs to be optimized. For this problem, the objective function can be defined as electricity production. The decision variables are the geographical coordinates of these four turbines: $(x_1, y_1; x_2, y_2; x_3, y_3; x_4, y_4)$. The production of electricity in this small array depends on the decision variables, because both the available wind energy resource, and the wake of each turbine (that may adversely affect power output for another turbine) are controlled by turbine locations. The feasible space or constraints can be defined as the leased area of the farm, because none of these turbines can be installed outside the allocated area. This can be mathematically formulated as $a \leq x_i \leq b$ and $c \leq y_i \leq d$, where a, b, c, and d define the geographical boundaries of this problem. This type of constraint is called *inequality constraint*. *Equality constraints* can also be applied. For the simple wind farm example, assume that, for aesthetic reasons, we want to install all four turbines in a straight line. Therefore, the geographical locations of all turbines must satisfy the equation of a line (i.e. $mx_i+ny_i+q = 0$). In general, an optimization problem can be formulated as

$$\begin{aligned} \text{minimize} \quad & f(x_1, x_2, \ldots, x_n) \\ \text{subject to} \quad & g_k(x_1, x_2, \ldots, x_n) \geq 0 \quad k = 1, 2, \ldots, N_{c_1}; \\ & h_l(x_1, x_2, \ldots, x_n) = 0 \quad l = 1, 2, \ldots, N_{c_2} \end{aligned} \tag{9.2}$$

where g_k and h_k represent inequality and equality constraints of the problem. N_{c_1} and N_{c_2} are the number of inequality and quality constraints, respectively; an optimization problem can have several inequality or equality constraints. It should be noted that applying too many constraints can make an optimization problem infeasible (e.g. if constraints are mutually contradictory), and so result in no solution.

The optimum solution of a problem can be local or global. Referring to Fig. 9.1, a function can have several minimums/optimums. The local optimum is a solution that is better than the neighbouring points, but not necessarily the best solution across the search space. If the objective function is *convex*

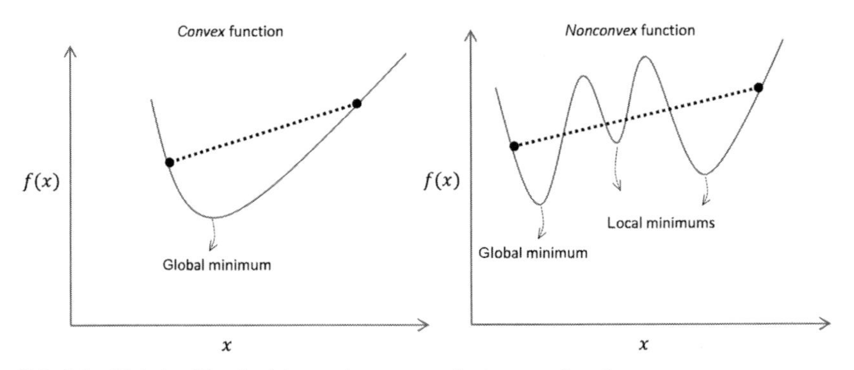

FIG. 9.1 Global and local minimums in convex and nonconvex functions.

(left plot in Fig. 9.1), there is only one minimum or optimal solution. By definition, a function is convex if the line segment connecting two points on the function always lies above it. This is not the case for the right-hand plot in Fig. 9.1, which is *nonconvex*. This definition can be generalized as a function with several variables. Although the majority of real life optimization problems are nonconvex, classical, simple optimization techniques can quickly find local optimums. Global optimization techniques are more complicated and, accordingly, more computationally expensive.

9.1.2 Search Methods and Optimization Algorithms

To briefly explain how an optimization algorithm works, the basic concepts of search techniques are explained here. An optimization technique searches for a point that minimizes the objective function. Many mathematical techniques are iterative. They start from an arbitrary initial point in the search space and update that point during each iteration:

$$\mathbf{x}_{k+1} = \mathbf{x}_k + \alpha_k \mathbf{g}_k \tag{9.3}$$

in which α is the step size, and $\mathbf{g}$ is the step direction. During each iteration, the decision variable vector is updated using Eq. (9.3); step size and step direction are computed based on the optimization algorithm (e.g. maximum gradient). Fig. 9.2 shows a 2D schematic of an optimization that starts at point $\mathbf{x}_1$ and moves iteratively towards the optimum point.

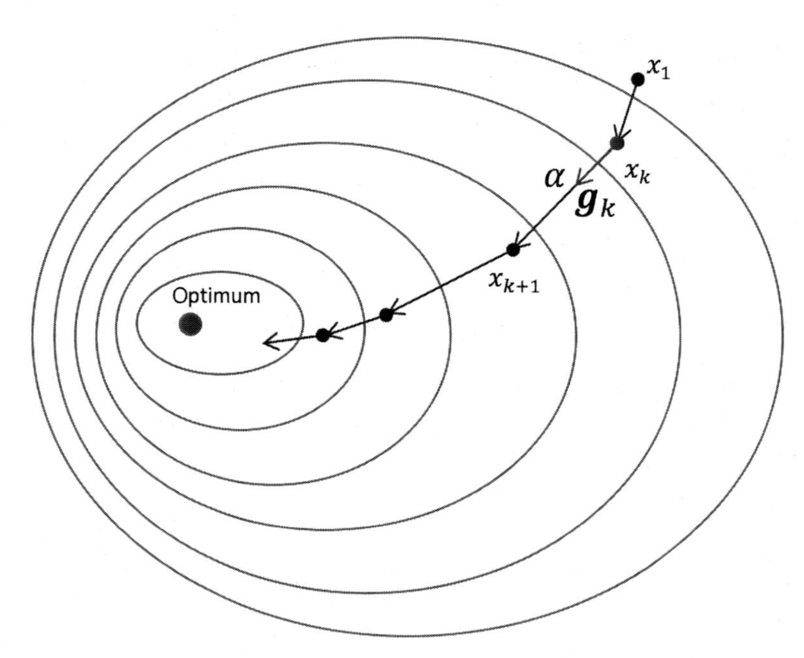

FIG. 9.2 Two-dimensional schematic of optimization using iterative gradient techniques.

A simple analogy to optimization is climbing towards the summit of a mountain (maximum elevation). Here, the objective function is elevation, and latitude and longitude are the decision variables. Assume that a helicopter leaves you at a random location on the mountain. Based on the best local information, you move towards a direction for a specific distance (step size), and then you change your direction. The distance that you move in each direction is not necessarily the same: the step size can also vary within each iteration. An obvious direction is going up, and in particular in the steepest ascent direction. The steepest ascent is mathematically the direction of the gradient, which can be written as

$$\nabla f = \left(\frac{\partial f}{\partial x_1}, \frac{\partial f}{\partial x_1}, \ldots, \frac{\partial f}{\partial x_n} \right) \tag{9.4}$$

and the method of steepest ascent becomes

$$\mathbf{x}_{k+1} = \mathbf{x}_k + \alpha_k \nabla f_k \tag{9.5}$$

This method almost guarantees that you reach a summit that is closest to your initial location but, as you can imagine, it does not guarantee that you will reach the highest summit of a mountain range. Many classical optimization techniques are based on gradient. The method of steepest ascent (or descent in minimization) is not very effective, especially near the solution. Referring to Fig. 9.1, it is clear that the gradient of the function (slope of the tangent) near the local and global minimums approaches zero. This causes some issues for methods that only use gradient for the direction. Other techniques are called conjugate gradient methods that use second derivatives (Hessian matrix) as well as first derivative (gradient) for direction.

9.2 INTRAARRAY OPTIMIZATION

Although understanding the natural (undisturbed) resource, and the operation of single devices, is essential, it is only when devices are installed in arrays that significant levels of electricity generation can be achieved. In this section, we discuss how devices within such arrays can be optimized for wind, tidal stream, and wave energy conversion.

9.2.1 Micro-Siting of Offshore Wind Farms

Referring to Section 4.4, marine spatial planning is the standard procedure used to find suitable and potential sites for the development of offshore wind farms in a region. Marine spatial planners, after a comprehensive study of technical, economic, environmental, sociopolitical, legal, and regulatory aspects, recommend areas that have minimum conflicts with other users of the ocean, in addition to a feasible energy resource. *Macro-siting* is the selection of a location for a wind

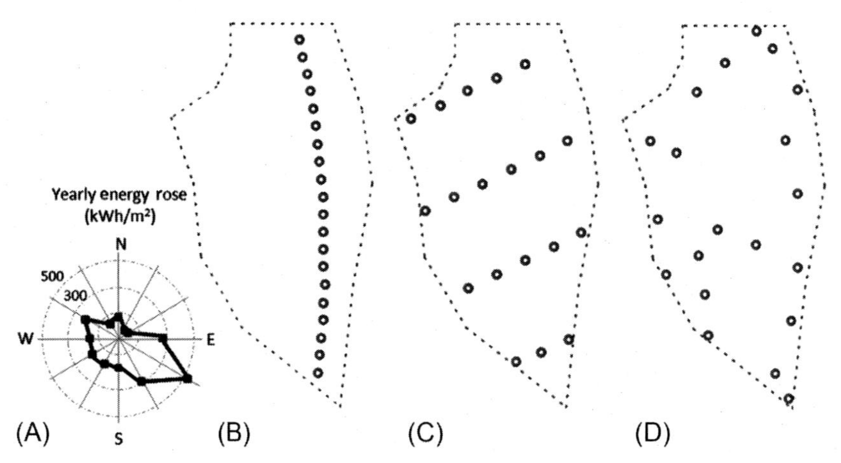

FIG. 9.3 Layout of the Middelgrunden offshore wind farm in various scenarios: (B) actual, (C) optimized with constraints on the arrangement, and (D) optimized without constraints on locations of the wind turbines [4,5]. (A) is the yearly wind energy rose.

farm within these recommended areas. Macro-siting can also be treated as an optimization problem, which is not discussed here.

After selection of a location/plot for an offshore wind farm, the best layout of turbines within a farm needs to be determined. The layout, in general, consists of the number of wind turbines, and the size and geographical location of each turbine. In many cases, the maximum investment, or the capacity of a farm, is decided in the first steps of a study as a constraint; therefore, if the size of the individual turbines is also established, the optimization problem reduces to finding the best/optimum geographical location for each turbine. This is referred to as *micro-siting* of offshore wind farms. Fig. 9.3 shows an example of wind farm optimization that can lead to around 5% increase in annual energy production (AEP; [4]).

Wake Effect

The layout of wind turbines in a wind farm affects AEP of the array. As we discussed in Chapter 4, wind has spatial variability; therefore, if the farm is large enough, in some places wind energy is higher than other places. If variability of the available resource was the only reason, we could just locate wind turbines in places that have the highest energy. A more complicated issue is the interaction of a wind turbine with neighbouring turbines. Fig. 9.4 shows how turbines can be located in the wake of each other under particular wind conditions. The wind speed in the wake of a turbine is significantly lower than the undisturbed wind speed (in the absence of the upwind turbine). This reduction of speed adversely affects total AEP. It is possible to place turbines very far from each other to avoid wake effects, but this can lead increased cost of cabling, electrical connections

FIG. 9.4 Visualization of the wake effect in a wind farm. *(Reproduced with kind permission of Vattenfall.)*

between turbines, maintenance cost, and a larger leased area. Therefore, finding the optimum layout is an optimization problem.

Estimating the reduction of wind speed in the wake of a turbine can be achieved by simplified methods, based on the distance of the turbines, hub height, and radius of turbines (e.g. [5,6]), or wakes can be more accurately simulated using CFD codes and high performance computing (e.g. [7,8]).

Decision Variables

In micro-siting of offshore wind farms, the layout of the farm, which consists of several variables, needs to be optimized. The simplest case is when the number and the capacity of turbines are given, in which the locations of the turbines are the decision variables. More complicated cases involve more decision variables, such as the number, size, type, and hub-heights of turbines, in addition to the locations of the wind turbines.

Objective Functions

As mentioned before, a renewable energy system can be optimized by a single or multiple objective functions. The simplest objective function is AEP (maximize). This objective function does not take into account the cost associated with AEP, which is a major factor in the decision-making process. Referring to the concept of levelized cost of energy (LCOE) introduced in Chapter 1, the ratio of the cost of energy to electricity production can be minimized, which is a more practical objective function. This objective function takes into account the increased cost of electrical cables and other elements of the farm when turbine spacing is increased:

$$\begin{aligned} &\text{minimize} \quad \frac{C_t}{E_p} \\ &\text{subject to} \quad \textit{constraints} \end{aligned} \tag{9.6}$$

where C_t can be chosen as the total cost of the project, and E_p is the total energy production during the lifetime of the project. Alternatively, the investment cost divided by AEP can be minimized, which does not take into account the interest rate, operation and maintenance cost, and the time value of money.

Constraints

Micro-siting is subject to several constraints. The most common constraint is the available area/plot for the farm (i.e. all turbines should be located inside the array). The minimum distance between turbines can also be a constraint due to technical issues or standards/regulations. Another constraint is the maximum available investment. Although a large project can produce reasonable and lower LCOE, the resources for financing a project are usually limited. The layout of a wind farm may be subjected to some constraints to reduce the visual impact or improve aesthetic design (e.g. Fig. 9.3B). Finally, forbidden areas in a farm can limit the search space. These areas may be excluded due to foundation issues, dedicated to electrical cable routes, etc.

Solution Techniques

Due to the complexity of the objective function and constraints of the wind farm optimization problem, classical optimization techniques that are often more suited to convex and continuous problems are less popular. Therefore, metaheuristic methods are more effective and common in this area. Metaheuristic approaches provide satisfactory solutions, but they do not necessarily find the theoretical optimum. Genetic algorithm, particle swarm optimization, the greedy algorithm, evolutionary algorithm, and ant colony are metaheuristic approaches applied in this area [5,9].

9.2.2 TEC Array Optimization

Interdevice Spacing

The 2009 EMEC guide the assessment of tidal energy resources [10] recommends that the lateral spacing between devices (the distance between axes) should be two-and-a-half times the rotor diameter (2.5D), and the downstream spacing should be 10D—both based on the assumption of horizontal axis turbines. Further, the EMEC guide states that devices should be positioned in an alternating downstream (i.e. staggered) arrangement (Fig. 9.5). Although the exact details of device spacings is device-specific and can be debated (and indeed no justification is given for the values 2.5D and 10D, although we can assume that such guidance has propagated through from the wind industry),

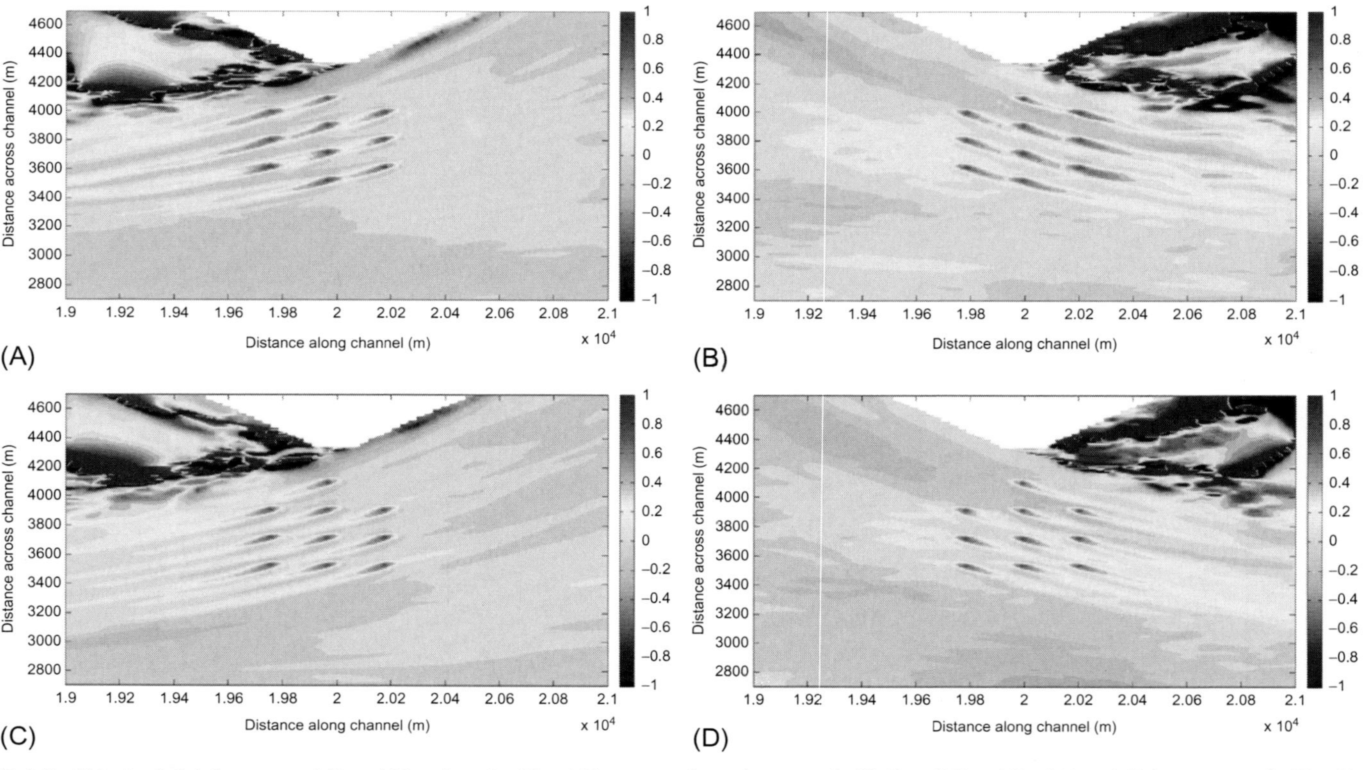

FIG. 9.5 Velocity deficit for staggered (A and B) and regular (C and D) array configurations at peak ebb (B and D) and flood (A and C) for a symmetrical headland with a constant bathymetry, representing an idealized Admiralty Inlet in Puget Sound. 'Staggered' in this sense refers to east/west velocities. *(Reproduced from T. Roc, D. Greaves, K.M. Thyng, D.C. Conley, Tidal turbine representation in an ocean circulation model: towards realistic applications, Ocean Eng. 78 (2014) 95–111, with permission from Elsevier.)*

the reasoning behind the spacings is introduced here. Each tidal stream device will generate a wake—a relatively narrow turbulent region downstream of the device where the velocity is below ambient. Clearly, placement of a subsequent device within this wake zone would lead to suboptimal device performance (e.g. increased turbulence), and so not exploiting the resource to its full potential (because the velocity in this region is below ambient). Therefore, the guidance of spacing devices 10*D* in the longitudinal direction is designed to minimize the wake effect. In addition, by staggering the devices, this will further prevent a device positioned downstream of another device from operating in its wake (Fig. 9.5). Myers and Bahaj [11] investigated the impact of lateral device spacing. They found that for very close lateral turbine spacings (0.5*D* measured between the innermost edges of the actuator disks, which is equivalent to 1.5*D* in the EMEC guidelines; based on flume experiments where the turbine rotors were represented by porous disks), the individual wakes generated by each of two devices merged by around 4*D* downstream. At increased lateral spacings (1.5*D*), a region of around 1*D* width accelerated flow between the two disks resulted, indicated by a negative velocity deficit[1] in Fig. 9.6. Therefore, for a final experiment, Myers and Bahaj [11] placed a third 'turbine' 3*D* downstream of the first row of disks in an attempt to exploit this region of accelerated flow. Although they did not perceive any significant negative changes to the efficiency of this third device, they found that the far wake region of the array had a relatively high velocity deficit due to the combined wakes, and so a third row of devices would need to be installed at a considerably increased longitudinal spacing to enable interception of flow speeds comparable to the

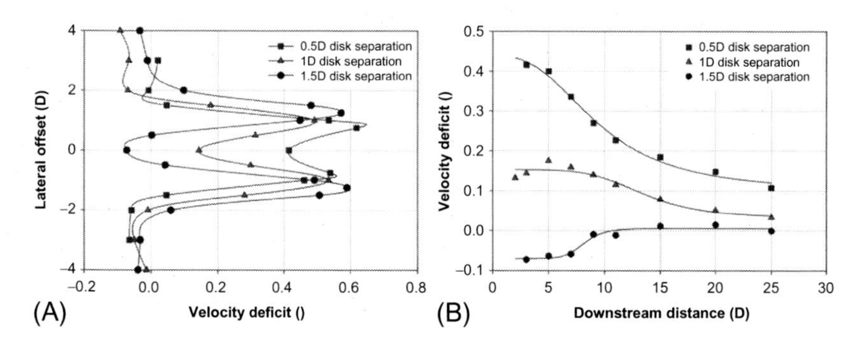

FIG. 9.6 Velocity deficit plots for dual actuator disk arrangements; (A) lateral centre-depth at 3*D* downstream and (B) longitudinal centreline. Note the region of accelerated flow (negative velocity deficit) for the 1.5*D* case. *(Reproduced from Myers and Bahaj L.E. Myers, A.S. Bahaj, An experimental investigation simulating flow effects in first generation marine current energy converter arrays, Renew. Energy 37 (1) (2012) 28–36, with permission from Elsevier.)*

1. Velocity deficit $U_{deficit} = 1 - U_W/U_0$, where U_W is the velocity at a point within the wake, and U_0 is the freestream velocity.

first row. Of course, flume experiments such as this are unidirectional, and so bidirectional (tidal) flows will require further consideration, for example, the 'first' row during the flood phase of the tidal cycle will become the 'last' row during the ebb.

Apart from potentially making use of regions of accelerated flow to strategically site devices (as discussed above), it is clearly advantageous, from both economic and practical perspectives, to minimize lateral and longitudinal extents of tidal energy arrays. Many leased tidal energy sites are situated in relatively narrow straits, where the principal flow dimension is much greater than the lateral channel dimension. Therefore, due to physical and navigational constraints, a more compact lateral array configuration (and hence spacing) is desirable. Excessive longitudinal spacing will significantly increase subsea cable lengths (e.g. the range of $5D$–$15D$ longitudinal spacing to avoid wake effects equates to 100–300 m for a 20 m diameter turbine), but such potentially increased cabling costs would depend on the design of the subsea connections and cable route between the devices and the substation (e.g. [12]).

Blockage

The theoretical upper limit to power extraction by a wind turbine is constrained by the *Betz limit* ($C_P = 0.59$), where C_P is the rotor power coefficient (see Section 3.13.1). However, in wind energy conversion, the cross-sectional area (CSA) of the turbine is very small in comparison to the CSA of the wind resource. By contrast, when tidal turbines are placed in a tidal channel, the turbine blockage ratio[2] increases, resulting in a theoretical C_P that can considerably exceed the Betz limit for array scales/configurations that are characterized by high blockage ratios [13]. However, the situation is complicated by the fact that by increasing the blockage ratio of a channel, there will be a corresponding reduction in the free-stream flow due to the increased drag resulting from tidal energy conversion. In other words, placing too many turbines in a tidal channel will merely block the flow.

To maximize turbine efficiency (i.e. the power available per turbine), tidal arrays should occupy the largest possible fraction of a channel's CSA [14]. However, practical constraints, such as navigation and the passage of marine life, will require gaps between the turbines in an array. In such a scenario, energy is dissipated in the wakes, and so the turbines are less effective than if they were deployed in a fence that spans the entire channel[3] [15].

An interesting extension of the work on simple tidal channels (e.g. [15]) is application to split channels [16]. Split channels are fairly common amongst tidal energy sites—for example, the Bay of Fundy, Puget Sound, and the

2. The ratio between the total swept area of the array to channel CSA.
3. The assumption of a complete fence is equivalent to assuming that the channel CSA is occupied by a single turbine [15]—a slightly unusual concept, but one that can be useful for analysis.

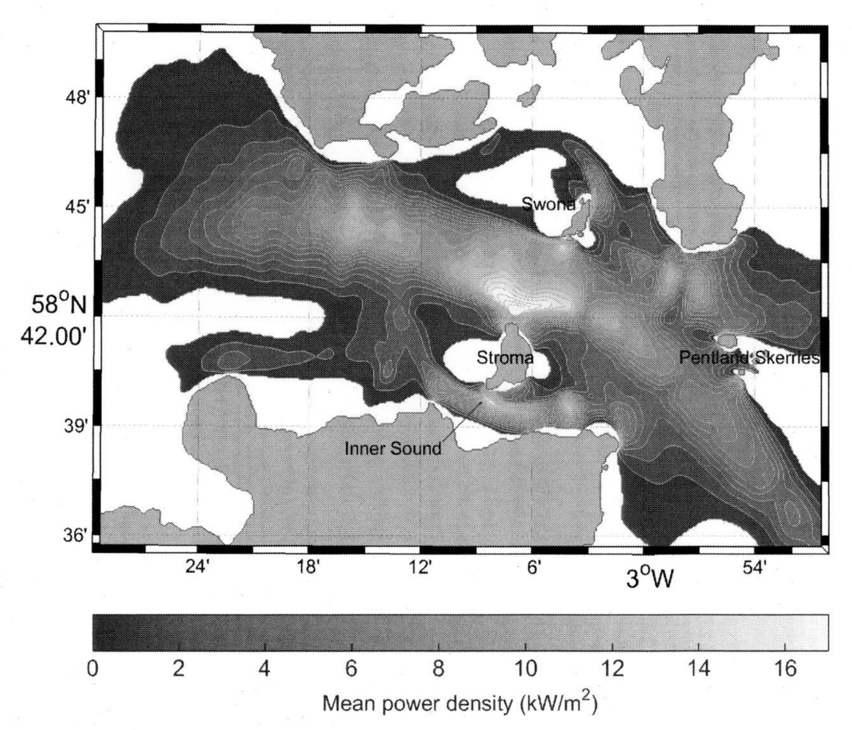

FIG. 9.7 Mean power density ($\geq 1\ \mathrm{kW/m^2}$) in the Pentland Firth over a spring tidal cycle. There are four leased tidal energy sites in this region, including the MeyGen project in the Inner Sound of Stroma. *(Reproduced from S.P. Neill, A. Vögler, A.J. Goward-Brown, S. Baston, M.J. Lewis, P.A. Gillibrand, S. Waldman, D.K. Woolf, The wave and tidal resource of Scotland, Renew. Energy 114 (2017) 3–17, under the Creative Commons Attribution License (CC BY).)*

Pentland Firth. How, for example, would large-scale tidal energy extraction in the Inner Sound influence the tidal energy resource within the Pentland Firth itself (Fig. 9.7)? In addition, for scenarios where there is a fairly equal channel division, devices could be strategically installed on one side of an island, leaving the opposite side clear for navigation [16]. Polagye and Malte [17] simulated a range of channel networks, including a split tidal channel (which they call a multiple-connected network). They found that transport in the impeded channel reduced, at maximum power, to around 50% of its magnitude compared with its undisturbed state. This is agreement with Cummins [18], who found, using an electric circuit analogue, that the transport in the impeded channel was reduced to 50–71%: a similar range to the much reported 58% calculated by Garrett and Cummins [19] for a single channel in the resistive limit. Polagye and Malte [17] found that maximum dissipation ($P_{\max}$) was 20% higher when extraction was evenly distributed between the two branches (Fig. 9.8), and that an even distribution minimizes far-field impacts. However, such an optimal arrangement

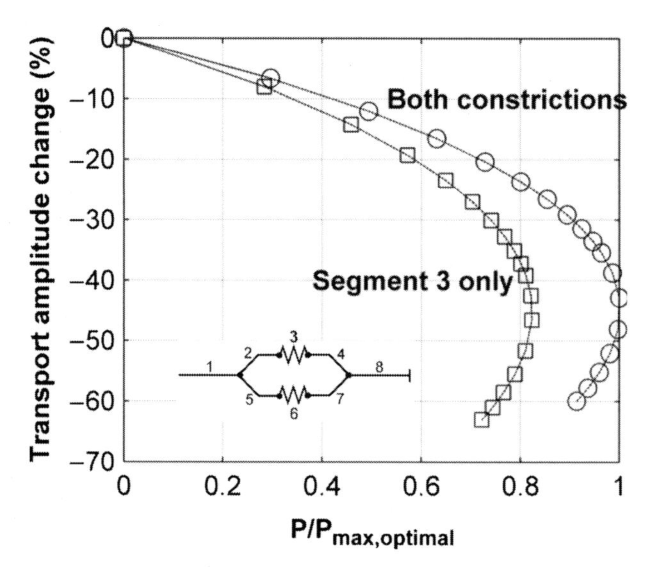

FIG. 9.8 Effect of energy extraction in a split channel. Change in transport amplitude is measured in segment 3 (upper constriction). □ denotes extraction only from segment 3 (upper branch). ○ denotes equal extraction from segments 3 and 6 (both branches). *(Reproduced from B.L. Polagye, P.C. Malte, Far-field dynamics of tidal energy extraction in channel networks, Renew. Energy 36 (1) (2011) 222–234, with permission from Elsevier.)*

may be neither desirable (e.g. it would preclude the dedicated navigational channel mentioned earlier) nor attainable (e.g. natural flow asymmetry between the two channel branches).

9.2.3 WEC Array Optimization

Park Effect

Since a renewable energy converter absorbs energy from its surrounding environment, the total available resource is reduced for neighbouring energy converters [20]. For a wave energy converter (WEC) array, the ‘park effect’ Q can be represented as the ratio of the power of the full array (P_{tot}) to the sum of the power for each isolated WEC (P_{isolated})

$$Q = \frac{P_{\text{tot}}}{\sum_{j=1}^{N} P_j^{(\text{isolated})}} \tag{9.7}$$

where N is the number of WECs in the array [21]. Although it is theoretically possible for some WEC array configurations to result in constructive wave interactions (i.e. $Q > 1$), in general $Q < 1$, and array designs tend to focus on limiting destructive interferences [20]. As the number of devices in a WEC array increases, the mean power per WEC reduces (Fig. 9.9B); but it is possible

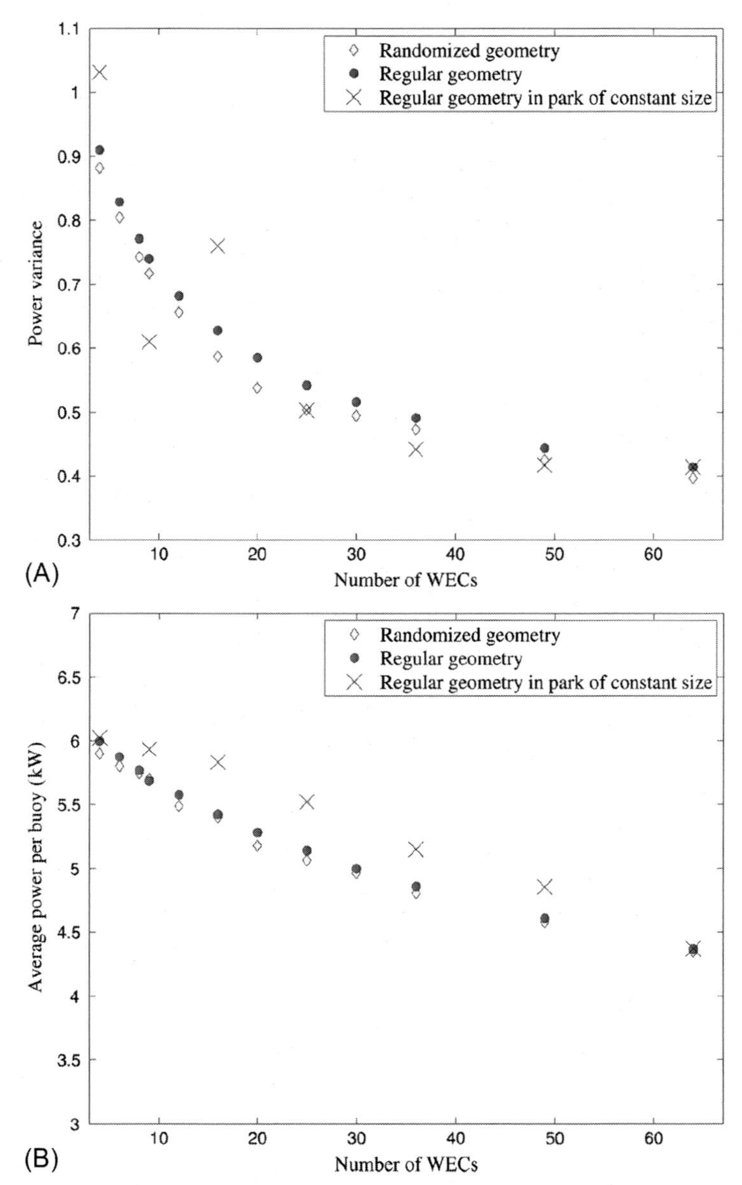

FIG. 9.9 (A) Variance and (B) mean power per WEC, as a function of the number of devices in a WEC array. The different symbols correspond to a range of WEC array configurations. *(Reproduced from M. Göteman, J. Engström, M. Eriksson, J. Isberg, M. Leijon, Methods of reducing power fluctuations in wave energy parks, J. Renew. Sustain. Energy 6 (4) (2014) 043103, with permission of AIP Publishing LLC.)*

to minimize such destructive interferences through optimization of various parameters such as array geometry and device spacing.

Smoothing Power Fluctuations

Wave energy comes in pulses, and so is unsuitable for direct conversion and transmission to the electricity grid [22]. The addition of devices to an array reduces the variance in power (Fig. 9.9A); however, at the expense of reduced average power per WEC due to the park effect (Fig. 9.9B). In sea trials with WECs, Rahm et al. [22] found that there was an 80% reduction in the standard deviation of electrical power output from three WECs, compared with the standard deviation from a single WEC. Based on model simulations of a larger number of WECs [21], normalized variance of power $\frac{Var}{\overline{P}^2}$ (where variance $Var = \sigma^2$ and σ is the standard deviation) reduced from ~0.9 to ~0.4 for 4 to 64 WECs, respectively (Fig. 9.9A).

Geometry of WEC Layout

It is possible to further smooth power fluctuations through careful consideration of the geometric WEC layout [22,23]. It has been shown that 'global' array layouts can increase power by 5% or reduce power by 30% [24]. Göteman et al. [25] simulated four global geometries, each with the same number of WECs (Fig. 9.10). Device spacings for the different configurations varied within the range 20–55 m, other than for the 'random' layout, where the minimal device spacing was set to 6 m. Because the incident waves propagate in the x-direction, clearly those devices at the rear of each array (increasing x coordinate) absorbed less power. However, it was not always the first row that absorbed the most power; for the rectangular configuration, it was the third row that absorbed most power, demonstrating a positive interference effect from the other WECs in the array. All four configurations resulted in similar energy absorption (Fig. 9.11A); however, power variance differed significantly between the four cases: 'rectangular' and 'random' layouts led to the highest and lowest variance, respectively (Fig. 9.11B). Engström et al. [26] investigated WEC array layouts that were only slightly randomized, that is, to represent uncertainty in mooring positions and temporal drift, again finding a reduction of variance in these more realistic 'randomized' layouts. This could be explained by considering regular WEC layouts that are aligned with the dominant wave direction (e.g. Fig. 9.10B). In this case, the electricity produced by all of the WECs distributed along the wave crest will generally be in phase for each WEC row. Staggering the devices slightly in the direction of wave propagation would introduce more phase diversity into the array, leading to more stable (less variable) aggregated power output. Minimizing the lateral dimension of the WEC array, and maximizing the longitudinal dimension, will have a similar effect, provided the longitudinal spacing is considered, in conjunction with the expected wavelengths. For example, a WEC array of 4×5 has lower variance if there are five rows along the direction of wave propagation, as compared to four [21].

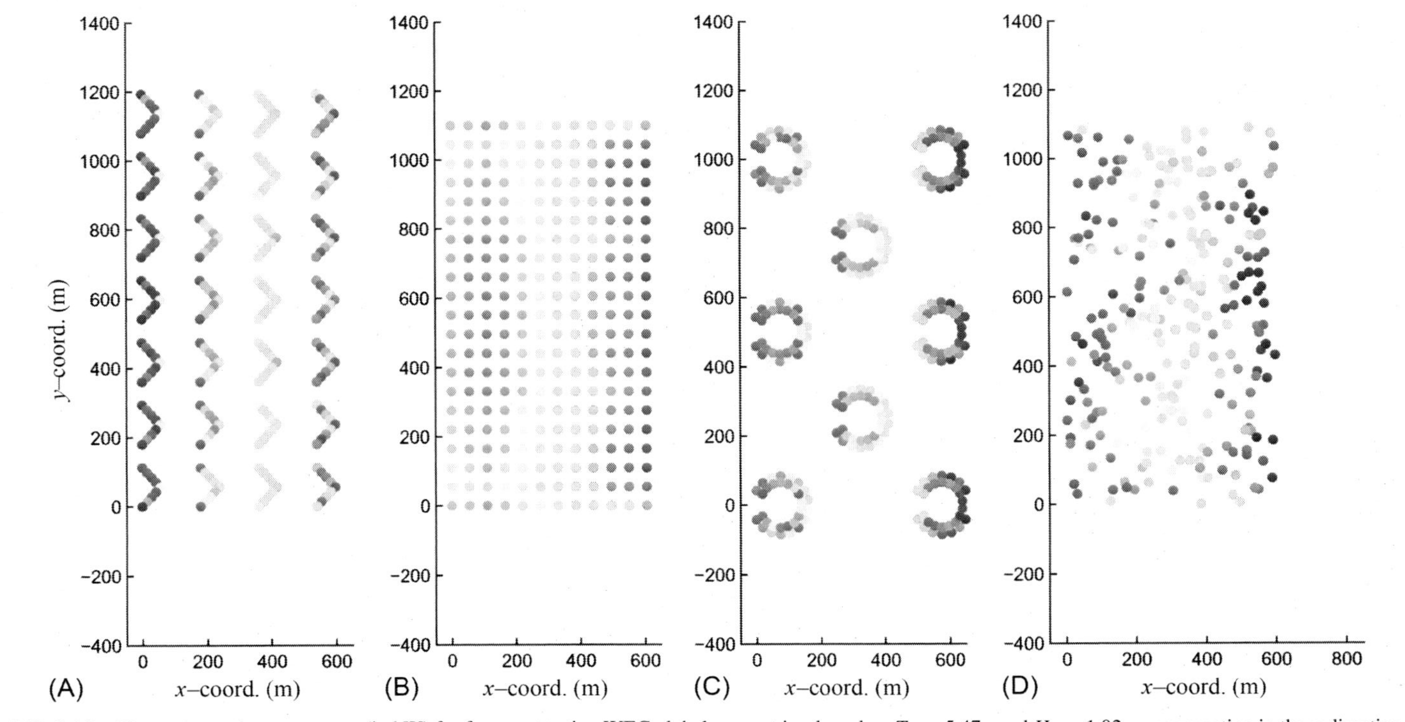

FIG. 9.10 Time-averaged power output (in kW) for four contrasting WEC global geometries, based on $T_e = 5.47$ s and $H_s = 1.02$ m, propagating in the x-direction. (A) wedge layout, (B) Rectangular layout, (C) Circle layout, and (D) Random layout. *(Reproduced from M. Göteman, J. Engström, M. Eriksson, J. Isberg, Optimizing wave energy parks with over 1000 interacting point-absorbers using an approximate analytical method, Int. J. Mar. Energy 10 (2015) 113–126, with permission from Elsevier.)*

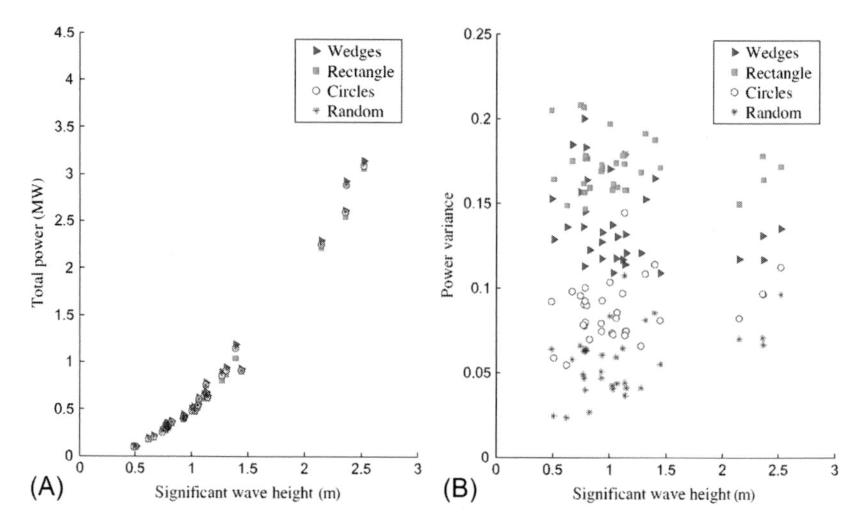

FIG. 9.11 (A) Total power output and (B) power variance for the four global WEC geometries shown in Fig. 9.10. *(Reproduced from M. Göteman, J. Engström, M. Eriksson, J. Isberg, Optimizing wave energy parks with over 1000 interacting point-absorbers using an approximate analytical method, Int. J. Mar. Energy 10 (2015) 113–126, with permission from Elsevier.)*

Device Spacing

The spacing between individual devices within a WEC array has been shown to influence variance in power output. As summarized by Göteman et al. [21], a 3×3 square wave array covering an area of $400\,m^2$ (10 m device spacing) has considerably lower variance than an array that covers $1600\,m^2$ (20 m device spacing). A relatively closely spaced WEC array, provided the 'park effect' is minimized, will have the added advantage of lower cabling costs, that is similar to cabling issues associated with the longitudinal and lateral spacing of tidal energy devices (Section 9.2.2). In a further study of a 3×3 array of point absorbers, Göteman et al. [25] found a localized peak in power variance associated with 18 m device spacing. Although this exact value is unique to this particular device, it demonstrates that careful design is required to avoid nonoptimal device spacings.

Size of WECs in an Array

One final issue in WEC array optimization is the size of individual devices within an array. Consider Fig. 9.10, for example, where larger WEC devices would be more suited to the 'front' of the array, and smaller devices to the 'rear'. Blending WEC sizes within an array can lead to higher power output for the array [21] and, it could be assumed, lead to more efficient use of each WEC within the array. Indeed, a concert of wave energy devices within a single array, each exploiting different parts of the wave energy spectrum, could be achieved, but only as a result of detailed knowledge of the wave climate, and extensive optimization.

9.3 INTERARRAY OPTIMIZATION

Examining the M2 (principal lunar semidiurnal) cotidal chart of the northwest European shelf seas (Fig. 9.12), Kelvin waves take a relatively long time to propagate along a coastline. For example, the southern part of the Irish Sea (around the Bristol Channel in Fig. 9.12) has a phase difference of around 3–4 h compared with the northern part of the Irish Sea (e.g. Holyhead). Because tides are predictable, it seems feasible that, with knowledge of the phase relationship between tidally energetic sites, we could strategically station a series of tidal power stations along a coastline, concurrently exploiting different parts of the same tidal wave, feeding the generated electricity into a unified electricity grid, and so reducing the variability that would result from a tidal power station operating in isolation.

9.3.1 Phasing of Tidal Stream Arrays

A 2005 study commissioned by The Carbon Trust [27] examined the patterns of energy availability from a range of tidal stream regions around the United Kingdom, with a focus on opportunities for diversification to reduce the variability of supply. The study is fairly comprehensive, but the results on optimizing variability can be summarized by reference to Fig. 9.13. In this figure, a range of development scenarios have been considered, from a relatively low level of development (10% of the UK tidal stream resource) up to an extreme scenario of 100% development. Within each of these development scenarios, optimization modelling was used to determine the contribution of each tidal stream site that resulted in the lowest average hourly variability (expressed as a percentage of maximum output). Development of 10% of the UK tidal stream resource across a range of sites would result in low variability. However, at larger scales of development, the synchronized output of larger sites (e.g. the Pentland Firth) dominates, increasing variability. For example, at 10% development, the Alderney Race (Channel Isles) accounts for 25% of all output compared with 10% for the Pentland Firth. However, at 100% development, the percentage contribution of the Alderney Race reduces to just 6%, whereas the Pentland Firth would account for 30% of the total output.

The issue of tidal phasing of tidal stream locations was revisited by Iyer et al. [28], who extended The Carbon Trust study [27] to include energetic sites in Wales: Anglesey (Skerries) and Ramsey Sound. Again, they concluded that there is insufficient diversity between the sites identified across the United Kingdom to be considered as a firm power source. In general, they found that the high-energy sites were in phase with one another, with the exception of Alderney Race in the Channel Isles. The problem was further investigated by Neill et al. [29,30] who, rather than constraining their analysis to a limited number of locations, considered the UK tidal energy system as a whole. By doing so, they did not limit their analysis solely to high energy sites, but considered a diverse range of current magnitudes in the analysis. In addition to seeking sites with high energy, they penalized site selection where the combination of

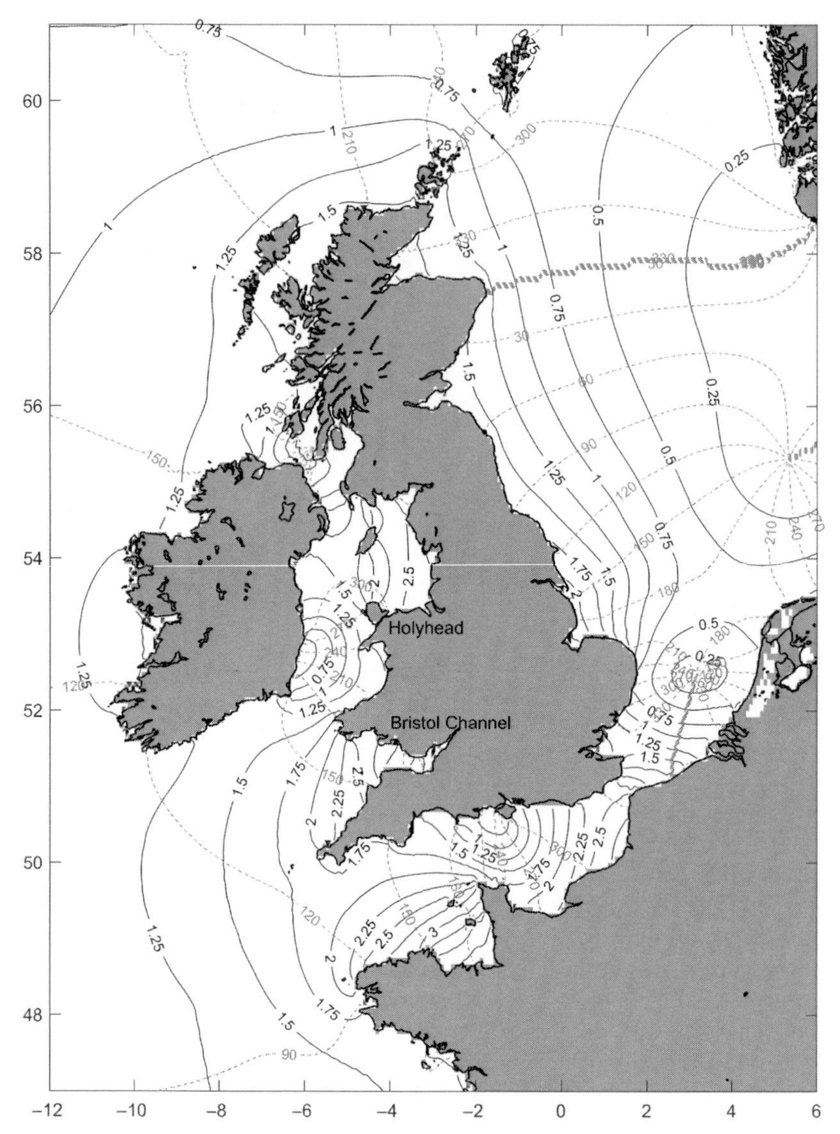

FIG. 9.12 M2 cotidal chart of the northwest European shelf seas. *Dashed contours* are cotidal lines (degrees), and *solid contours* are coamplitude lines (in metre). *(Adapted from S.P. Neill, J.D. Scourse, K. Uehara, Evolution of bed shear stress distribution over the northwest European shelf seas during the last 12,000 years, Ocean Dyn. 60 (5) (2010) 1139–1156, and reproduced with the permission of Springer.)*

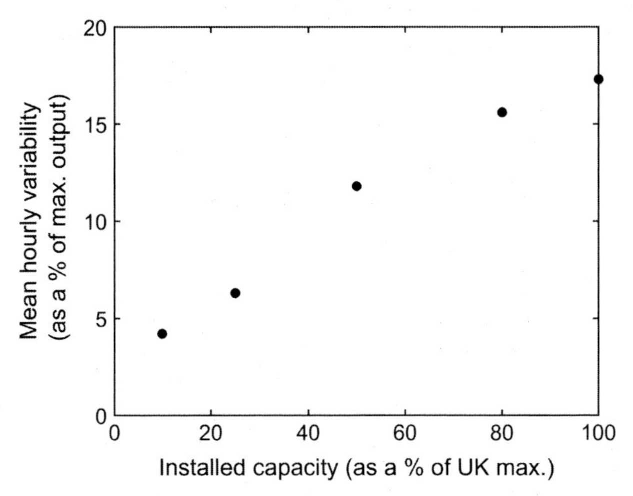

FIG. 9.13 Influence of level of development of the UK tidal stream resource on variability. *(Data from G. Sinden, Variability of UK Marine Resources, The Carbon Trust, London, 2005.)*

multiple sites led to a reduction in phase diversity. A summary of the findings of Neill et al. [30] is presented in Fig. 9.14. Because the high-energy tidal stream sites are generally in phase with one another, there is high variability in the resulting (aggregated) power—in agreement with previous studies [27,28]. In contrast, there is considerably more phase diversity amongst less energetic sites. Therefore, an energy policy that seeks to minimize variability in the developed tidal stream resource would be advised to include lower-energy sites in the energy mix. Optimization, in this case the greedy algorithm with penalty function, led to a proposed road map for the evolution of a UK tidal stream industry (Fig. 9.15), with the key assumption that the Pentland Firth is the first region to host an array—an assumption that has now been proven to be valid with development of the MeyGen array. It is noted, in common with the Carbon Trust report [27], that at very high levels of tidal stream development, limitations on sea space would lead to an aggregated electricity signal that is dominated by relatively few high tidal stream locations, with a particular focus on the Pentland Firth. Incidentally, none of these studies included feedback between energy extraction and the resource—a topic that is covered in Chapter 10. In reality, at such large scales of exploitation, these feedback effects would be significant, and so there is a gap in knowledge of how best to optimize at interarray scale.

9.3.2 Phasing of Multiple Tidal Range Power Plants

Few studies have considered multiple lagoon operation, never mind multiple lagoon optimization. However, Yates et al. [31] considered (including feedbacks) the operation of five tidal *barrages* in the Irish Sea, all operating in ebb-only mode. The results show that all of the proposed tidal barrages in the northern part of the Irish Sea (Dee, Morecambe, Solway, Mersey) are in phase with one

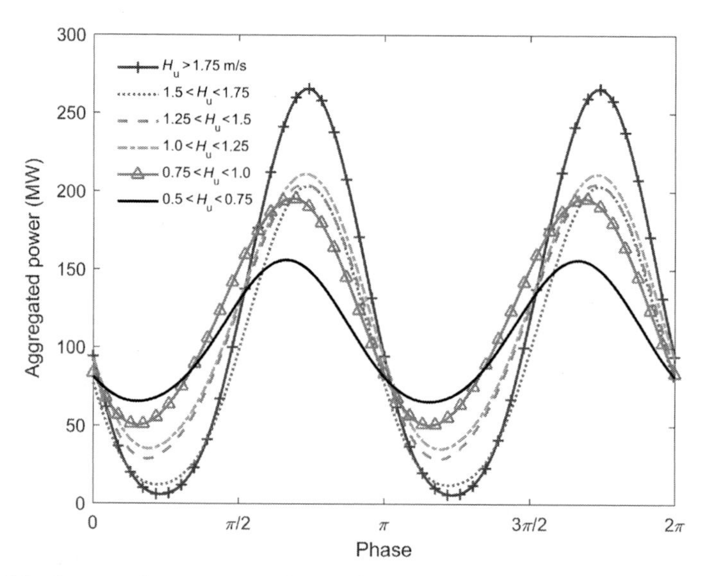

FIG. 9.14 Aggregated power generated over a tidal cycle for various M2 tidal current amplitude ranges throughout the NW European shelf seas. *(Reproduced from S.P. Neill, M.R. Hashemi, M.J. Lewis, Tidal energy leasing and tidal phasing, Renew. Energy 85 (2016) 580–587, under the Creative Commons Attribution License (CC BY).)*

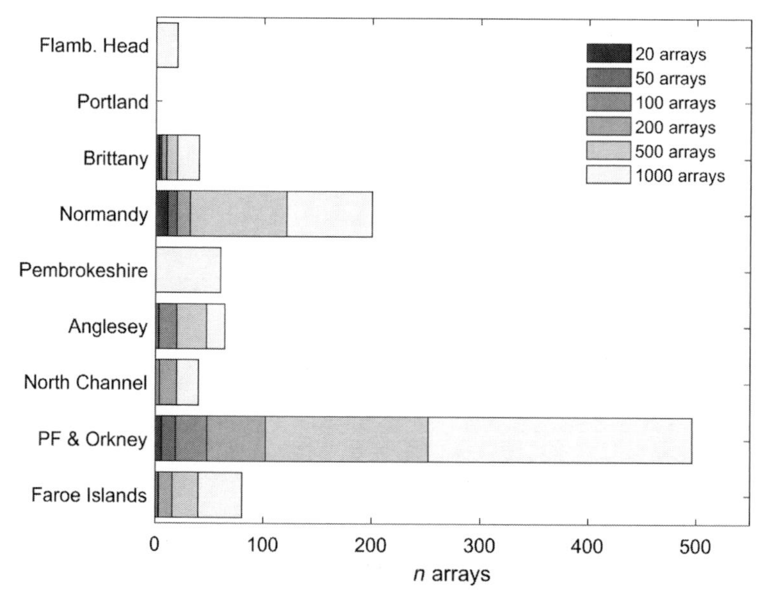

FIG. 9.15 Optimized temporal evolution of tidal stream development over the NW European shelf seas. An 'array' is a unit of 10×2 MW devices. *(Reproduced from S.P. Neill, M.R. Hashemi, M.J. Lewis, Optimal phasing of the European tidal stream resource using the greedy algorithm with penalty function, Energy 73 (2014) 997–1006, under the Creative Commons Attribution License (CC BY).)*

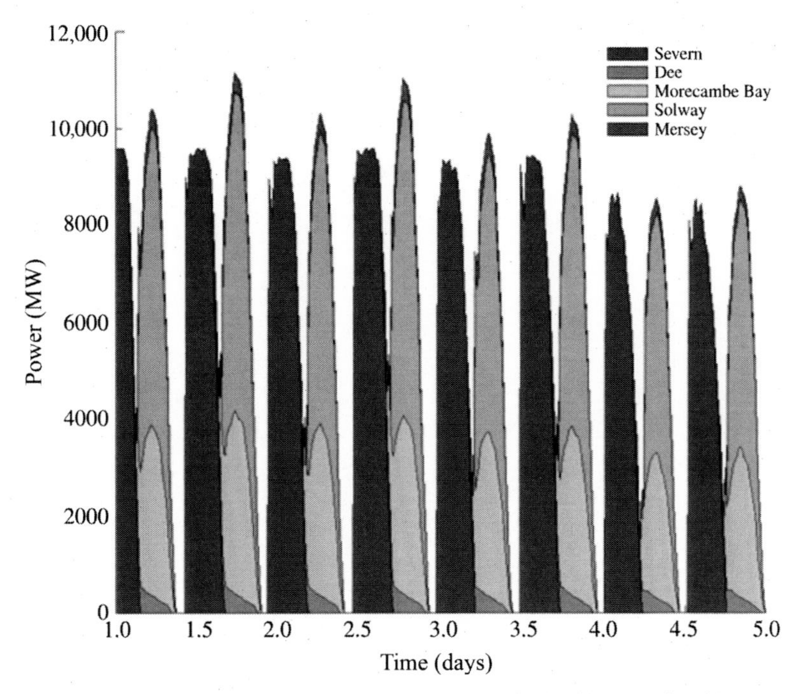

FIG. 9.16 Electricity generation from five barrages in the Irish Sea during spring tides. *(Reproduced from N. Yates, I. Walkington, R. Burrows, J. Wolf, Appraising the extractable tidal energy resource of the UK's western coastal waters, Philos. Trans. R. Soc. A 371 (1985) (2013) 20120181, with permission from the Royal Society.)*

another, whereas a Severn barrage in the southern Irish Sea is approximately 180 degrees out of phase (Fig. 9.16). This is perhaps not surprising if we examine the M2 cotidal chart of the region (Fig. 9.12). Because the northern part of the Irish Sea is a standing wave system (Section 3.4), there are few cotidal lines in this region (cotidal lines are all 300–360 degrees in Fig. 9.12), indicating that tidal elevations throughout this region are more or less in phase. The cotidal lines in the Severn Estuary have a phase of around 180 degrees in Fig. 9.12, explaining the 180 degrees change in power output between power plants in the southern and northern Irish Sea (Fig. 9.16). However, further optimization (flood-only, ebb-only, and dual model) could be performed on this set of tidal barrages to further smooth power output. Similar optimization could also be performed on multiple tidal lagoon power plants, as considered, in part, by Angeloudis and Falconer [32], who optimized annual electricity production for the Bristol Channel and Severn Estuary. Although they did not optimize to minimize variability of power output, they advise that multiple projects should also consider the environmental consequences of changing the mode of operation; for example, flood-only and ebb-only operation may lead to greater loss of intertidal area and far-field effects, compared with dual-model operation.

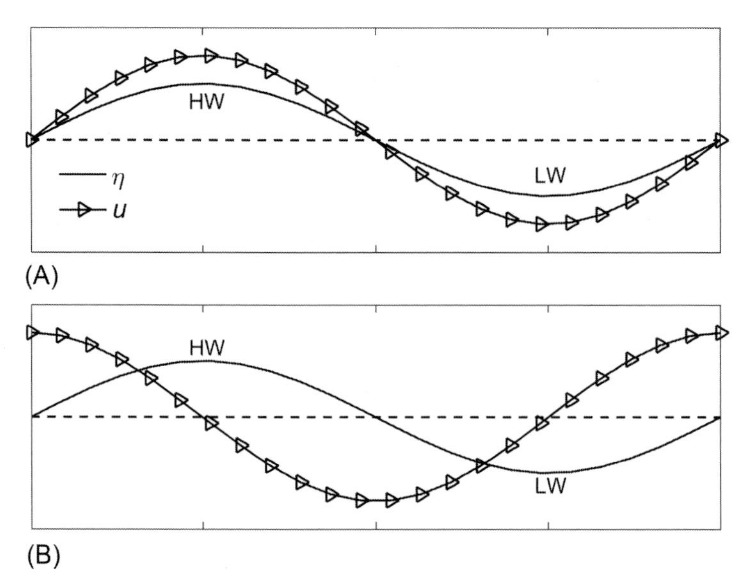

FIG. 9.17 (A) Progressive and (B) standing wave systems over a single tidal cycle, where η is elevation, and u is current. In (A), peak currents coincide with HW and LW. In (B), HW and LW coincide with slack water.

9.3.3 Combined Tidal Range and Tidal Stream Phasing

For locations that exhibit an energetic tidal stream *and* tidal range resource, it is interesting to consider the nature of the tides, as either progressive or standing wave systems. In a progressive wave, tidal elevations and currents are in phase with one another (Fig. 9.17A). High water (HW) is typically a time of *standing* for a tidal range power plant (Section 3.14), when no electricity would be generated; whereas this is the time associated with peak tidal streams in a progressive wave. Therefore, there is scope for phase diversity between tidal elevations and currents within the same region for a progressive wave system. In a standing wave system (Fig. 9.17B), peak tidal currents occur mid-way between HW and LW, at a stage when a tidal lagoon power plant would be at peak generation (Section 3.14); therefore there is less potential for phase diversity between currents and elevations in a standing wave system. Unfortunately, due to resonance (Section 3.5), most promising tidal range sites will be standing wave systems, hence minimizing the potential for developing combined tidal range and tidal stream power plants at the same location.

9.4 OPTIMIZATION OF HYBRID MARINE RENEWABLE ENERGY PROJECTS

There are many advantages in colocating or combining marine renewable energy power plants at a single location, such as shared grid infrastructure, common substructure/foundation systems, and smoothing the power output when combining multiple renewable resources [33]. Although one colocated combination was briefly discussed in Section 9.3.3 (tidal range power plants and tidal stream arrays), in this section we focus on the colocation of wind and wave arrays.

9.4.1 Combined Wind-Wave Projects

One obvious example where multiple renewable energy resources could be combined at a single location is wave and offshore wind. The offshore wind and wave industries face the same hostile marine environment, and face similar administrative and technological barriers [33]. From a resource perspective, although wind fields lead to the generation of wind waves, often there will be a phase difference or a lag between wind power and wave power, for example, consider the spread of data in Fig. 9.18 calculated for WaveHub, UK. In particular, swell waves are independent of the local wind climate, and so in many ways the wind and wave resource are complementary (although optimization would be required to select the best sites for colocation).

Stoutenburg et al. [35] examined the temporal correlation (Pearson's) between wind power and wave power at the same and different wave buoys off the coast of California (Fig. 9.19). Lower values (i.e. weakly correlated sites) indicate increased diversity, and so good potential for colocation of wave and

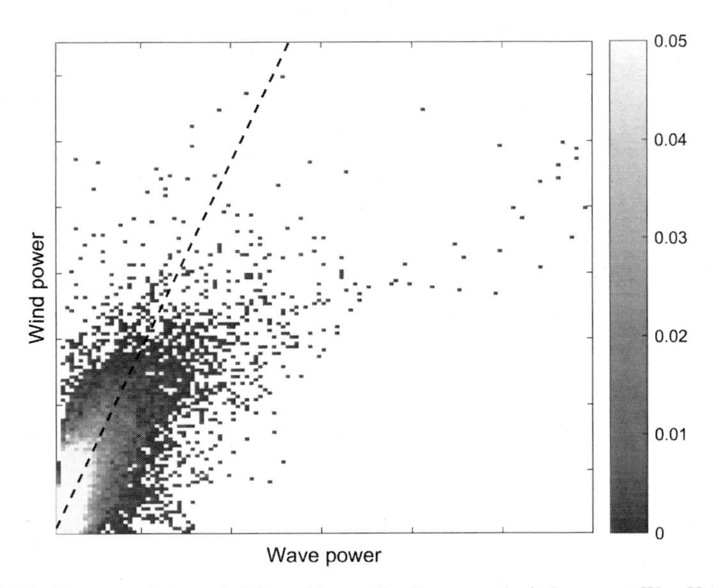

FIG. 9.18 Percentage joint probability of (normalized) wave and wind power at WaveHub, UK. Wind speed is from the ECMWF ERA-Interim reanalysis dataset, and wave data is from a modified version of the wave model presented by Neill and Hashemi [34] forced with ERA-Interim data. Time series analysed is 10 years.

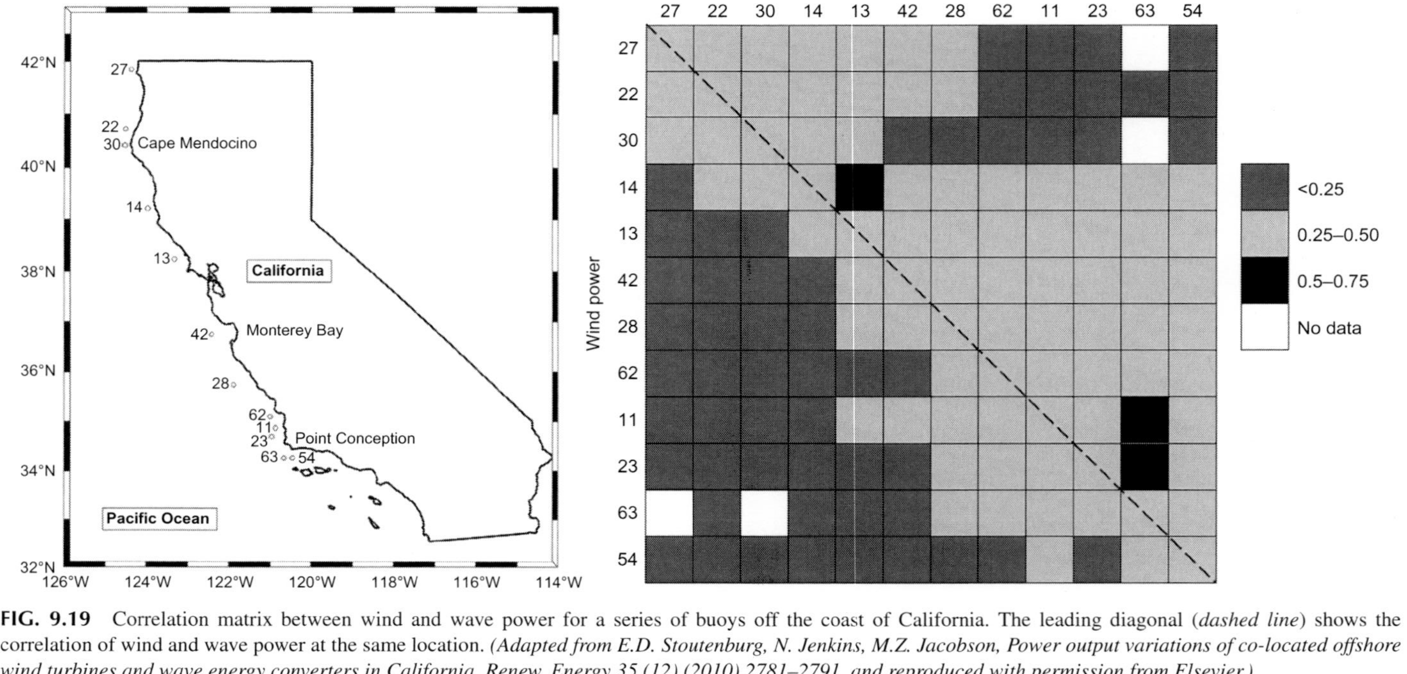

FIG. 9.19 Correlation matrix between wind and wave power for a series of buoys off the coast of California. The leading diagonal (*dashed line*) shows the correlation of wind and wave power at the same location. *(Adapted from E.D. Stoutenburg, N. Jenkins, M.Z. Jacobson, Power output variations of co-located offshore wind turbines and wave energy converters in California, Renew. Energy 35 (12) (2010) 2781–2791, and reproduced with permission from Elsevier.)*

wind arrays. Aggregate power from a colocated wind and wave farm achieves reductions in variability equivalent to aggregating power from two offshore wind farms approximately 500 km apart, or two wave farms approximately 800 km apart [35].

In general, a trend of energetic wind and wave activity in winter months coincides with an increased demand for electricity for heating and lighting (e.g. see Fig. 10.1 in the next chapter). However, with significant interannual variability in the wind and wave resources, it is a high-risk strategy to put too much reliance on these stochastic forms of energy conversion. In Fig. 1.12C, we looked at how demand for electricity varied throughout the day, with well-defined peaks at around 08:00 and 18:00. It would therefore be useful if a wind/wave energy mix could be optimized to match these peaks in demand. Cradden et al. [36] considered wind/wave energy mixes in Orkney (at the EMEC wave test site) in the ratios 100% wind:0% wave, 75% wind:25% wave, etc. down to 0% wind:100% wave. Considering only time periods when electricity demand exceeded 90% of peak, the frequency distribution of capacity factor for these different wind/wave energy mixes was calculated (Fig. 9.20). At 100%

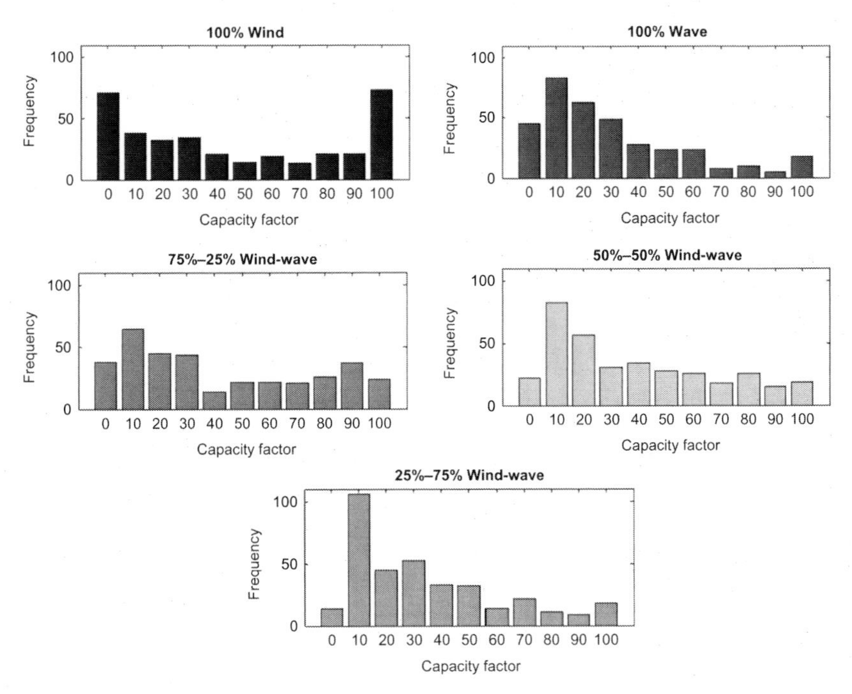

FIG. 9.20 Frequency distribution of capacity factor for a range of wind/wave energy mixes, calculated when demand is >90% of peak for a 2-year time series from the EMEC wave test site. *(Reproduced from L. Cradden, H. Mouslim, O. Duperray, D. Ingram, Joint exploitation of wave and offshore wind power, in: Proceedings of the 9th European Wave and Tidal Energy Conference (EWTEC), Southampton, UK, 2011, pp. 1–10.)*

wind power, there are relatively high frequencies of zero and maximum power, but these two extremes are much less frequent in the 100% wave scenario, indicating that the wave resource is less extreme, but more likely to be present (i.e. more reliable) than the wind resource. A more even distribution across all capacity factors results from all of the combined wind/wave scenarios. However, very specific optimization would be required to determine which of these scenarios is advantageous from both generation and demand perspectives. Further, colocation of wind and wave farms can lead to more complexity in the structural design of the devices due to loading and fluid-structure interactions.

9.5 OPTIMIZATION TOOLS

9.5.1 HOMER

Whilst a number of tools are available to optimize distributed and hybrid renewable energy systems, we selected Hybrid Optimization Model for Electric Renewable (HOMER; www.homerenergy.com) as an example to describe some of the capabilities of these tools. HOMER is an example of a computer-based optimization tool developed by National Renewable Energy Laboratory of the US for hybrid renewable energy systems. The main components of this model are simulation, optimization, and sensitivity analysis.

Simulation

Assume that you have designed a configuration for a hybrid renewable energy system. This system has several components such as one or several wind turbines, a storage system (e.g. batteries, hydrogen tank), and diesel generators (for hybrid systems). HOMER simulates the performance of the system every hour for the duration of a year. It calculates the energy produced and compares it with electricity demand. The hourly time series of demand should be provided as an input to the model. HOMER calculates the surplus or deficit at each time step, and tries to address these differences by storing energy during surplus periods, and using stored energy, diesel generators, or grid imports, in times of deficit. Several criteria can be imposed by the user as constraints to determine whether the performance of a system/configuration is acceptable; for example, minimum percent of the time that energy demand should be met, the share of the power supplied by renewable energy (in a hybrid system), and a limit on the emissions can be imposed. In the simulation phase, HOMER also calculates total net present cost of the system (i.e. all future costs are discounted to the present by a discount/interest rate). The costs include initial capital cost, the O&M cost during the lifetime of the project, fuel, and power purchased from the grid. The revenue from selling the power is then subtracted from the total cost to compute the net present cost. LCOE can also be computed given total cost and energy produced during a year.

Optimization

Finding the best configuration of a renewable energy system is the objective of the optimization step. HOMER simulates the performance of several configurations. It selects those that meet the constraints (demand, and minimum percentage of renewable energy contribution in a hybrid system). The best configuration is a solution/configuration that has minimum net present cost. Sample independent (decision) variables that can be changed by HOMER include the number and type/size of wind turbines, the size of the storage system (number of batteries), and the size of diesel generators. A search space that contains all possible scenarios for a component (e.g. various battery sizes) is provided by the user.

Sensitivity Analysis

An optimized system is calculated based on certain assumptions about a number of uncertain parameters; examples are fuel price, grid power price, discount rate, life time of the project, and demand. By performing sensitivity analysis on these variables, a user can deal with uncertainty and understand how/if the optimized system changes with these parameters. For instance, what is the minimum lifetime for a project? What interest rates are feasible for a renewable energy system?

HOMER has been applied in several studies to find the optimum energy system. For instance, in a case study in a remote village in India, four energy sources (hydropower, solar, wind, and bio-diesel generators) were combined [37]. The optimal off-grid system was identified and compared with the alternative of grid extension. This study showed that the hybrid off-grid system is cost-effective, compared with grid extension.

Some studies have optimized energy systems at much larger scales. As an example, Budischak et al. [38] considered a large grid in the eastern United States (Fig. 9.21) with a capacity of 72 GW (PJM Interconnection; www.pjm.com). By combining several renewable energy resources distributed across the region (onshore wind, offshore wind, and solar), and storage systems (batteries and fuel cells), an alternative system based on renewable energy was simulated and optimized. This study proposed an energy system that can meet the electricity demand of this large region based on 90% to 99.99% renewable sources, and with a cost comparable to conventional energy systems by 2030. Notably, Budischak et al. [38] recommended excessive energy production as the least cost option; the renewable system should produce almost three times the electricity demand, on average, to compensate for intermittency, and to avoid high costs associated with energy storage. Offshore wind energy had nearly the same contribution as onshore wind in the energy mix for the 99.99% renewable case in 2030. Both HOMER and another tool RREEOM (Regional Renewable Electricity Economic Optimization Model) were used, whilst PREEOM was recommended for larger grids like PJM.

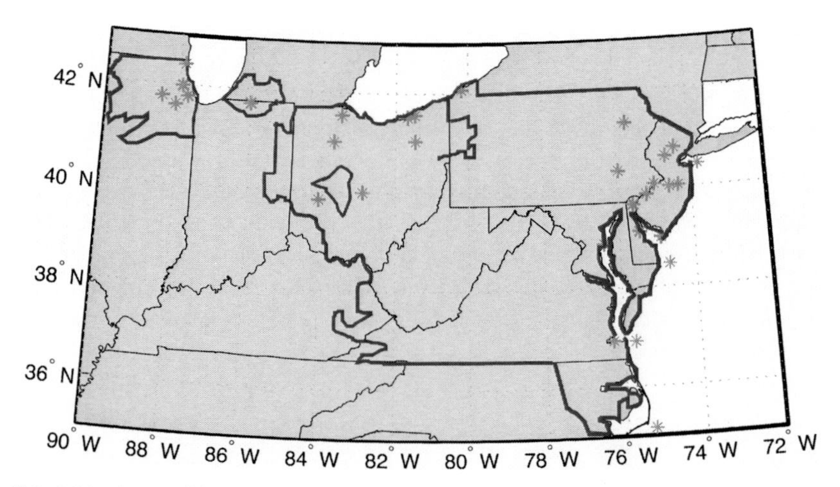

FIG. 9.21 A map of the current PJM system (*thick lines*). The meteorological stations (inland and offshore) for the wind energy assessments are shown as *asterisks*. *(Reproduced from C. Budischak, D. Sewell, H. Thomson, L. Mach, D.E. Veron, W. Kempton, Cost-minimized combinations of wind power, solar power and electrochemical storage, powering the grid up to 99.9% of the time, J. Power Sources 225 (2013) 60–74, with permission from Elsevier.)*

9.5.2 OpenTidalFarm

OpenTidalFarm is an open source software designed to simulate and optimize arrays of tidal energy converters (TECs; http://opentidalfarm.org). Optimization of TEC arrays generally requires the running of computationally expensive 2D or 3D models, and so only a limited number of options can be explored; for example, Divett et al. [39] explored 'only' four TEC array configurations using their 2D shallow water equation model to maximize global array power output. Funke et al. [40] presented an automated procedure for TEC array optimization using the adjoint technique, based on a 2D shallow water equation model. In contrast to less efficient gradient-free optimization algorithms (such as the greedy algorithm presented in Section 9.3.1, which optimizes the objective function, e.g. power output, as a black box), the adjoint method is a gradient-based method. Gradient-based optimization algorithms update the position in parameter space at each iteration using derivatives of the objective function. However, as mentioned before, this involves differentiating through the solution of a partial differential equation, which can be challenging in the case of complex models. In the tidal energy case, we wish to optimize a single variable—power output—with respect to many input parameters, and 'adjoint linearization' is a method that can efficiently compute the derivative of a single output with respect to all inputs.

OpenTidalFarm has, amongst other applications, been used to optimize the siting of 256 turbines in the Inner Sound of the Pentland Firth [40], and to minimize the cost of cabling in the same region [12] (Fig. 9.22).

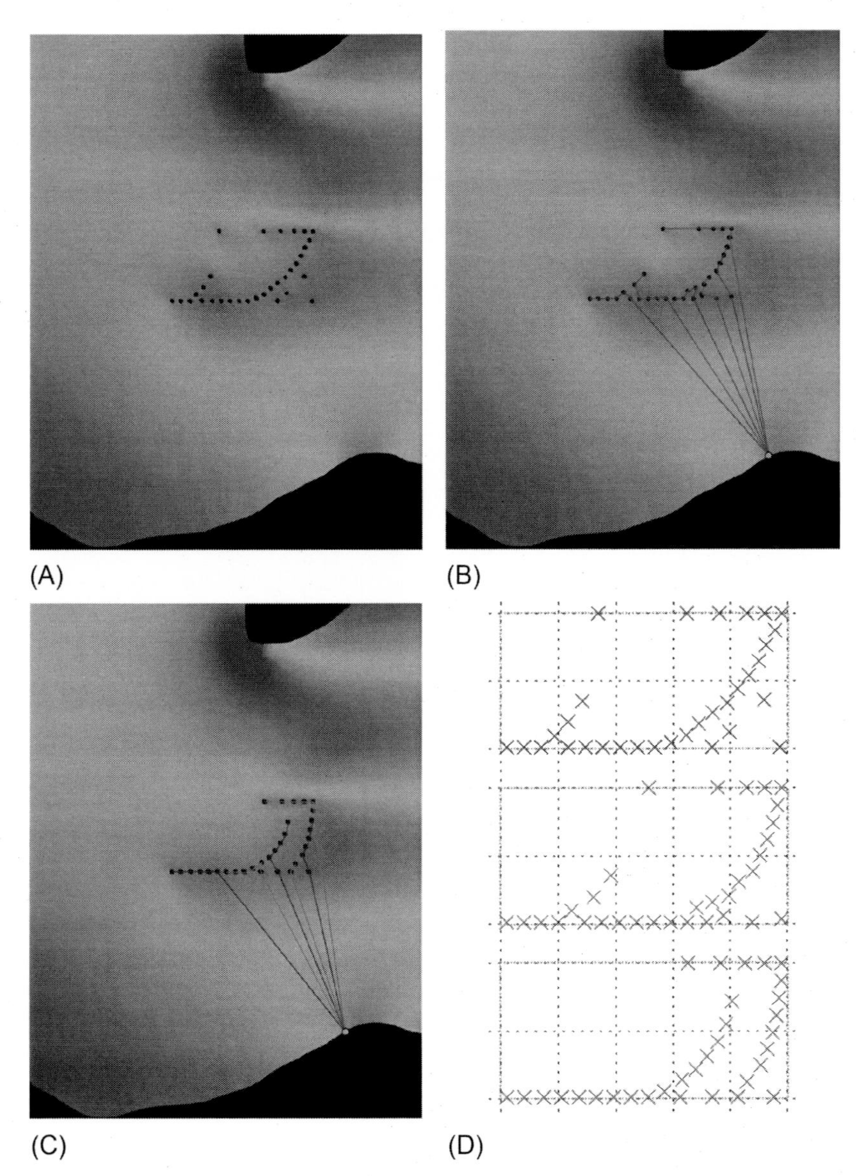

FIG. 9.22 Application of OpenTidalFarm to the Inner Sound of the Pentland Firth, with optimized micro-siting for (A) maximization of power only and (B) maximization of financial return (showing cable routing). (C) shows a sensitivity scenario where the cost of cabling doubles, and (D) shows a comparison of the three scenarios, with (A) in *blue* (*top*), (B) in *green* (*middle*), and (C) in *red* (*bottom*). *(Reproduced from D.M. Culley, S.W. Funke, S.C. Kramer, M.D. Piggott, Integration of cost modelling within the micro-siting design optimisation of tidal turbine arrays, Renew. Energy 85 (2016) 215–227, under the Creative Commons Attribution License (CC BY).)*

REFERENCES

[1] W. Su, J. Wang, J. Roh, Stochastic energy scheduling in microgrids with intermittent renewable energy resources, IEEE Trans. Smart Grid 5 (4) (2014) 1876–1883.

[2] S.S. Rao, S.S. Rao, Engineering Optimization: Theory and Practice, John Wiley & Sons, New York, NY, 2009.

[3] S. Boyd, L. Vandenberghe, Convex Optimization, Cambridge University Press, Cambridge, 2004.

[4] A. Neubert, A. Shah, W. Schlez, Maximum yield from symmetrical wind farm layouts, in: Proceedings of DEWEK, 2010.

[5] J.S. González, M.B. Payán, J.M.R. Santos, F. González-Longatt, A review and recent developments in the optimal wind-turbine micro-siting problem, Renew. Sustain. Energy Rev. 30 (2014) 133–144.

[6] S. Frandsen, On the wind speed reduction in the center of large clusters of wind turbines, J. Wind Eng. Ind. Aerodyn. 39 (1–3) (1992) 251–265.

[7] S. Chowdhury, J. Zhang, A. Messac, L. Castillo, Unrestricted wind farm layout optimization (UWFLO): investigating key factors influencing the maximum power generation, Renew. Energy 38 (1) (2012) 16–30.

[8] M.M. Yelmule, E.A. Vsj, CFD predictions of NREL phase VI rotor experiments in NASA/AMES wind tunnel, Int. J. Renew. Energy Res. 3 (2) (2013) 261–269.

[9] R. Banos, F. Manzano-Agugliaro, F.G. Montoya, C. Gil, A. Alcayde, J. Gómez, Optimization methods applied to renewable and sustainable energy: a review, Renew. Sustain. Energy Rev. 15 (4) (2011) 1753–1766.

[10] C. Legrand, Assessment of Tidal Energy Resource: Marine Renewable Energy Guides, European Marine Energy Centre, Orkney, UK, 2009.

[11] L.E. Myers, A.S. Bahaj, An experimental investigation simulating flow effects in first generation marine current energy converter arrays, Renew. Energy 37 (1) (2012) 28–36.

[12] D.M. Culley, S.W. Funke, S.C. Kramer, M.D. Piggott, Integration of cost modelling within the micro-siting design optimisation of tidal turbine arrays, Renew. Energy 85 (2016) 215–227.

[13] R. Vennell, Exceeding the Betz limit with tidal turbines, Renew. Energy 55 (2013) 277–285.

[14] R. Vennell, Tuning turbines in a tidal channel, J. Fluid Mech. 663 (2010) 253–267.

[15] C. Garrett, P. Cummins, The efficiency of a turbine in a tidal channel, J. Fluid Mech. 588 (2007) 243–251.

[16] J.F. Atwater, G.A. Lawrence, Power potential of a split tidal channel, Renew. Energy 35 (2) (2010) 329–332.

[17] B.L. Polagye, P.C. Malte, Far-field dynamics of tidal energy extraction in channel networks, Renew. Energy 36 (1) (2011) 222–234.

[18] P.F. Cummins, The extractable power from a split tidal channel: an equivalent circuit analysis, Renew. Energy 50 (2013) 395–401.

[19] C. Garrett, P. Cummins, The power potential of tidal currents in channels, Proc. Roy. Soc. 461 (2005) 2563–2572.

[20] A. Babarit, On the park effect in arrays of oscillating wave energy converters, Renew. Energy 58 (2013) 68–78.

[21] M. Göteman, J. Engström, M. Eriksson, J. Isberg, M. Leijon, Methods of reducing power fluctuations in wave energy parks, J. Renew. Sustain. Energy 6 (4) (2014) 043103.

[22] M. Rahm, O. Svensson, C. Boström, R. Waters, M. Leijon, Experimental results from the operation of aggregated wave energy converters, IET Renew. Power Gener. 6 (3) (2012) 149–160.
[23] J. Cruz, R. Sykes, P. Siddorn, R.E. Taylor, Wave farm design: preliminary studies on the influences of wave climate, array layout and farm control, in: Proceedings of the 8th European Wave and Tidal Energy Conference (EWTEC), Uppsala, Sweden, 2009, pp. 736–745.
[24] B.F.M. Child, J. Cruz, M. Livingstone, Development of a tool for optimising arrays of wave energy converters, in: Proceedings of the 9th European Wave and Tidal Energy Conference (EWTEC), Southampton, UK, 2011, pp. 5–9.
[25] M. Göteman, J. Engström, M. Eriksson, J. Isberg, Optimizing wave energy parks with over 1000 interacting point-absorbers using an approximate analytical method, Int. J. Mar. Energy 10 (2015) 113–126.
[26] J. Engström, M. Eriksson, M. Göteman, J. Isberg, M. Leijon, Performance of large arrays of point absorbing direct-driven wave energy converters, J. Appl. Phys. 114 (20) (2013) 204502.
[27] G. Sinden, Variability of UK Marine Resources, The Carbon Trust, London, 2005.
[28] A.S. Iyer, S.J. Couch, G.P. Harrison, A.R. Wallace, Variability and phasing of tidal current energy around the United Kingdom, Renew. Energy 51 (2013) 343–357.
[29] S.P. Neill, M.R. Hashemi, M.J. Lewis, Optimal phasing of the European tidal stream resource using the greedy algorithm with penalty function, Energy 73 (2014) 997–1006.
[30] S.P. Neill, M.R. Hashemi, M.J. Lewis, Tidal energy leasing and tidal phasing, Renew. Energy 85 (2016) 580–587.
[31] N. Yates, I. Walkington, R. Burrows, J. Wolf, Appraising the extractable tidal energy resource of the UK's western coastal waters, Philos. Trans. R. Soc. A 371 (1985) (2013) 20120181.
[32] A. Angeloudis, R.A. Falconer, Sensitivity of tidal lagoon and barrage hydrodynamic impacts and energy outputs to operational characteristics, Renew. Energy 114 (2017) 337–351.
[33] C. Perez-Collazo, D. Greaves, G. Iglesias, A review of combined wave and offshore wind energy, Renew. Sustain. Energy Rev. 42 (2015) 141–153.
[34] S.P. Neill, M.R. Hashemi, Wave power variability over the northwest European shelf seas, Appl. Energy 106 (2013) 31–46.
[35] E.D. Stoutenburg, N. Jenkins, M.Z. Jacobson, Power output variations of co-located offshore wind turbines and wave energy converters in California, Renew. Energy 35 (12) (2010) 2781–2791.
[36] L. Cradden, H. Mouslim, O. Duperray, D. Ingram, Joint exploitation of wave and offshore wind power, in: Proceedings of the 9th European Wave and Tidal Energy Conference (EWTEC), Southampton, UK, 2011, pp. 1–10.
[37] R. Sen, S.C. Bhattacharyya, Off-grid electricity generation with renewable energy technologies in India: an application of HOMER, Renew. Energy 62 (2014) 388–398.
[38] C. Budischak, D. Sewell, H. Thomson, L. Mach, D.E. Veron, W. Kempton, Cost-minimized combinations of wind power, solar power and electrochemical storage, powering the grid up to 99.9% of the time, J. Power Sources 225 (2013) 60–74.
[39] T. Divett, R. Vennell, C. Stevens, Optimization of multiple turbine arrays in a channel with tidally reversing flow by numerical modelling with adaptive mesh, Philos. Trans. R. Soc. A 371 (1985) (2013) 20120251.
[40] S.W. Funke, P.E. Farrell, M.D. Piggott, Tidal turbine array optimisation using the adjoint approach, Renew. Energy 63 (2014) 658–673.

FURTHER READING

[1] T. Roc, D. Greaves, K.M. Thyng, D.C. Conley, Tidal turbine representation in an ocean circulation model: towards realistic applications, Ocean Eng. 78 (2014) 95–111.
[2] S.P. Neill, A. Vögler, A.J. Goward-Brown, S. Baston, M.J. Lewis, P.A. Gillibrand, S. Waldman, D.K. Woolf, The wave and tidal resource of Scotland, Renew. Energy 114 (2017) 3–17.
[3] S.P. Neill, J.D. Scourse, K. Uehara, Evolution of bed shear stress distribution over the northwest European shelf seas during the last 12,000 years, Ocean Dyn. 60 (5) (2010) 1139–1156.

Chapter 10

Other Aspects of Ocean Renewable Energy

In this book, we have introduced and discussed many aspects of ocean renewable energy relating to tidal, wind, and wave energy resource characterization, including modelling, observation, and optimization. In this chapter, we introduce more contemporary aspects of ocean renewable energy—state-of-the-art research topics, and issues that are not yet fully resolved. These include the variability of multiple ocean renewable energy resources (and how these relate to other forms of renewable energy), and how the resource is likely to vary in the future due to global warming. We also discuss uncertainty in resource characterization (e.g. due to feedbacks between energy extraction and the resource) and wave-tide interaction. After a consideration of the development of an ocean renewable energy project from site selection and device design through to grid connection and commissioning, we finally discuss how tidal energy conversion influences sediment dynamics—a process, via morphodynamics, that can influence the resource itself.

10.1 RESOURCE VARIABILITY

We have seen in Chapters 3–5 that there is considerable spatial and temporal variability in all ocean renewable energy resources. For example, the tidal range resource is amplified in regions that are in resonance (i.e. standing wave systems), and tidal currents are much stronger within the confines of narrow straits and around rocky headlands. Further, tides in most regions of the world are either diurnal or semidiurnal, with the latter being characterized by the fortnightly spring-neap cycle. By contrast, the wave resource tends to be characterized by seasonal cycles (although with significant interannual variability) and is greatest in those regions that are exposed to long fetches, for example, the west coast of Ireland and Scotland (exposed to the North Atlantic), and the west coast of the United States. Similarly, the wind resource is greatest in exposed regions and again exhibits strong interannual and intraannual variabilities. Therefore, for any future energy mix that includes significant levels of generation from multiple ocean renewable energy resources, it is important to consider temporal variability *between* resources, from both supply and demand

Fundamentals of Ocean Renewable Energy. https://doi.org/10.1016/B978-0-12-810448-4.00010-0

perspectives. Further, within the context of global warming, it is appropriate to consider how ocean renewable energy resources will vary in the future, particularly the tidal range resource, since the embankments of tidal range power plants are expected to have a lifespan (> 100 years) that will witness evolution in the Earth's climate. Due to sea-level rise, tidal dynamics are expected to change in the future.

10.1.1 Timescales of Multiple Ocean Renewable Energy Resources

The temporal variability of various proxies for a selection of UK ocean renewable energy resources (tidal elevation, wind speed, and wave height) is shown in Fig. 10.1A–C for a typical year—2012. UK demand for electricity over the same time period is also shown (Fig. 10.1D). Fig 10.1A demonstrates clearly the spring-neap cycle that is characteristic of many tidal regions throughout the world, and indeed is one of the main challenges that the tidal energy industry faces: time periods with high potential for electricity generation (springs), followed by time periods with minimal potential for generation (neaps). By contrast, neither winds nor waves suffer from such fortnightly variability. However, beyond seasonal variability, wind and wave resources are stochastic, and so not predictable, other than at relatively short (e.g. 24–28 h) timescales (i.e. based on forecasts generated by well-constrained operational models). Rarely in Fig. 10.1A, B, or C does the variability of individual ocean renewable energy resources track the pattern of demand for electricity (Fig. 10.1D).

Widén et al. [1] presented a useful seasonal/diel comparison across a range of renewable energy resources (solar, wind, wave, tide) at a variety of contrasting locations (Fig. 10.2). It is important to note, in this figure, that each of the resources has been normalized by the 98th percentile at each location, and so the magnitudes between resources and locations should not be compared directly. At high latitudes (e.g. Sweden in this example), there is strong seasonal variability in the solar resource—something that is absent at low latitudes. The solar resource is of course strongly linked to the diel cycle (because incoming solar radiation peaks at noon), and so can map conveniently to the 24 h demand for electricity. However, it should be noted that at relatively high latitudes, the seasonal trend in the solar energy resource (summer peak, Fig. 10.2) tends to be 180 degrees out of phase with the seasonal demand for electricity (winter peak, Fig. 10.1D). By contrast, although there is a moderate seasonal correlation between the wave resource and demand for electricity, the wave resource is independent of the time of day. The two tidal examples shown are for semidiurnal and diurnal sites at Pennsylvania and Mississippi, respectively.[1] Although the time of HW (or LW), peak flood (or ebb), etc. does not correspond

1. Example time series for semidiurnal and diurnal sites are plotted in Fig. 3.15.

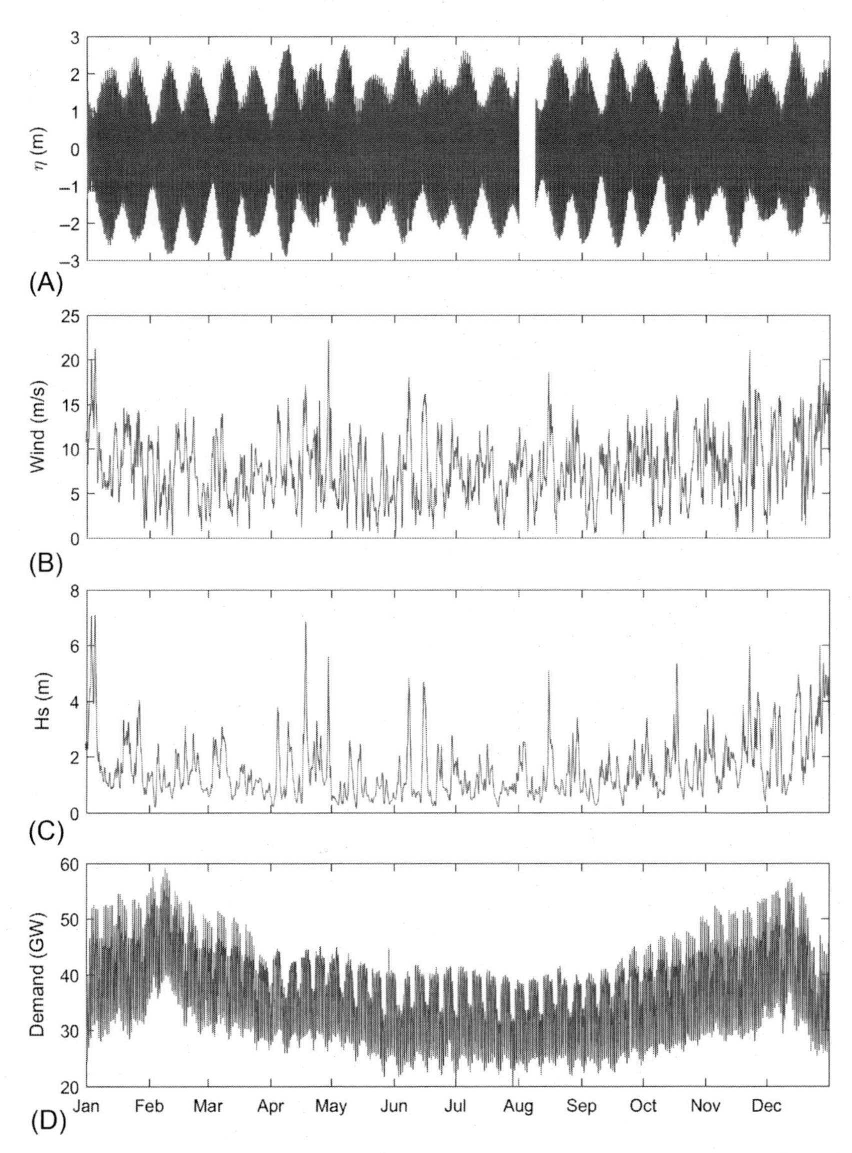

FIG. 10.1 The 2012 time series of (A) tidal elevation (at Newlyn tide gauge), (B) wind speed (at WaveHub), (C) significant wave height (at WaveHub), and (D) UK electricity demand. The gap in tidal elevation data during the first week of August is due to instrument failure. Tidal data are from the National Tidal and Sea Level Facility, provided by the British Oceanographic Data Centre and funded by the Environment Agency. *(Wind speed is from the ECMWF ERA-Interim reanalysis dataset, wave data are from a modified version of the wave model presented by S.P. Neill, M.R. Hashemi, Wave power variability over the northwest European shelf seas, Appl. Energy 106 (2013) 31–46, forced with ERA-Interim data, and electricity data are from http://www.gridwatch.templar.co.uk/.)*

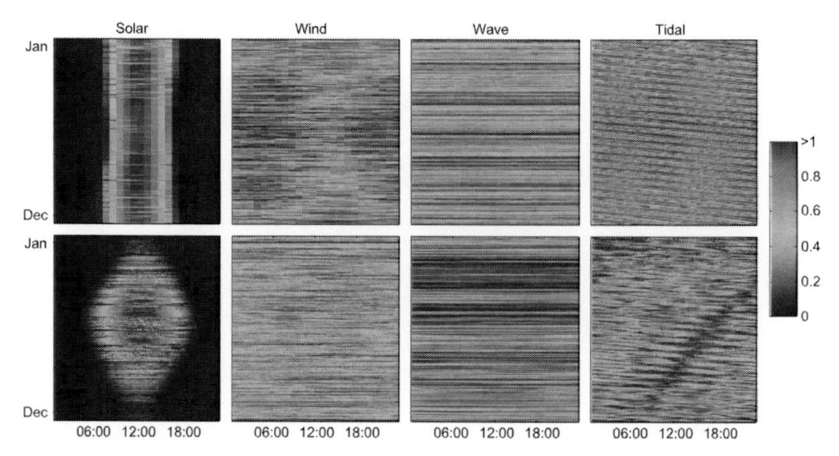

FIG. 10.2 Variability of renewable energy resources at contrasting locations. Results are normalized to the 98th percentile measured for each site. The solar sites are Botswana (*top*) and Sweden (*bottom*); wind sites are South Africa (*top*) and Sweden (*bottom*); wave sites are Hawaii (*top*) and Sweden (*bottom*); tidal sites are Pennsylvania—semidiurnal (*top*) and Louisiana—diurnal (*bottom*). *(Reproduced from J. Widén, N. Carpman, V. Castellucci, D. Lingfors, J. Olauson, F. Remouit, M. Bergkvist, M. Grabbe, R. Waters, Variability assessment and forecasting of renewables: a review for solar, wind, wave and tidal resources, Renew. Sustain. Energy Rev. 44 (2015) 356–375, with permission from Elsevier.)*

to either seasonal or diel demand for electricity, there is an underlying trend. Because we measure time based on the Earth's rotation (and hence the S2 constituent that has a period of 12 h), this means that at any one location, spring (or neap) tides—which are a combination of the M2 and S2 tidal constituents—will occur at the same time every lunar cycle. For example, in semidiurnal regions, when the S2 phase is close to 0 degrees, then maximum values of S2 will occur near the time of solar transit (i.e. noon or midnight) [2]. Therefore, mean high water spring (MHWS) will occur near noon and midnight, at times when the S2 and M2 constituents are in phase. However, if the S2 phase is near 180 degrees, then S2 will have a minimum value near noon and midnight, and mean low water spring (MLWS) will occur at these times. Therefore, particular stages within the spring neap-neap cycle are linked to time of day, and could be considered from an electricity demand perspective. Examining the S2 cotidal chart for the northwest European shelf seas (Fig. 10.3), we can determine the time of day that HW (or LW) spring tides will occur, and similar charts can be used to determine at what time of day peak spring flood (or ebb) currents occur (e.g. [3]).

Relating to variability, one issue that is shared by the majority of ocean renewable energy resources is, although they have a high energy density and may be predictable, they are not *dispatchable*. For example, if there is a sudden increase in demand for electricity, then a pumped storage plant such as Dinorwic (Electric Mountain) in North Wales can attain full power, from standby, in under

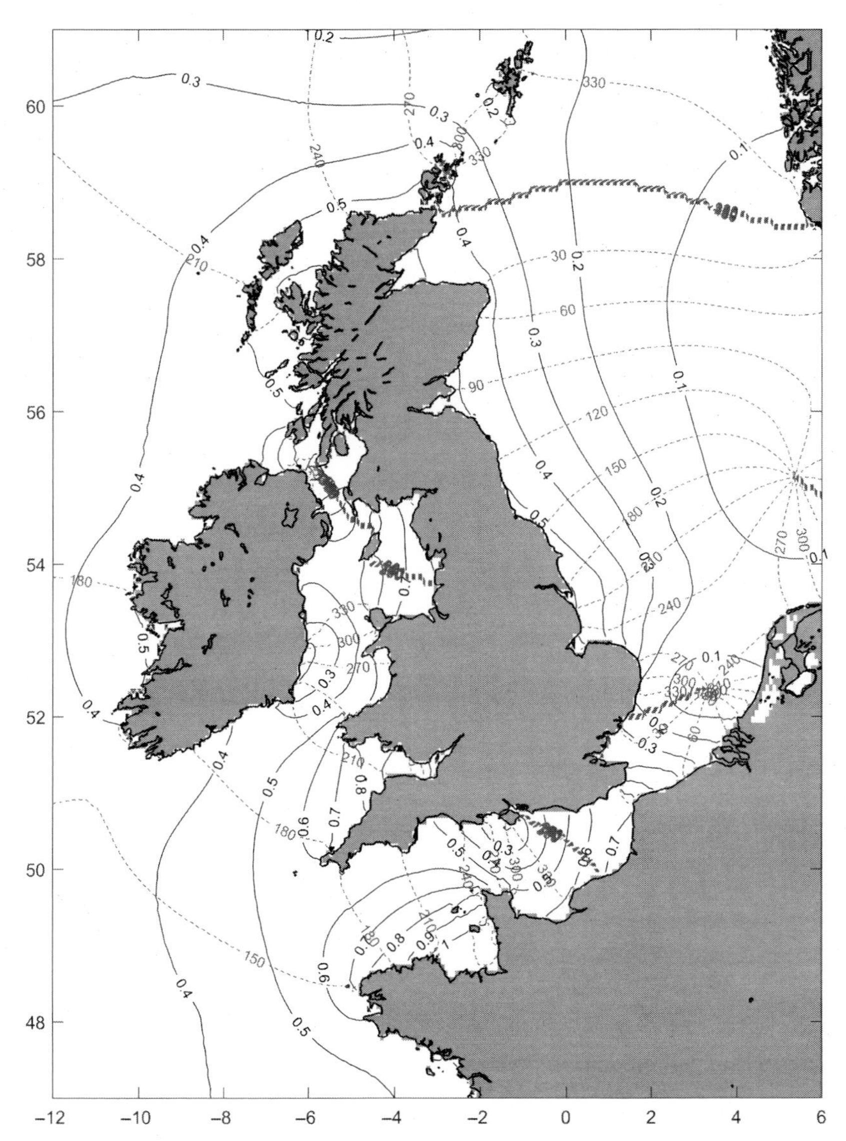

FIG. 10.3 S2 cotidal chart for the northwest European shelf seas (compare with M2 cotidal chart—Fig. 9.12). At locations where the S2 phase equals 0 degrees (*thick dashed contours*), MHWS will always occur at noon and midnight. *(Reproduced from S.P. Neill, J.D. Scourse, K. Uehara, Evolution of bed shear stress distribution over the northwest European shelf seas during the last 12,000 years, Ocean Dyn. 60 (5) (2010) 1139–1156, with permission from Springer.)*

20 s (i.e. it is said to be *dispatchable*). This is a valuable asset for any renewable energy resource, especially when a resource is also predictable. One ocean energy resource that could meet both of these conditions is tidal range energy. As discussed in Section 3.14, tidal range power plants make use of an impoundment to store potential energy. Although there will be many constraints, release of this potential energy for conversion into electricity could be controlled. For example, if the timing of tides on a particular day is such that electricity can only be generated via a tidal range power plant between 03:00–06:00 (when demand is low), this may not be as useful for the electrical grid system as electricity that can be generated between 06:00–09:00 when a peak in demand occurs (see Fig. 1.12). However, if the potential energy could be temporarily stored in the impoundment and then released some time later, this would inject some much needed flexibility into the grid, from a predictable/renewable resource that has significant capacity potential. Another option is to pump water into the lagoon during periods of low power demand (preferably using other renewable energy resources) to optimize generation [4]. Tidal range power plants, therefore, almost uniquely amongst ocean renewable energy resources, offer some degree of flexibility/dispatchability.

An effective method that could resolve the short-timescale variability of renewable energy sources, when supply and demand are not in phase, is energy storage. Conventional fossil fuel-based energy systems have a major advantage over renewables, because the power output from fossil fuel-based power plants can be varied to meet demand, that is, they are controllable (but not necessarily dispatchable). Therefore, energy storage technologies have not evolved at the same pace as conventional energy technologies in the past. However, as renewable energy systems gradually displace fossil fuel-based systems, many energy storage solutions are emerging to help balance supply and demand. Some of these methods include mechanical (pumped-storage, compressed air), electrochemical (rechargeable batteries), and chemical (hydrogen storage).

10.1.2 Long Timescale Changes to the Resource

It is now accepted that global warming will lead to a future change in the statistical distribution of weather patterns. Global warming will, therefore, influence the future marine renewable energy resource, and this can manifest in several ways. First, globally coordinated efforts to reduce carbon emissions, in addition to dwindling fossil fuel reserves, will increase the proportion of renewables in the future energy mix (e.g. see Section 1.2). Second, climate change will, by its very definition, lead to a change in weather patterns that will directly influence wave and offshore wind resources. Further, global warming leads to sea-level rise, and this will alter the tidal range and tidal stream resources; for example, by increasing mean water depth, the resonance characteristics of ocean basins will be affected (e.g. see Section 3.5), leading to a change in the magnitude and distribution of regions of high tidal range.

In addition, with generally increasing global temperatures and a blurring of the distinction between seasons, the intraannual pattern of demand for electricity may vary significantly in the future within the context of global warming. For example, hotter summers could lead to more demand for electricity to provide cooling during summer months (generally a period of lower demand for electricity at relatively high latitudes—see Fig. 10.1D), yet milder winters would lead to lower demand for electricity for heating during what is currently the time of year that is associated with the highest demand for electricity.[2] Here, we focus on how marine renewable energy resources (rather than demand for electricity) are likely to change in the future.

Wind Energy

There have been several studies that examine how the wind resource is likely to vary in the future, although these studies generally make no distinction between the onshore and offshore wind resource. It is important to note that there is already large interannual variability in the wind resource, and so future trends are often difficult to distinguish in such analysis. In general, studies are based on the outputs of global climate models (GCMs) through either dynamical or statistical downscaling [5]. Within Europe, the general trend is for an increase, towards the end of the 21st century, in the autumn and winter wind energy resource in northern and central Europe, with a corresponding reduction of the resource across the Mediterranean (with the exception of the Aegean) [6–8], and this trend also generally applies to annual mean wind speeds [5]. These findings are consistent with a progressive tendency towards the positive phase of the North Atlantic Oscillation (see Section 5.5), which is strongly linked to winter wind speeds in northern Europe. Changes in the annual mean wind energy density are within the range ±10%, but the change during winter months alone is around ±20% towards the end of the 21st century, with increases up to 25% predicted for the United Kingdom [8]. In the period 2081–2100, Cradden et al. [9] calculated, based on analysis of a range of climate scenarios, that in the United Kingdom, wind speeds will generally increase by around 5–8% in autumn/winter, and reduce by 5–8% during summer months (Fig. 10.4). However, they found that, although there were spatial and temporal differences in the future UK resource, in general the annual capacity factor from two offshore wind farms is almost unchanged in the future (2081–2100), compared with a 1961–90 baseline, under a range of Intergovernmental Panel on Climate Change Fourth Assessment Report (IPCC AR4) climate scenarios (Table 10.1). Similar studies for Ireland [10] and southern Africa [11] also demonstrate significant seasonal changes, but little change in the annual mean wind resource up to the middle of the 21st century. Fant et al. [11] reported that the largest change in wind speeds in the middle of the 21st century across southern Africa

2. Although the demand for lighting will not be affected by climate change.

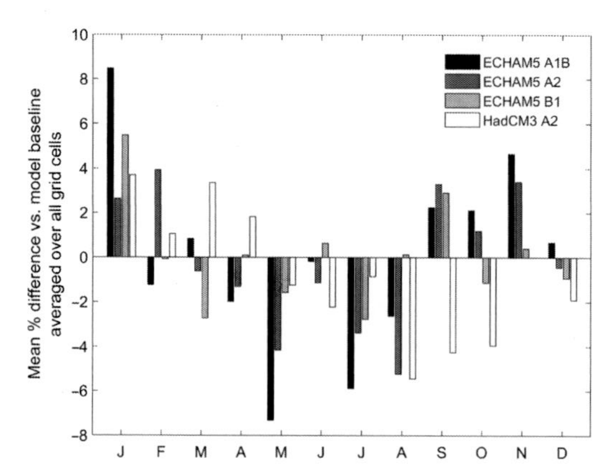

FIG. 10.4 Future changes in spatially averaged monthly wind speeds over the United Kingdom for a range of climate scenarios. For IPCC AR4 climate change scenarios (A1B, A2, B1), see Table 10.1. *(Based on a figured presented in L.C. Cradden, G.P. Harrison, J.P. Chick, Will climate change impact on wind power development in the UK?, Clim. Change 115 (3–4) (2012) 837–852.)*

TABLE 10.1 Annual Capacity Factor for Two UK Wind Farms for Baseline (1961–90) and Future (2081–2100) IPCC AR4 Scenarios

		Annual Capacity Factor (%)	
Scenario	**Description**	**North Hoyle**	**Kentish Flats**
Baseline	1961–90	41.6	35.0
A2	A very heterogeneous world, with an emphasis on self-reliance and preservation of local identities. Economic development and technological change are fragmented	41.7	33.9
A1B	Rapid economic growth and rapid introduction of new and more efficient technologies, with a balance between fossil-intensive and nonfossil energy sources	41.7	34.3
B1	Rapid change in economic structures towards a service and information economy, with reductions in material intensity and introduction of clean and resource-efficient technologies	41.3	34.2

Source: Data from Cradden et al. [9].

will occur offshore, with typical increases in the median wind speed of around 0.6 m/s, and with significant seasonal trends.

Wave Energy

In a relatively early study, Harrison and Wallace [12] used a simple model (based on the Pierson-Moskowitz wave energy spectrum) to investigate the sensitivity of the wave energy resource to the west of Scotland to changes in the annual mean wind speed (a proxy for climate change). A 20% reduction in annual mean wind speed lowers available wave power in this region by 67%, whereas an equivalent increase raises the available power by 146%. Although it is clear that changes in wind speed will significantly alter the wave energy resource, it is important to know how the resource is likely to change in the future, and we can learn something about the future wave resource by examining past trends. A study of wave height variations to the west of Norway from 1881 to 1999 demonstrated that there is a positive trend in mean wave height from 1960 to 1999 [13], but a similar study shows that the southern California coast witnessed a negative trend in wave heights, and hence wave energy, towards the end of the 21st century [14].

There have been relatively few studies that examine how wave heights, and particularly wave power, are likely to vary in the future. In a multimodel ensemble of wave-climate projections, Hemer et al. [15] found a projected decrease (2070–2100 time period compared with 1979–2009 baseline) in annual mean significant wave height (H_S) over 26% of the global ocean area, and a projected increase in annual mean H_S over 7% of the global ocean, but mainly concentrated in the Southern Ocean (Fig. 10.5). Within the context of wave energy conversion, Janjić et al. [16] found an overall reduction in wave energy flux along the European coast towards the end of this century, and this is in agreement with Reeve et al. [17] who found a 2–3% (depending on the IPCC AR4 climate scenario) reduction in available wave power at the Wave Hub site, UK, but stress that such changes are relatively small compared with natural variability of the wave climate at this location. Further, it is noted that variance of wave climate projections associated with study methodology dominates other sources of uncertainty (e.g. climate scenario and model uncertainties) [15]. Climate model uncertainty (i.e. intermodel variability) is significant nearly globally, and its magnitude is comparable or greater than that of the common signal (the signal simulated by multiple model ensembles of H_S simulations) in all areas, with the exception of the eastern tropical Pacific [18].

Other issues associated with global warming that can influence the wave energy resource are reduction in sea ice extent and sea-level rise. A reduction in the extent of sea ice will increase fetch lengths (e.g. Fig. 10.6). In northern Europe, for example, northerly winds would generate larger waves in this region due to a reduction in Arctic sea ice extent (and hence a longer fetch), and this effect needs to be considered in wave models that consider the future climate. In addition, sea-level rise would lead to increased water depths in coastal waters,

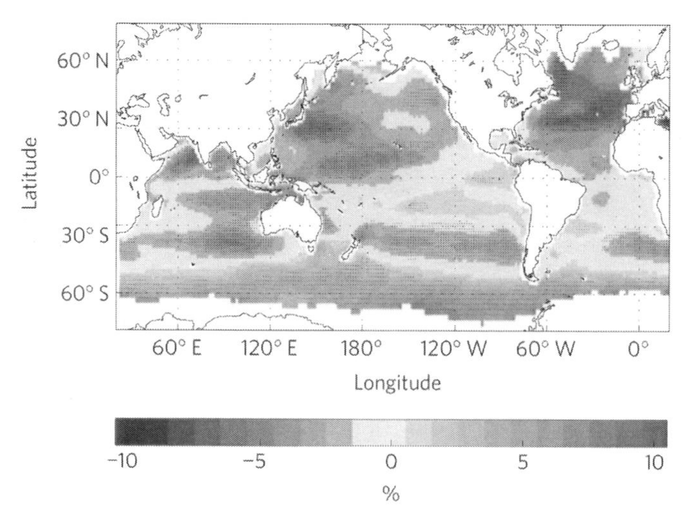

FIG. 10.5 Projected future change in annual significant wave height for 2070–2100, compared with a 1979–2009 baseline. *(Reproduced from M.A. Hemer, Y. Fan, N. Mori, A. Semedo, X.L. Wang, Projected changes in wave climate from a multi-model ensemble, Nat. Clim. Change 3 (5) (2013) 471–476, with permission of Nature Publishing Group.)*

and this would influence the distribution of wave properties along coastlines due to (a) wave shoaling (Section 5.2.1) and (b) wave refraction (Section 5.2.2).

Tidal Range

Sea levels increased at an average rate of 1.7 mm/year over the 20th century [19] (e.g. see Fig. 1.9C).[3] The most recent Intergovernmental Panel on Climate Change Fifth Assessment Report (IPCC AR5) considered four future greenhouse gas concentration trajectories (Representative Concentration Pathways or RCPs) (i.e. four possible future climates) known as RCP2.6, RCP4.5, RCP6.0, and RCP8.5, representing radiative forcing values in the year 2100 relative to preindustrial levels (+2.6, +4.5, +6.0, and +8.5 W/m^2, respectively). Projections of global mean sea-level rise (SLR) across the four scenarios, relative to 1986–2005, are shown in Fig. 10.7. The solid line shows the median projections, and the *likely*[4] range is indicated by the shaded area. Global mean sea level is therefore likely to rise in the range 0.28–0.98 m by the end of the 21st century, with a median value across the four scenarios of 0.44–0.74 m.

Several studies consider how a future rise in sea level would influence tidal dynamics. Although some studies consider a sea-level rise of at least 2 m—considerably in excess of the likely range reported by IPCC AR5

3. Although there is considerable spatial variability in this rate.
4. Defined by the IPCC as a probability of at least 66%.

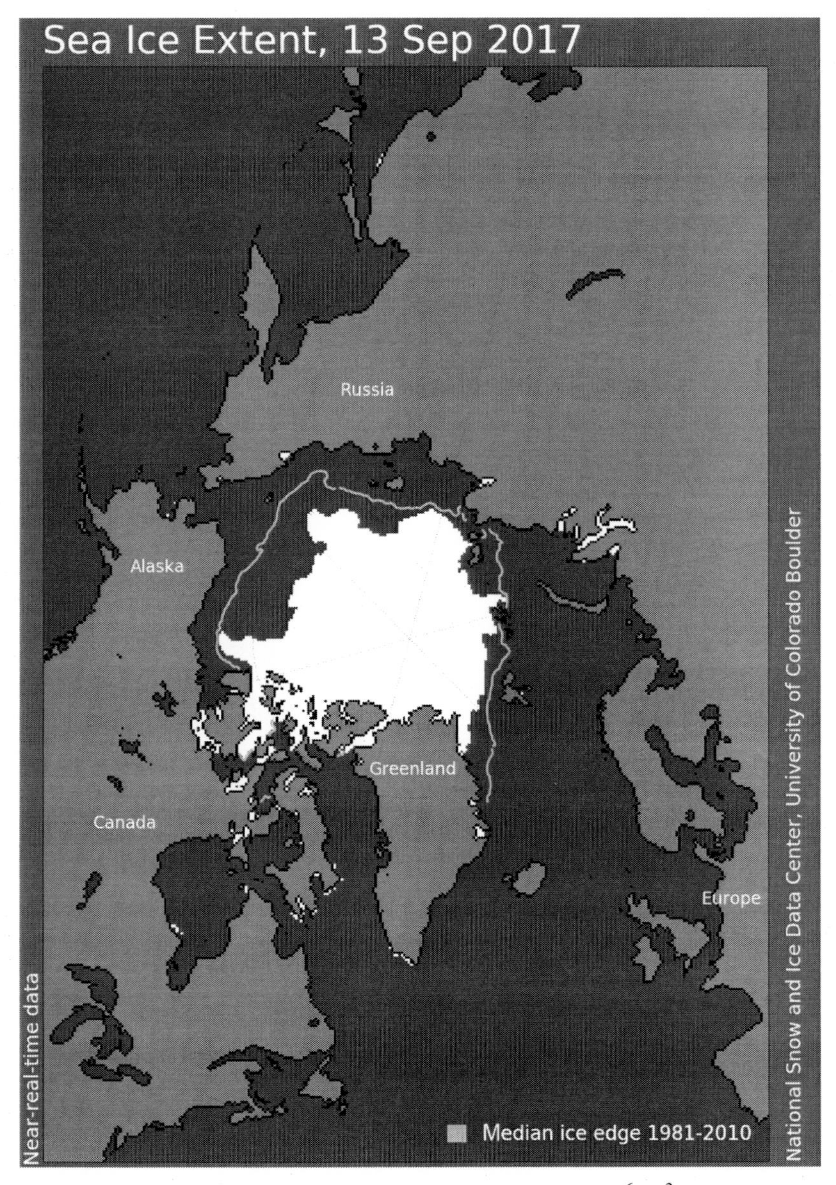

FIG. 10.6 Arctic sea ice extent on September 13, 2017 was 4.64×10^6 km^2, the eighth lowest in the 38-year satellite record. The *contour line* shows the 1981–2010 average extent for that day. *(Image courtesy of National Snow and Ice Data Center.)*

(Fig. 10.7)—we here restrict our discussion to those studies that considered SLR more in line with the above values (i.e. *likely* values of SLR towards the end of this century). With a spring tidal range of 16 m, the Bay of Fundy has the largest tidal range in the world, and hence a vast tidal range resource. The Bay

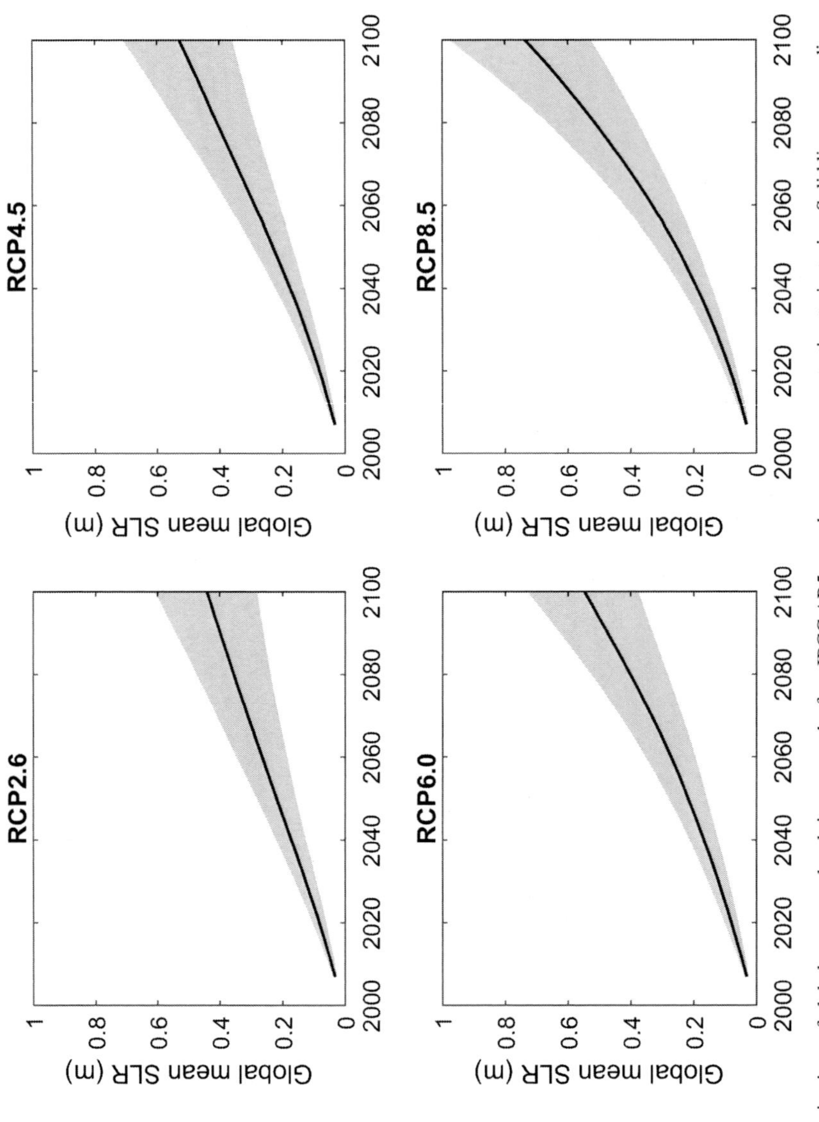

FIG. 10.7 Projections of global mean sea-level rise across the four IPCC AR5 greenhouse gas concentration trajectories. Solid lines are median projects, and shaded regions are the likely range. *(Data Provided by IPCC AR5, T. Stocker, Climate Change 2013: The Physical Science Basis: Working Group I Contribution to the Fifth Assessment Report of the Intergovernmental Panel on Climate Change, Cambridge University Press, Cambridge, 2014.)*

of Fundy is in resonance at the semidiurnal frequency, and so the water depth (in comparison to the channel length) is expected to be sensitive to changes in mean sea level—see Section 3.5. Pelling and Green [20] simulated the response in tidal dynamics that a 1 m SLR would have in this region. This 1 m SLR led to an increase in the M2 tidal amplitude of around 0.1 m throughout the Bay of Fundy, and this is attributed to the basin approaching true resonance as sea level (i.e. the mean water depth in the basin) increases. In a study of the northwest European shelf seas, 1 m of SLR led to significant changes in the tidal dynamics [21]. Again, in regions that are in resonance such as the northern Irish Sea, tidal amplitude increased (by around 0.05 m), but there were many regions that reduced in amplitude, and one important region is the Bristol Channel (which hosts Swansea Bay—proposed location of a tidal lagoon power plant), where the M2 amplitude reduced by around 0.05 m. The M2 amplitude in Swansea Bay is around 3 m (Fig. 3.6). Since theoretical tidal range power is related to tidal range squared (e.g. Section 3.14), a reduction of 0.05 m in the M2 amplitude would therefore represent a 3% reduction in the mean tidal range resource.

Finally, although the IPCC AR5 is a highly respected source of projected SLR, there are other studies that consider the possibility of even greater SLR by the end of the century. It is useful to consider more extreme future scenarios since, although they are highly unlikely to occur, they are associated with very high levels of impact. The UK Met Office UKCP09 includes a high-plus-plus (H++) scenario, that is associated with SLR of up to 1.9 m by 2100. The response of the M2 amplitude across the world's oceans to an SLR of such magnitude is plotted in Fig. 10.8 [22], demonstrating a complex pattern of significant regional increases/decreases in the tidal range resource. For example, the changes in the Irish Sea discussed earlier are further amplified under such a scenario, and there is a significant reduction in the tidal range resource in the Gulf of St Malo (where La Rance barrage is situated)—see Pickering et al. [22] (which is open access) for more details, including a consideration of the additional tidal constituents S2, K1, and O1.

Tidal Stream

There are surprisingly few published studies that investigate how the tidal stream resource is likely to vary in the future. In an appendix included in Pickering et al. [22], it is noted that Manning's equation expresses the depth-averaged velocity as a function of the square root of the hydraulic slope, and so a 10% change in elevation amplitude would result in ~3% change in current amplitude. Under a scenario of 1 m SLR, a typical change in M2 elevation amplitude of 0.1 m was reported in the previous section for the Bay of Fundy (where the typical M2 elevation amplitude is in the range 2–5 m [20]), and this equates to a 2–5% change in tidal elevation. Therefore, currents in the Bay of Fundy would change by 1.4–2.2% which, for a typical M2 current amplitude of 1 m/s would lead to an increase in current speed of 1.4–2.2 cm/s. This change is likely to be within the bounds of natural variability (e.g. consideration of nonastronomical

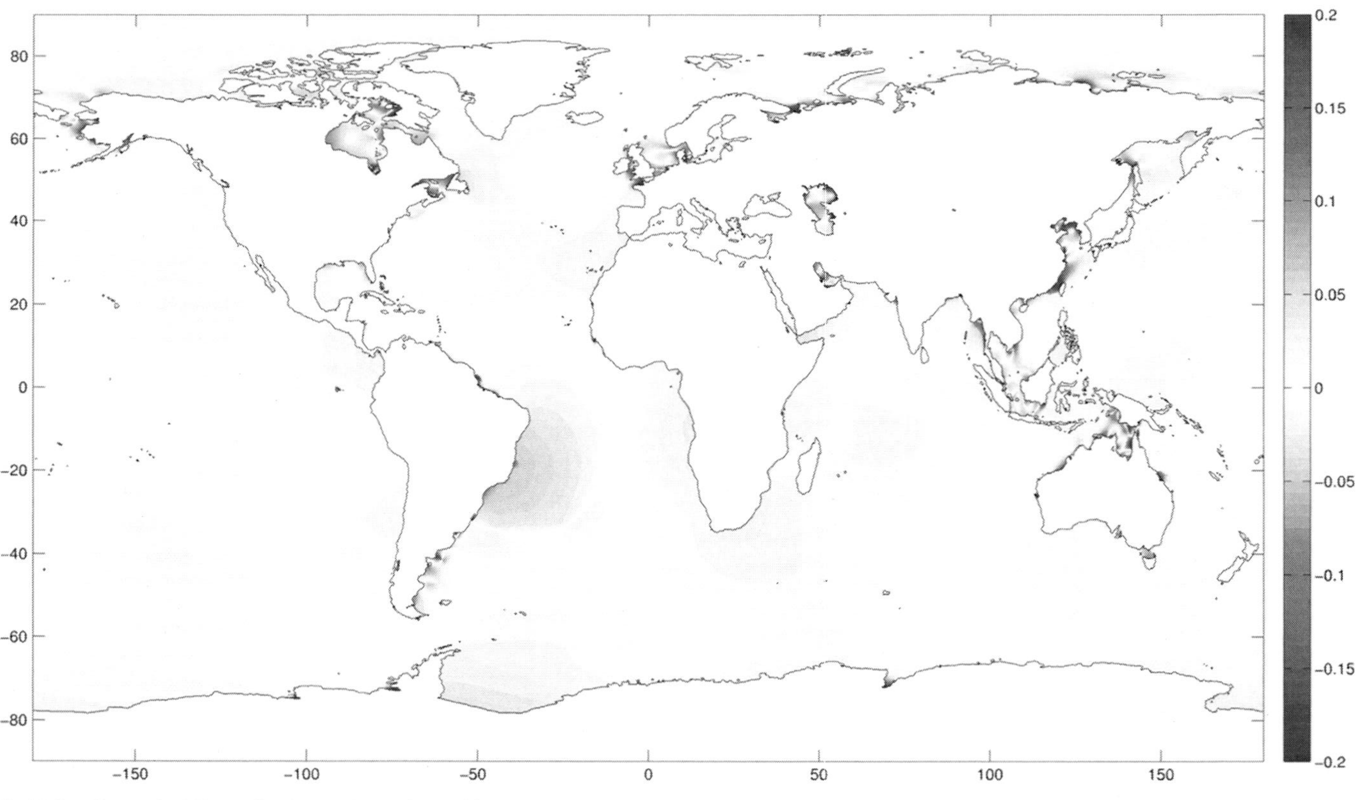

FIG. 10.8 Change in M2 amplitude (m) due to 2 m uniform global sea-level rise [22]. *(This figure is available under the Creative Commons Attribution License (CC BY).)*

effects such as wind-driven currents), and less than the uncertainties associated with both numerical models and observations (which tend to be around 5–10%). However, tidal power is related to velocity cubed, and so these relatively small changes in current speed would be amplified in the power signal, leading to a 4–7% increase in instantaneous power. Although they incorporate large-scale tidal energy extraction in their simulations, De Dominicis et al. [23] similarly show changes of order 1–2 cm/s in the middle of the 21st century across many regions of the northwest European shelf seas due to RCP8.5 (Fig. 10.7).

10.2 INTERACTIONS OF ENERGY RESOURCES AND ENERGY DEVICES (RESOURCE UNCERTAINTY)

Marine renewable energy resources interact with one another through wave-tide interaction—a process in which short waves interact with long waves. In addition, the extraction of energy from the marine environment will influence the resource itself, and this can be an important consideration for large levels of energy extraction. It is possible to account for the processes of wave-current interaction and feedbacks between resource exploitation and the resource itself in numerical models of the marine renewable energy resource. Both issues, combined under the common theme of 'resource uncertainty', are discussed in this section.

10.2.1 Wave-Tide Interaction

Some regions of the world (e.g. the northwest European continental shelf seas) concurrently experience both energetic wave climates and tidal current regimes. The interaction of waves and tides can affect resource characterization, device performance, and maintenance. In general, we can investigate how waves affect a tidal energy site, or conversely: how tides can affect a wave energy site.

Wave properties such as group velocity, wave height, and wave power are altered in the presence of tides. For instance, wave frequency/period is affected by the Doppler shift. When waves oppose tidal currents, wave height, and wave steepness increase. Strong opposing tidal currents can even lead to wave breaking, and a complete blockage of wave energy propagation.

Previous studies have shown that the presence of waves in relatively shallow coastal environments leads to an enhancement of the (apparent) bed roughness. This enhancement of bottom friction can reduce the speed of tidal currents, and lead to a reduction of the tidal energy resource. Further, extreme waves can damage tidal energy devices; therefore, tidal energy devices do not operate during these extreme events. If these events happen frequently, they will significantly reduce the annual energy yield from tidal devices.

Therefore, in sites where wave-tide interactions are significant, further assessment is required to consider these interactions. Here, we provide a brief summary of the physical mechanisms of wave-tide interaction, and tools that can be used to investigate them. More details can be found in other resources (e.g. [24–26]).

In general, tides are long waves with periods in the range of hours and days, whereas wind-generated waves are short waves with periods in the order of seconds; therefore, wave-tide interaction can be regarded as the interaction between long and short waves.

Effect of Tidal Currents on Wave Energy

Considering a wave that propagates in the presence of currents, the frequency of the wave changes by an ambient current as follows (i.e. Doppler shift):

$$\omega = \sigma + ku \tag{10.1}$$

where σ is the relative wave frequency, ω is the absolute wave frequency, u is the ambient current velocity, and k is the wave number. The *absolute* wave frequency can be measured by a stationary observer (i.e. the wave frequency that would be experienced by a wave energy device), but the *relative* frequency can be measured by an observer moving at the same speed as the currents. Equations derived from linear wave theory are valid for an observer moving with the currents [27].

As mentioned before (Section 5.1.3), wave energy propagates at the group velocity, C_g. Due to the Doppler shift, we can write

$$C_g = \frac{d\sigma}{dk} = \frac{d(\omega - ku)}{dk} = \frac{d\omega}{dk} - u \rightarrow C_g = C_g^* - u \tag{10.2}$$

where C_g^* is the absolute group velocity (stationary observer). Referring back to Chapter 5, the wave power for monochromatic waves is given by

$$P = \left(\frac{1}{8}\rho g H^2\right) C_g = EC_g \tag{10.3}$$

where P is the wave power, H is the wave height, and E is wave energy averaged over the wave period. Referring to Eq. (10.2), we can conclude that

$$C_g^* = C_g + u \rightarrow EC_g^* = EC_g + uE \tag{10.4}$$

EC_g^* is the wave power in the presence of currents; it is a combination of the wave power originating from waves (EC_g), and the wave energy transport by tidal currents (uE). The interaction of tidal currents and waves leads to further complications for wave energy resource assessment. When waves propagate in the presence of currents, the principle of wave energy conservation is no longer valid, due to the energy exchange between wave and current fields. This is the reason why spectral wave models such as SWAN do not use conservation of wave energy in their formulation. Instead of wave energy, wave action (i.e. E/σ) is conserved in the presence of ambient currents (Section 8.4.1).

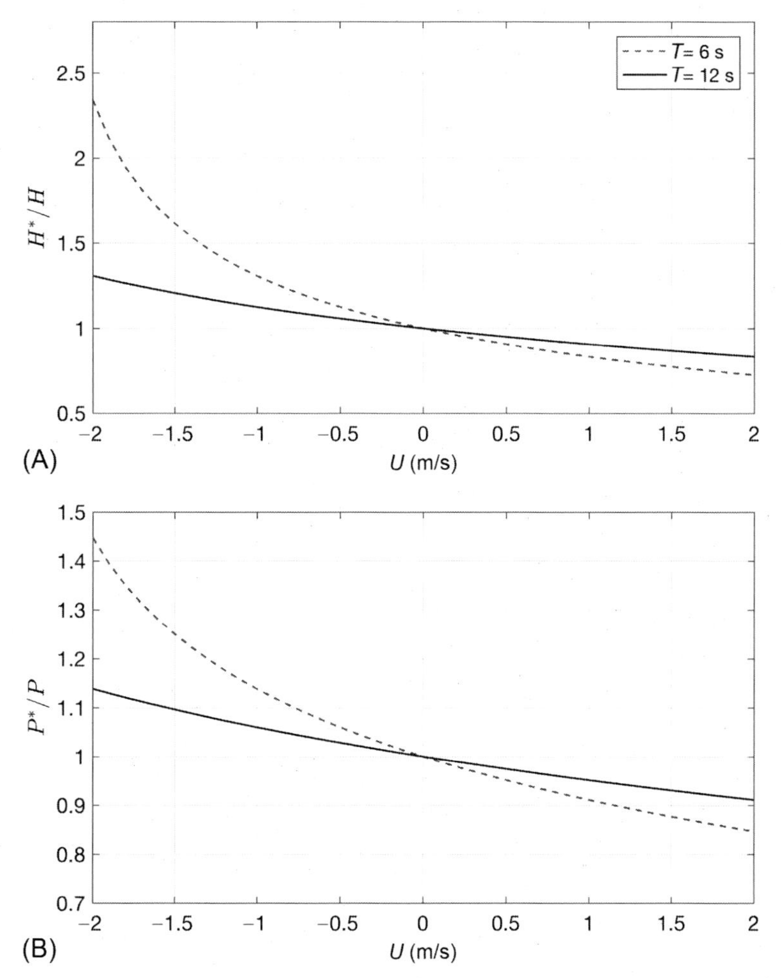

FIG. 10.9 Effect of tidal currents on wave height and wave power using a simplified method [25]. Note that opposing currents have negative speed. (A) Effect of currents on wave height. (B) Effect of currents on wave power.

Using the conservation of wave action, we can assess the effect of tidal currents and tidal elevations (water depth) on wave properties. For instance, for linear monochromatic waves, and assuming the deep water approximation for the dispersion relation (i.e. $C = g/\omega$), the effect of currents on wave height can be estimated as [25]

$$\frac{P^*}{P} = \frac{\sigma}{\omega}\left[1 + \frac{2u}{C}\frac{\sigma}{\omega}\right]^{-1}; \quad \frac{\sigma}{\omega} = 2\left[1 + \left(1 + 4\frac{u}{C}\right)^{1/2}\right]^{-1} \tag{10.5}$$

where P^* is the wave power in the presence of currents. Fig. 10.9 demonstrates how wave height and wave power are affected by opposing or following currents based on Eq. (10.5). In particular, opposing currents can increase wave height and wave steepness. This can lead to wave breaking. Referring to Eq. (10.4), the wave group velocity will be reduced by opposing currents. If tidal currents are strong enough, the group velocity approaches zero, which means that waves can be completely blocked by currents.

The discussion above was based on a simplified method to help understand the concepts. In real applications, several ocean models offer coupled wave and tide modelling capability. These models can be used to simulate the interactions of waves and tides. For instance, SWAN has been coupled with both ADCIRC and ROMS models (introduced in Chapter 8). SWAN can include the effect of tidal currents on wave power by importing the tidal current and elevation fields from a tidal model (e.g. ADCIRC or ROMS).

Effect of Waves on the Tidal Energy Resource

Energetic waves can alter tidal currents and tidal elevations. For instance, waves add additional momentum/force to the tidal flow (i.e. wave radiation stresses). Further, the interaction of wave orbital velocities and the bottom boundary layer (of currents) leads to an increase in the roughness felt by currents. The enhancement of bed roughness due to wave-current interaction has been estimated as [28]

$$k_a = k_s \exp\left(\Gamma \frac{U_w}{u}\right) < 10, \quad \Gamma = 0.80 + \phi - 0.3\phi^2 \tag{10.6}$$

in which k_a is the apparent roughness, k_s is the physical roughness, u is the current velocity, U_w is the near-bed wave-induced orbital velocity, and ϕ is the angle between wave and current directions (in radians). As this equation demonstrates, the apparent bed roughness can be much higher (up to 10 times) than the physical bed roughness. In tidal models, friction coefficients such as drag or Manning's are used to represent the bottom friction rather than the bed roughness. For the drag coefficient, it can be shown that [26,29]

$$\gamma = \frac{C_D^*}{C_D} = \left[1 + 1.2\left(\frac{\lambda}{1+\lambda}\right)^{3.2}\right] < 2.2, \quad \lambda = \frac{\tau_w}{\tau_c} \tag{10.7}$$

where C_D and C_D^* are the drag coefficients in the absence and presence of waves (respectively) averaged over the wave period. τ_c is the bed shear stress due to currents only, and τ_w is the bed shear stress due to waves only. These shear stresses can be determined based on the current velocity and the near-bed wave orbital velocity. Fig. 10.10 shows sample calculations for the increase in the drag coefficient. The increase in bottom friction is proportional to the near-bed orbital velocity, and physical roughness (k_s). This graph has been generated for a tidal current of 1 m/s.

Eq. (10.6) or (10.7) can be embedded in a tidal model to estimate the increased bottom friction, and also identify whether the increase in bottom

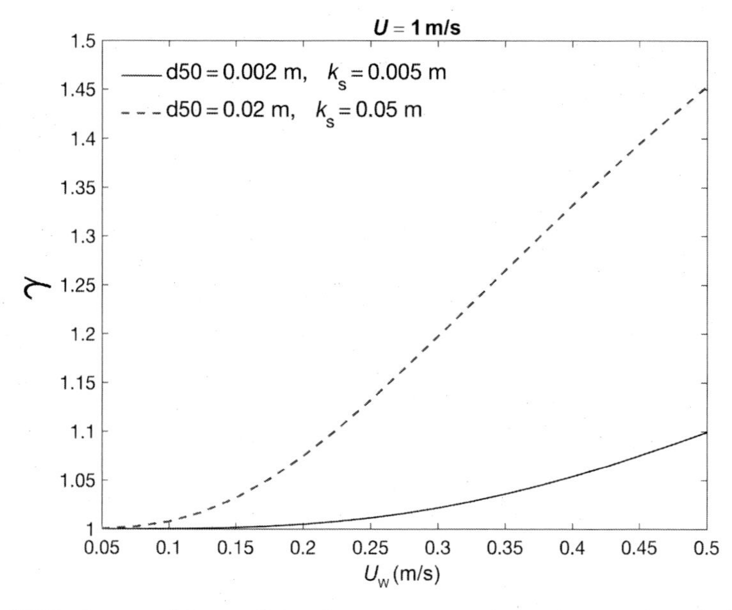

FIG. 10.10 Increase of drag coefficient due to wave-current interaction, assuming a 1 m/s tidal current. U_W is the near-bed wave orbital velocity; γ represents the ratio of the drag coefficient in the presence and absence of waves.

friction can significantly affect the tidal energy resource over long-time periods. At some sites, up to 20% reduction in tidal energy resources has been reported [26,30].

10.2.2 Accounting for Ocean Energy Devices in Numerical Models

Up until now, we have only considered *undisturbed* wave and tidal energy resources. However, the act of harvesting energy from these resources will disturb the underlying hydrodynamics (e.g. [31]). By contrast, only at very large levels of energy extraction would there be significant feedback between wind energy conversion and the wind resource [32], hence wind energy feedbacks are not considered here. In this section, we consider the implications of (and methods for) extracting wave and tidal energy within numerical models.

Tidal Models

Tidal energy extraction can influence both near- and far-field hydrodynamics. Locally, energy extraction can lead to the generation of wakes and changes in eddy patterns, and far-field effects can lead to tidal flow diversion from the project site, and changes in tidal phasing [31]. When assessing the tidal energy resource, it is therefore necessary to consider the impact of various levels of energy harvesting. In general, the IEC Technical Specification for Tidal Energy Resource Assessment and Characterization [31] recommends that energy extraction be accounted for in ocean models when the installed capacity is greater than 10 MW, or the proposed level of energy extraction exceeds 2% of the theoretical tidal energy resource.

Energy extraction can be incorporated in ocean models by modifying the governing equations of flow, and so the turbine is represented in the model by including the mechanics of the energy extraction process, rather than the physical structure of the turbine itself [33]. At the scales of relevance to ocean models, generally additional friction, or a *momentum sink* approach is used to represent turbines, but the *actuator disc* model is also gaining popularity [33]. Actuator disc theory was explained in detail in Section 3.13.

In an early paper on tidal energy extraction, Bryden and Couch [34] presented a relatively simple 1D case of energy extraction in a rectangular cross-section channel, based on an additional bed roughness term. Their finding that 10% of the energy flux can be extracted before there is a significant change in the flow characteristics, remained a popular 'rule-of-thumb' for the limits of energy extraction for many years. Neill et al. [35] applied a similar technique of an additional quadratic friction term in the momentum equation to a 1D study of the Severn Estuary/Bristol Channel under various extraction scenarios, finding that a relatively small amount of energy extraction (particularly at critical points, relating to tidal symmetry) can lead to a significant impact on the sediment dynamics of a channel, including far-field effects. However, in common with Bryden and Couch [34], their 1D model did not account for lateral effects, and both studies could, perhaps, have overstated the effect of energy extraction. If turbines occupy only a fraction of the cross-section, the available power drops below that obtainable from a complete fence,[5] since energy is lost in downstream merging of the wake and the surrounding fluid [36]. For example, in a 2D model the flow can also by-pass the turbine, and so efficiency of energy extraction will reduce significantly. Other research makes use of a momentum sink approach in either 2D (depth-averaged) [37] or 3D [38] models, but it is the use of 3D models that is of particular interest. With increasing access to supercomputing facilities and increased power of desktop computers, it is becoming increasingly common for resource characterization studies to make use of 3D models [39]. There is on-going research about whether a turbine (or an array of turbines) needs to be parameterized in these 3D models at the actual depth of energy extraction (i.e. over the swept area of the rotor), or whether a simpler 'depth-averaged' turbine representation is sufficient. It is important to note in such discussions that the support structure of the turbine will, in the majority of device designs, extend over a greater portion of the water column

5. Rather than installing devices individually within arrays, a tidal fence consists of a 'continuous' row of turbines, extending across the width of a channel.

than the rotor, possibly even the full water column. Drag on the support structure serves to reduce the flow without generating any power [36]. Whereas a depth-averaged representation of the support structure may be sufficient in a 3D model, the rotor itself may require a 3D parameterization in the model (e.g. [40]).

Although it is possible to tune modelling methodologies for tidal energy extraction, datasets are only available for scaled laboratory experiments of turbines at present. Once large-scale arrays are commissioned, and data are made publicly available, it will be possible to validate these techniques against appropriate field data.

Wave Models

In contrast to simulating the tidal energy resource (e.g. [31]), the corresponding IEC technical specification for wave resource assessment (IEC 62600-101 TS) [41] does not provide specific guidance on when to consider WEC device/array feedbacks in models that simulate the wave energy resource. Rather, the academic community tend to focus on how wave energy extraction could influence the nearshore wave climate from an environmental (impact) perspective. However, IEC 62600-101 TS does state that when it is considered appropriate to include the effects of the WEC array on wave propagation in the numerical model, any modifications made to the numerical model to account for the effects of a WEC array should be documented and justified.

In a fairly preliminary study, Neill and Iglesias [42] made use of a 1D cross-shore wave model (UNIBEST-TC) to simulate the nearshore impacts of wave energy extraction, with a focus on subtidal bars. They used wave buoy observations coincident with the model boundary to create a joint probability distribution of wave period and wave height, then simply calculated the 'deep water' wave power, and reduced this power by an appropriate amount (e.g. 10%) to represent the presence of a WEC array, simulated by a proportional reduction in wave height and period at the model boundary. Although 1D cross-shore wave models can be insightful, it is more common to make use of 2D (spectral) wave models for resource assessment, such as the SWAN model that was described in Section 8.4.

SWAN has an OBSTACLE command that can be used to input characteristics of a (line of) subgrid obstacles through which waves are transmitted/reflected [43]. By application of this OBSTACLE command, several studies have attempted to account for the presence of WEC arrays in the wave model (e.g. [44]). The transmission coefficient (K_t) is defined as the ratio of H_{st} (wave height in the lee of the wave array) to H_{si} (incident wave height)

$$K_t = \frac{H_{st}}{H_{si}} \tag{10.8}$$

and acts as an energy sink in the wave action balance equation (8.39), extracting a fraction of the incident wave energy to represent a wave energy converter. An example K_t measured under wave tank conditions for the WaveCat device

TABLE 10.2 Implementation of WEC Module in SNL-SWAN: Description of Obstacle Cases

Obcase	Description
0	Baseline SWAN formulations using constant K_t, specified by user
1	WEC power matrix used to calculate K_t, applied as a constant value across all frequencies
2	WEC RCW used to calculate K_t, applied as a constant value across all frequencies
3	WEC power matrix used to calculate K_t, applied as a unique value at each binned frequency
4	WEC RCW used to calculate K_t, applied as a unique value at each binned frequency

is in the range 0.51–0.81, with a mean value across a range of experiments of $K_t = 0.74$. Abanades et al. [44] used these values to parameterize an array of 11 WaveCat devices in the SWAN model, simulating changes in the wave conditions between the array and the coastline. However, the problem with applying a constant transmission coefficient is that the reduction in wave energy is applied across all wave frequencies. In reality, a WEC will extract wave energy as a function of wave frequency. To address this, Smith et al. [45] developed a frequency-dependent transmission coefficient in SWAN, which has now been formalized in a version of SWAN known as SNL[6]-SWAN [46]. This version of SWAN incorporates a WEC module with five different options (referred to as OBCASE) that modify the baseline SWAN OBSTACLE formulation (Table 10.2). Power matrices for Wave Energy Converters are a familiar concept from Chapter 5 (Fig. 5.17), and relative capture width (RCW) is a dimensionless performance parameter that is used extensively in the WEC industry—it can be thought of as the proportion of incident wave energy (for a crest length equal to the nominal WEC dimension) captured by the device [47]. An illustrative comparison between Obcase 2 (RCW applied as a constant across all frequencies) and Obcase 4 (RCW applied as a unique value at each binned frequency) is provided in Fig. 10.11.

10.3 OCEAN ENERGY PROJECT DEVELOPMENT

Whilst some renewable energy sources such as wind and solar are cost-competitive and operate at commercial scales, the ocean renewable energy sector is generally at the research and development (R&D) stage. Nevertheless,

6. Sandia National Laboratories.

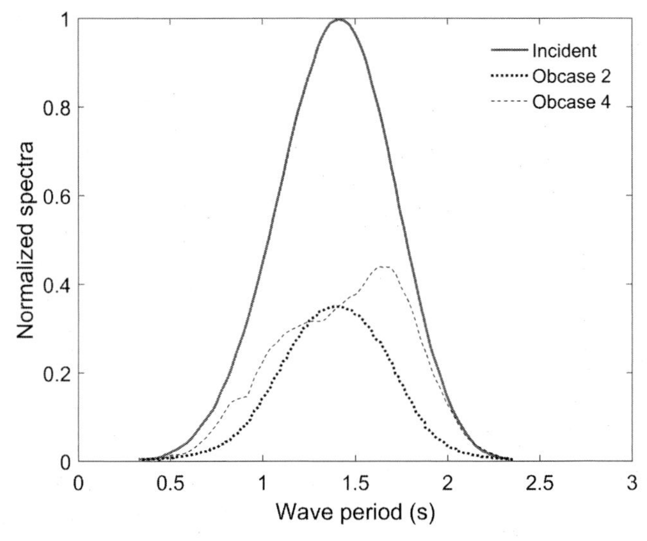

FIG. 10.11 Conceptual comparison between constant transmission coefficient applied across all frequencies (Obcase 2) and applying unique values at each binned frequency (Obcase 4). *(Data from K. Ruehl, A. Porter, C. Chartrand, H. Smith, G. Chang, J. Roberts, Development, verification and application of the SNL-SWAN open source wave farm code, in: The 11th European Wave and Tidal Energy Conference: EWTEC 2015, 2015.)*

the ocean is a vast and untapped resource of renewable energy, and with sufficient investment in R&D, it can potentially become a major competitive industry in the future.

Development of a marine renewable energy project is a complicated and iterative process, involving significant risks and unknowns. Several models of project development have been proposed in the literature (e.g. see Refs [48,49] for further details). Development of a marine renewable energy project usually consists of the following elements: site, market, permits, technology, project team, and capital.

A suitable site is usually chosen/leased by marine spatial planning with the objective of finding the best technical location (maximum energy, minimum construction difficulty, closest to the consumer, minimum cost), with minimum conflicts amongst other users of the ocean, and considering regulations. The energy project at the site should also have minimum environmental impacts. An important step of siting is resource characterization—a topic that has been covered extensively in previous chapters of this book.

Another essential element of any project is the market: the user of the generated electricity should be identified. The buyer of energy should be secured by contracts such as power purchase agreements. Depending on the type of site (grid-connected or off-grid), other costs such as cabling, transmission, and some infrastructure may need to be added to the total cost.

In addition to leasing of a site, required permits for construction, operation, and maintenance of a project should be applied and issued. Permitting is often integrated into marine spatial planning, because it involves close interactions with local and governmental authorities.

In terms of technology, depending on the purpose of a project, various energy devices at different stages of development can be deployed at a site. Some sites are dedicated to demonstration where new devices can be evaluated, whilst at commercial sites, devices that have already been tested at prototype scales (proven technologies) are deployed.

Project development is a team effort that requires effective and close collaboration of technical, legal, financial, and business experts with authorities and stakeholders (e.g. local communities, fishers).

If a project is well designed, it can potentially attract investors and financial resources. As the levelized cost of energy for marine renewable energy projects is still significantly higher than other conventional (e.g. thermal power stations) and renewable (e.g. onshore wind) sources, policymakers create incentives to encourage investments in this sector. This is a key issue in the success of marine renewable energy projects.

Operation and maintenance of a project is the next step after construction of a project. At this stage, monitoring of device performance and their impacts on the environment can provide valuable data for other projects.

10.4 IMPACT OF TIDAL ENERGY EXTRACTION ON SEDIMENT DYNAMICS

Previous studies into the environmental impacts of marine renewable energy extraction have focussed on issues that do not directly influence the resource itself, for example, collision risk of marine mammals and the effects of underwater noise (e.g. [50]). However, there is one aspect of the environmental impacts of marine renewable energy that can have a direct influence on the resource—changes to sediment dynamics and associated morphodynamics [51]. These issues are introduced briefly in this section, covering changes to offshore sand banks (due to tidal stream energy conversion), and changes in sedimentation as a result of tidal lagoon operation. We also discuss how tidal stream array operation can alter sediment transport pathways for varying degrees of tidal asymmetry. However, it is important to emphasize that at present, prior to the existence of and long-term monitoring of lagoons or significant scale tidal energy conversion, feedback between energy extraction and sediment dynamics/morphodynamics remains a research question, and one that is yet to be observed in situ. Although it is likely that large-scale wave energy arrays will lead to changes in sediment dynamics, the change is likely to occur between the array and the coastline, and is therefore not likely to influence the incident wave climate (e.g. [44]).

10.4.1 Impact of TEC Arrays on Sediment Dynamics

Neill et al. [35] first introduced the possibility of nonlocalized, long timescale, changes in morphodynamics as a result of tidal array operation. Prior to this research, it had been assumed that either (a) highly energetic tidal sites were devoid of mobile sediment or (b) any impacts would be localized, i.e. scouring of sea bed sediments around the base of tidal energy devices would be the main issue. However, even in highly energetic environments such as the Pentland Firth (Scotland), the predominant bedrock will be interspersed with regions of sand (e.g. [52]). Such pockets of mobile sediment are important habitats for fisheries, and important repositories of sediment that exchange material with neighbouring beaches over a range of timescales [53]. In addition, larger offshore sand banks, such as headland sand banks generated by eddies associated with strong tidal flow past headlands [54], have a significant and important influence on the flow field [55]. Therefore, any changes in the morphodynamics of offshore sand banks due to tidal energy extraction will likely have a significant impact on the tidal energy resource.

Tidal Asymmetry

Since astronomical tides are generated by the combined gravitational forces of the Sun and the Moon, the frequencies of tidal constituents in the deep oceans directly relate to lunar or solar days, and can be expressed in terms of diurnal and semidiurnal components. The propagation of (barotropic) tides in the deep ocean is primarily governed by linear processes, where their interactions generate subharmonic tides [56]. For instance, the combination of the principal semidiurnal lunar (M_2) and solar (S_2) tidal constituents describes the spring neap cycle (Section 3.9).

Over continental shelves, and particularly in shallow coastal waters, other nonlinear forces and processes such as friction, advection (due to advective inertia forces), and diffusion (due to turbulence) become increasingly responsible for the dynamics of the tides. As a result, the tidal signal is more complex in such regions, and can no longer be represented by simple linear superposition of semidiurnal and diurnal components. Using the concept of Fourier series, by combining higher-frequency tidal components or superharmonic tides, any nonlinear tidal signal can be reconstructed. Unlike astronomical tides, superharmonic tidal components are generated by localized shallow water forces. Accordingly, the nonlinear interaction of an astronomical tidal component with itself and other tidal components generates overtides and compound tides, respectively, with higher frequencies, for example,

$$\mathrm{M}_4(2\omega_{\mathrm{M}_2}), \mathrm{M}_6(3\omega_{\mathrm{M}_2}), \mathrm{S}_6(3\omega_{\mathrm{S}_2}), \mathrm{MS}_4(\omega_{\mathrm{M}_2}+\omega_{\mathrm{S}_2}), \mathrm{MN}_4(\omega_{\mathrm{M}_2}+\omega_{\mathrm{N}_2}), \ldots \tag{10.9}$$

Overtides and compound tides are the main causes of tidal asymmetry, and their role in understanding and accurately simulating tides is very important in some regions [57]. Using simple mathematics, it can easily be shown that the nature of tidal asymmetry is related to the phase difference between semidiurnal and quarterdiurnal tidal constituents. Further, Speer et al. [58] showed that

$$2\phi_{M_2} - \phi_{M_4} \approx \phi_{M_2} + \phi_{S_2} - \phi_{MS_4} \approx \phi_{M_2} + \phi_{N_2} - \phi_{MN_4} \tag{10.10}$$

It is generally the phase relationship between the principal semidiurnal tidal current (M_2) and its first harmonic (M_4) that dominates tidal asymmetry [59,60]. Although the combination of M_2 and M_4 tidal currents in Fig. 10.12A results in a distorted tide (Fig. 10.12B), the flood and ebb tides are equal in magnitude, as is the net power (a function of velocity cubed) generated during the flood and ebb phases of the tidal cycle. By combining M_2 and M_4 tidal currents as in Fig. 10.12C, however, the flood tide is stronger than the ebb (Fig. 10.12D). Although there is no net residual flow, the integrated cube of the velocity (U^3) is greater during the flood phase of the tide. Hence, there will be a strong bias of power production in favour of the flood phase of the tidal cycle. In the case where the flood and ebb currents are equal

$$2\phi_{M_2} = \phi_{M_4} + 90\,\text{degrees} \tag{10.11}$$

and where there is a maximum asymmetry

$$2\phi_{M_2} = \phi_{M_4} \tag{10.12}$$

where ϕ_{M_2} and ϕ_{M_4} are the phases (in degrees) of the M_2 and M_4 tidal currents, respectively. Hence, tidal asymmetry can be quantified by

$$2\phi_{M_2} - \phi_{M_4} \tag{10.13}$$

Calculations of peak bed shear stress due to M_2 and M_4 tidal interactions can be used to infer sediment transport pathways (e.g. Fig. 10.13). Regions where bed shear stress vectors diverge are known as bed load partings, for example, the region in the middle of the Irish Sea (between Ireland and Wales) and in the English Channel (between England and France).

Morphological evolution can be simulated by application of the Exner equation

$$\frac{\partial z}{\partial t} = -\frac{1}{1-p}\left\{\frac{\partial q_x}{\partial x} + \frac{\partial q_y}{\partial y}\right\} \tag{10.14}$$

where z is the change in bed level, p is the bed porosity, and q_i is the transport of sediment in the i direction. Therefore, in regions where there is a *divergence* in sediment transport (i.e. a bed load parting), z will reduce. This will occur in a region of tidal symmetry, since a maximum in bed-level change is associated with zero residual sediment transport (Fig. 10.14). By contrast, in the case of maximum residual sediment transport (either flood or ebb dominant), zero bed-level change results. Neill et al. [35] examined the contrasting impact on large-scale sediment transport and morphodynamics for a tidal stream array located in regions of (a) tidal symmetry (either bed load parting or convergence),

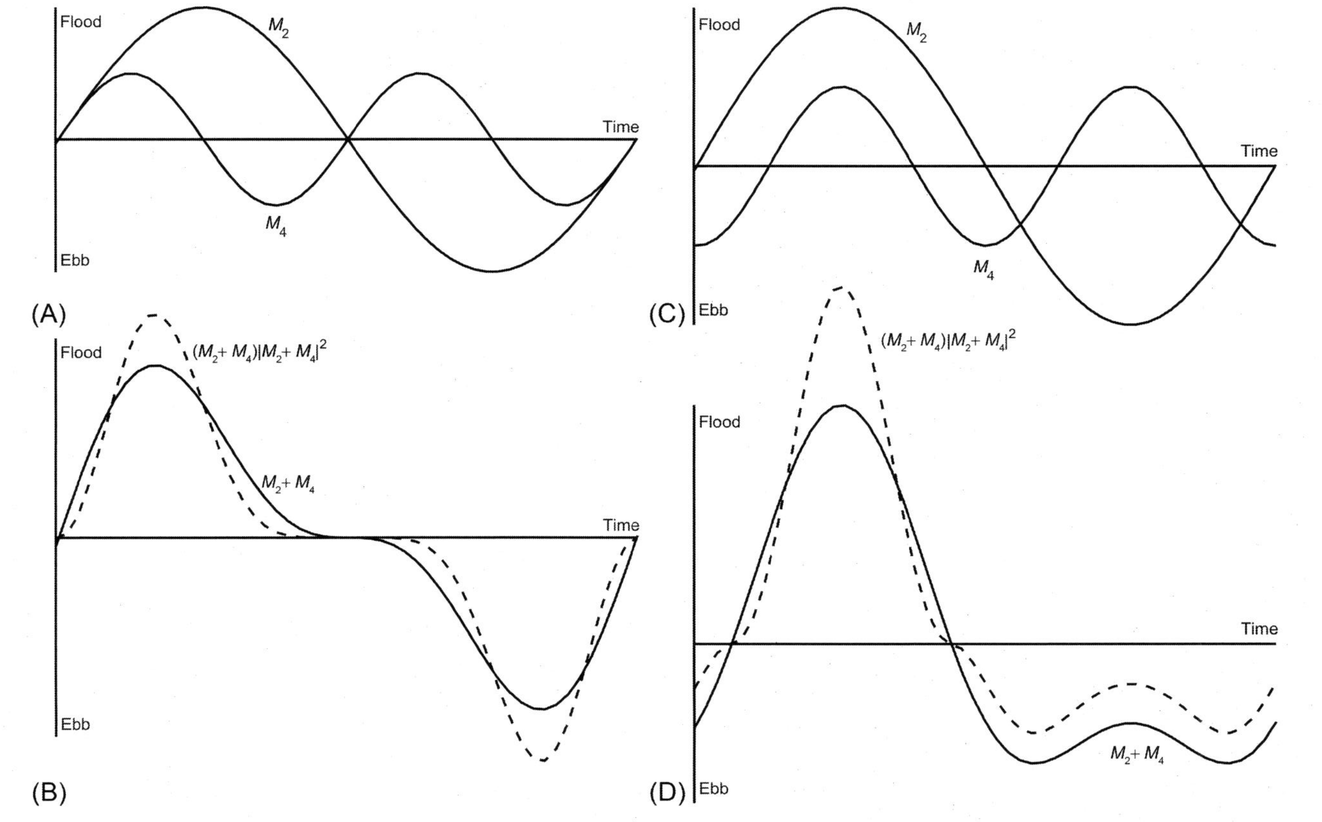

FIG. 10.12 Contributions of M_2 and M_4 tidal constituents leading to (B) a distorted but symmetrical tide and (D) tidal asymmetry. In (A), the phase lag between M_2 and M_4 is 0 degrees, and in (C) the phase lag is 90 degrees.

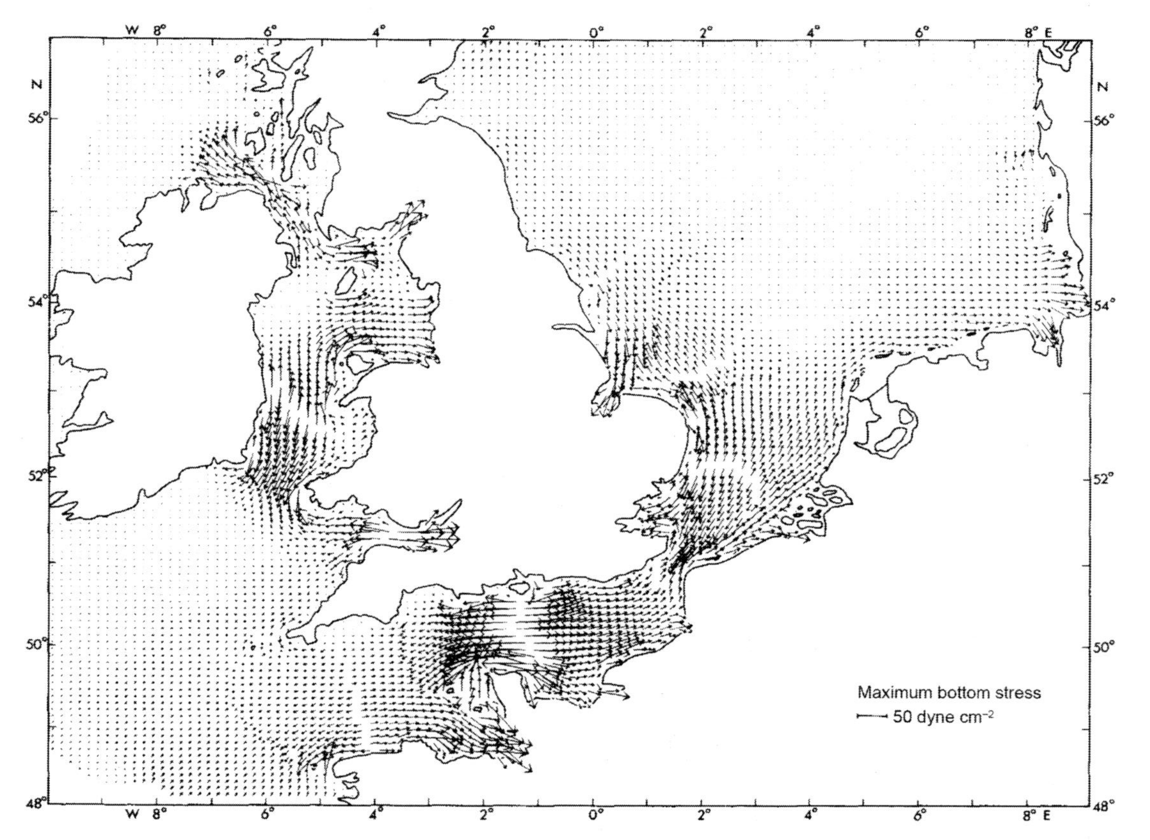

FIG. 10.13 Peak bed shear stress vectors due to M_2 and M_4 interactions. *(Reproduced from R.D. Pingree, D.K. Griffiths, Sand transport paths around the British Isles resulting from the M_2 and M_4 tidal interactions, J. Mar. Biol. Assoc. UK 59 (1979) 497–513, with permission from Cambridge University Press.)*

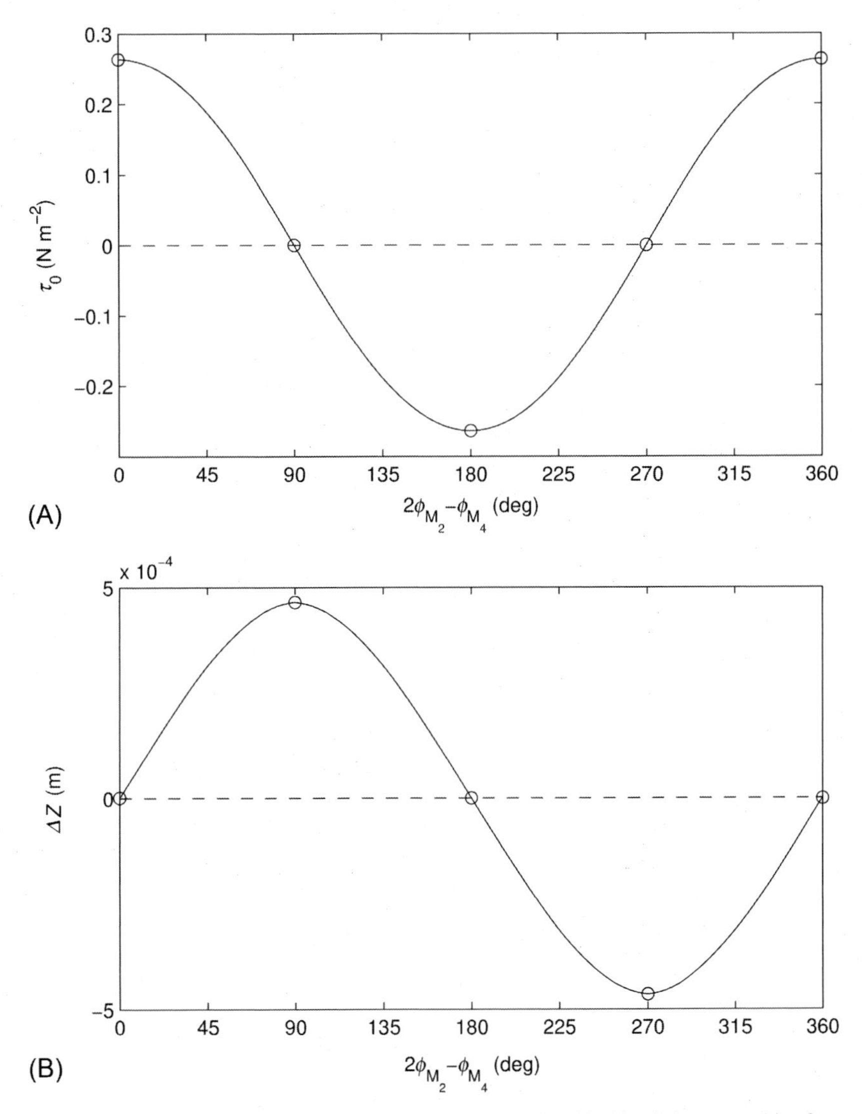

FIG. 10.14 (A) Tidal residual bed shear stress and (B) associated bed level change resulting from varying phase relationships between M_2 and M_4 tidal constituents. *(Reproduced from S.P. Neill, E.J. Litt, S.J. Couch, A.G. Davies, The impact of tidal stream turbines on large-scale sediment dynamics, Renew. Energy 34 (12) (2009) 2803–2812, with permission from Elsevier.)*

and (b) asymmetry. Using the Bristol Channel as a model case study, they concluded that TEC arrays sited in regions of tidal asymmetry had considerably more impact on sediment dynamics, since energy conversion alters residual sediment transport in regions of asymmetry. By contrast, TEC array operation in regions of tidal symmetry did not influence net sediment transport, and had less impact on morphodynamics. It is therefore recommended, from both impact and resource perspectives (since regions of symmetry lead to balanced power generation between flood and ebb phases of the tidal cycle), that where possible, regions of tidal symmetry should be favoured for siting TEC arrays. In addition, Neill et al. [35] found that the impacts of tidal energy conversion influenced far-field sediment dynamics and morphodynamics, of order 50 km from the array, and the impacts were not just confined to the site of energy extraction, as assumed previously.

Offshore Sand Banks

Offshore sand banks are common features of many coastlines around the world. They represent repositories of sediment that exchange material with neighbouring beaches over various timescales [53]. In addition, they are an important natural form of coastal defence, since the shallow waters associated with offshore sand banks reduce the energy of storm waves via depth-induced wave breaking. Offshore sand banks are also an important source of aggregate for the construction industry [61], and can act as nursery grounds for fisheries. For the above reasons, it is important that tidal energy extraction does not interfere with the natural cycle of sand bank evolution, and it is this influence that we investigate here. In particular, many regions of strong tidal flow that are suitable for tidal energy conversion are associated with offshore sand banks, and one common category is the headland sand bank.

Strong tidal flow past a headland leads to the generation of eddy systems, with an opposite sense of vorticity between the flood and ebb phases of the tidal cycle (Fig. 10.15). Within each eddy system, the outward-directed centrifugal force (CF) is balanced by an inward-directed pressure gradient force (PG). Since velocities are weaker close to the sea bed (e.g. Fig. 3.19), the centrifugal force is weaker at the sea bed, and stronger higher in the water column. The inward-directed pressure gradient force (which is the same at all depths) can therefore exceed the centrifugal force at the sea bed, and this leads to flow towards the centre of the eddy system at the sea bed (balanced by an outward directed flow at the sea surface, where CF $>$ PG). In fact, the result is actually a ‘spiralling’ flow, since the system is 3D (e.g. consider this secondary flow superimposed on the 2D horizontal eddy systems shown in Fig. 10.15). The inward-directed flow at the sea bed leads to net movement of sediment (e.g. via bed load transport) towards the centre of the eddy system. Over time, this net transport of sediment leads to the formation of ‘headland sand banks’—one on either side of the headland (Fig. 10.15). Evolution over long timescales results in a system that is in equilibrium,[7] with the sand bank influencing the flow field, and vice versa. However, headlands are attractive regions for siting tidal energy arrays

7. Although the tides are relatively constant, sand bank variability will still be influenced by wave action, for example, at storm and seasonal timescales.

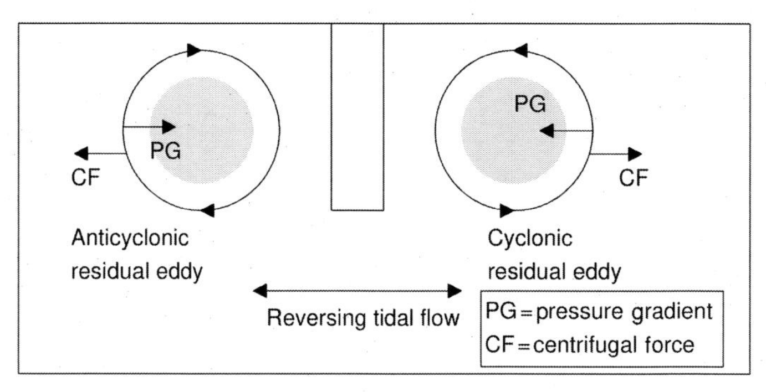

FIG. 10.15 Mechanism for headland sand bank formation. *(Reproduced from S.P. Neill, The role of coriolis in sandbank formation due to a headland/island system, Estuar. Coast. Shelf Sci. 79 (3) (2008) 419–428, with permission from Elsevier.)*

(e.g. [62]). It is therefore useful to examine the influence that tidal energy conversion in the vicinity of headlands would have on the evolution (and maintenance) of headland sand banks.

Neill et al. [38] initially developed an idealized model to demonstrate that a ~300 MW TEC array sited in the vicinity of a headland could have a significant influence on headland sand bank formation (Fig. 10.16). Subsequent application to a case study—flow past the island of Alderney in the Alderney Race—showed how tidal energy conversion influenced the evolution of a neighbouring headland sand bank. They also found that siting the array close to a headland significantly increased flow between the array and the coastline, hence potentially increasing coastal erosion. A modest offset in the array position, siting it to one side of the headland, led to a significant reduction on the hydrodynamic and sedimentary impact. A similar study of the same region, although focussed more on the French side of the Alderney Race, again demonstrated the significant impact of a 300 MW array on sediment dynamics, and how careful and localized siting of the array can minimize such impacts [63]. However, offshore sand banks are characterized by interannual and intraannual variabilities [61]. To determine the influence of tidal energy extraction, it is therefore necessary to compare the magnitude of this impact to natural variability. Robins et al. [37] attempted to do so for a modest sand bank known as Langdon sand bank that is located in the vicinity of Anglesey Skerries—a leased tidal energy site that was once under active development by Marine Current Turbines. By combining outputs from tidal and wave models, they showed that energy extraction had little influence on the evolution of Langdon sand bank, perhaps because its maintenance is not strongly controlled by flow through the Skerries where the hypothetical array is located. However, their study did lead to one interesting finding. Local change in bed shear stress due to energy extraction was greater than natural variability when the TEC array exceeded 87 MW (Fig. 10.17). However, this result was valid only for

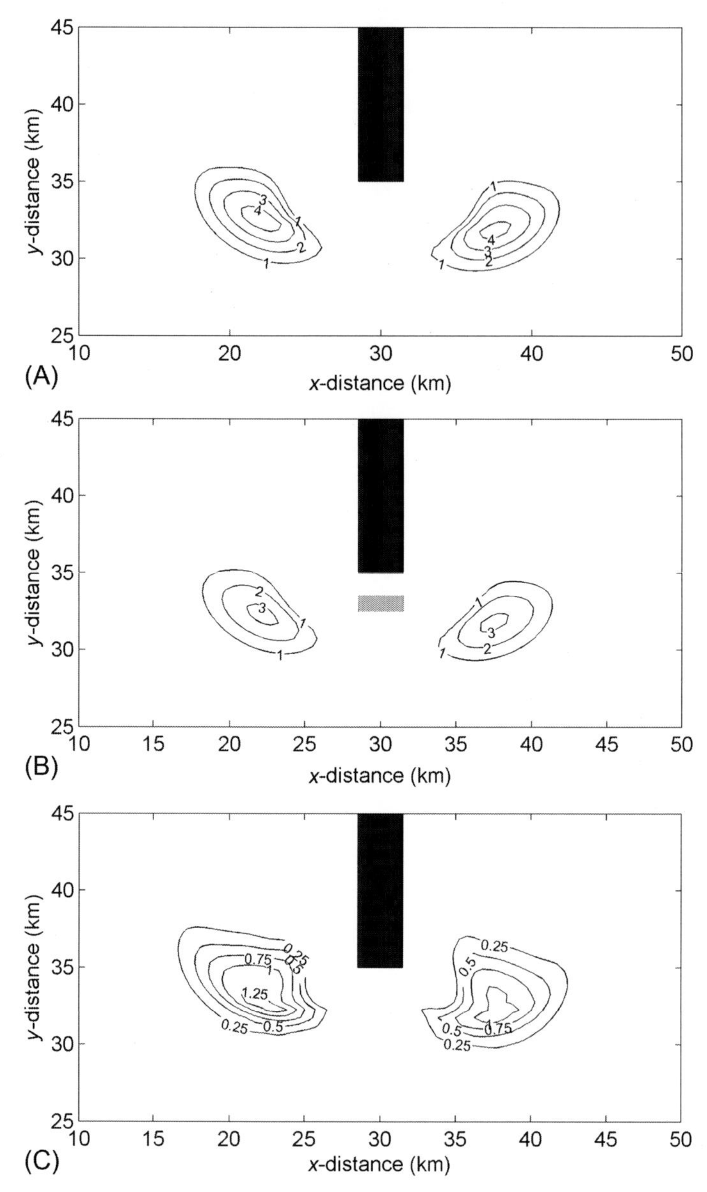

FIG. 10.16 Idealized headland sand bank formation based on a sediment grain size of 300 μm. Contours (in centimetres) show the magnitude of deposition after 50 tidal cycles. (A) is the baseline case, (B) is the 300 MW energy extraction scenario, and (C) is the difference between (A) and (B). *(Reproduced from S.P. Neill, J.R. Jordan, S.J. Couch, Impact of tidal energy converter (TEC) arrays on the dynamics of headland sand banks, Renew. Energy 37 (1) (2012) 387–397, with permission from Elsevier.)*

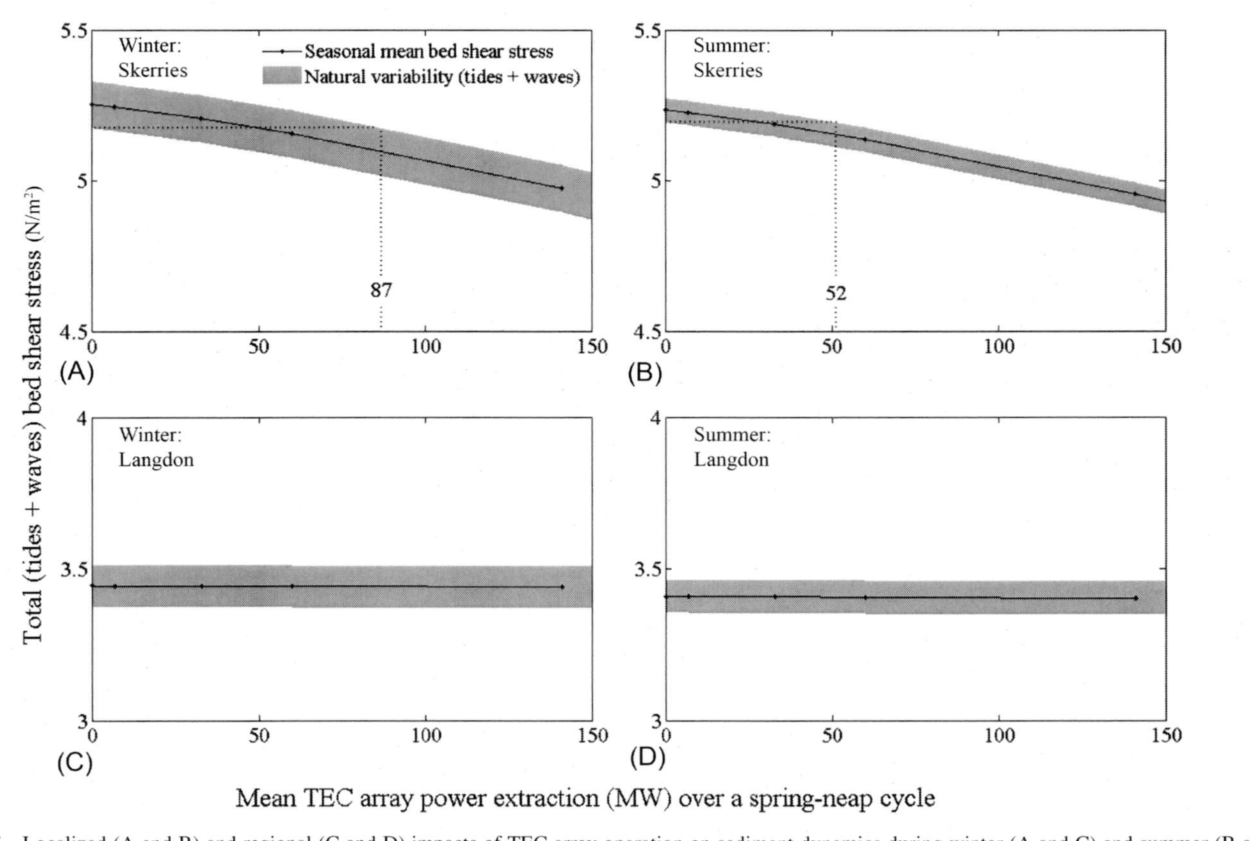

FIG. 10.17 Localized (A and B) and regional (C and D) impacts of TEC array operation on sediment dynamics during winter (A and C) and summer (B and D) months. *(Reproduced from P.E. Robins, S.P. Neill, M.J. Lewis, Impact of tidal-stream arrays in relation to the natural variability of sedimentary processes, Renew. Energy 72 (2014) 311–321, with permission from Elsevier.)*

winter months, when there was significant natural variability due to an energetic, yet highly variable, wave climate. In summer months, there was less natural variability, and so a smaller array (52 MW) was sufficient to exceed natural variability. It is therefore important to consider the timescale that is used to measure the impact of a TEC array on sedimentary processes, when assessing how the impact compares to natural variability. This has implications for time periods when instruments suitable for monitoring are deployed, especially since winter deployments tend to be more challenging than summer deployments.

All of the studies before considered single TEC arrays. Many regions, such as the Pentland Firth in Scotland, are expected to accommodate several discrete arrays. Therefore, the cumulative impact of arrays on sediment dynamics should be considered in such regions. Fairley et al. [52] demonstrated that the impacts of four leased sites in the Pentland Firth, somewhat unexpectedly, combined linearly (i.e. the sedimentary impacts of each array could be considered in isolation), and then combined. However, the authors comment that this may not necessarily be the case for very large scales of energy conversion. In addition, it is noted that the commercial model used (MIKE3) was not explicit about the energy extraction term, and so it is difficult to compare the methodology against other studies that used open access code (e.g. [37,38,63]).

10.4.2 Tidal Lagoons

There is very little published research that examines the impact of tidal lagoon power plants on sediment dynamics. However, there is research into indirect impacts (e.g. changes to hydrodynamic flow fields) that can be used to infer (but perhaps not quantify) the possible impacts of lagoons on sediments [51].

There are two physical environments where the sediment dynamics will be affected by tidal lagoon power plants—the region that is impounded by the lagoon, and the region that remains outside of the impoundment. In both of these environments, the operating mode (e.g. ebb-only, flood-only, or flood and ebb generations) will have a significant influence on the sediment dynamics. For example, counter-rotating eddy systems may form in the turbine jets [64], leading to localized changes in sediment transport: an impact that can be minimized by optimizing the operating mode, and evenly distributing the turbines along the lagoon structure [65]. Inside the lagoon, and away from the power house, there will be a significant reduction in energy, both in terms of the wave climate and tidal currents. In particular, in the absence of any real wave climate (other than for very short-period waves generated within the severely fetch-limited local confines of the impoundment), there will be negligible interannual and intraannual variabilities in sediment dynamics and associated morphodynamics. Since sediment exchange between the lagoon and the regional ocean would be so drastically altered, so too would be the sediment dynamics in the nearshore environment (e.g. beach processes). In addition, processes driven by waves (e.g. wave radiation stresses leading to longshore transport) will be altered by the presence of a tidal lagoon power plant, with significant consequences for

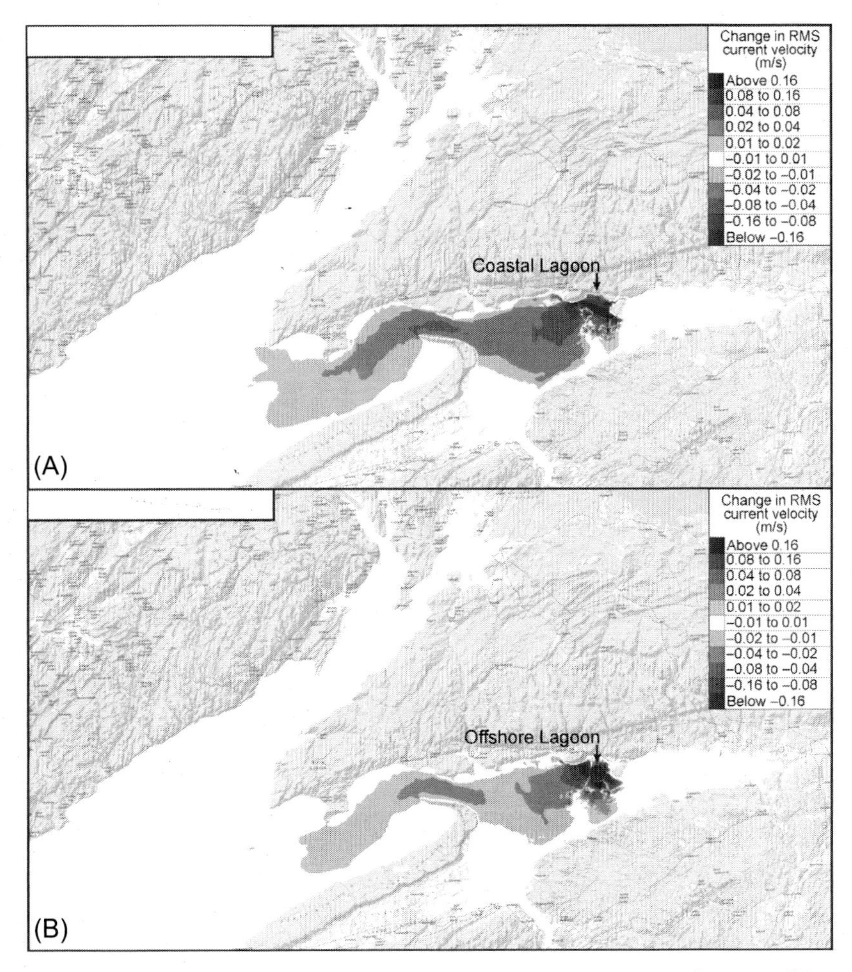

FIG. 10.18 Change in RMS current speed in the Bay of Fundy due to (A) coastal and (B) offshore tidal lagoon power plants. (A) S1: coastal lagoon. (B) S2: offshore lagoon. *(Reproduced from A. Cornett, J. Cousineau, I. Nistor, Assessment of hydrodynamic impacts from tidal power lagoons in the Bay of Fundy, Int. J. Mar. Energy 1 (2013) 33–54, with permission from Elsevier.)*

long-term erosion and deposition of sediment at regional scale. Changes in current speed outside of a lagoon (e.g. Fig. 10.18) will also lead to significant changes in sediment dynamics and morphodynamics. Although peak changes will occur close to the lagoon itself (e.g. changes in U_{RMS} of over 0.1 m/s in Fig. 10.18), there will also be a wider region that is characterized by a significant change in flow speeds. The impacts of tidal range power plants on sediment dynamics are an issue that needs to be investigated further through numerical simulations, prior to construction, to minimize impacts through optimization of (a) shape of the embankment and impounded area, (b) location of power house, and (c) operating mode.

REFERENCES

[1] J. Widén, N. Carpman, V. Castellucci, D. Lingfors, J. Olauson, F. Remouit, M. Bergkvist, M. Grabbe, R. Waters, Variability assessment and forecasting of renewables: a review for solar, wind, wave and tidal resources, Renew. Sustain. Energy Rev. 44 (2015) 356–375.

[2] D.T. Pugh, Tides, Surges and Mean Sea-Level (Reprinted With Corrections), John Wiley & Sons Ltd, New York, NY, 1996.

[3] S.P. Neill, M.R. Hashemi, M.J. Lewis, Tidal energy leasing and tidal phasing, Renew. Energy 85 (2016) 580–587.

[4] N. Yates, I. Walkington, R. Burrows, J. Wolf, The energy gains realisable through pumping for tidal range energy schemes, Renew. Energy 58 (2013) 79–84.

[5] S.C. Pryor, R.J. Barthelmie, Climate change impacts on wind energy: a review, Renew. Sustain. Energy Rev. 14 (1) (2010) 430–437.

[6] S.C. Pryor, R.J. Barthelmie, E. Kjellström, Potential climate change impact on wind energy resources in northern Europe: analyses using a regional climate model, Clim. Dyn. 25 (7–8) (2005) 815–835.

[7] A. Bloom, V. Kotroni, K. Lagouvardos, Climate change impact of wind energy availability in the Eastern Mediterranean using the regional climate model PRECIS, Nat. Hazards Earth Syst. Sci. 8 (6) (2008) 1249–1257.

[8] H. Hueging, R. Haas, K. Born, D. Jacob, J.G. Pinto, Regional changes in wind energy potential over Europe using regional climate model ensemble projections, J. Appl. Meteorol. Climatol. 52 (4) (2013) 903–917.

[9] L.C. Cradden, G.P. Harrison, J.P. Chick, Will climate change impact on wind power development in the UK?, Clim. Change 115 (3–4) (2012) 837–852.

[10] P. Nolan, P. Lynch, R. McGrath, T. Semmler, S. Wang, Simulating climate change and its effects on the wind energy resource of Ireland, Wind Energy 15 (4) (2012) 593–608.

[11] C. Fant, C.A. Schlosser, K. Strzepek, The impact of climate change on wind and solar resources in southern Africa, Appl. Energy 161 (2016) 556–564.

[12] G.P. Harrison, A.R. Wallace, Climate sensitivity of marine energy, Renew. Energy 30 (12) (2005) 1801–1817.

[13] F. Vikebø, T. Furevik, G. Furnes, N.G. Kvamstø, M. Reistad, Wave height variations in the North Sea and on the Norwegian Continental Shelf, 1881–1999, Cont. Shelf Res. 23 (3) (2003) 251–263.

[14] D.R. Cayan, M. Tyree, M.D. Dettinger, H. León, G. Hugo, T. Das, E.P. Maurer, P. Bromirski, N. Graham, R. Flick, Climate change scenarios and sea level rise estimates for California 2008 Climate Change Scenarios Assessment, Paper from: California Climate Change Research Center, Sacramento, 2009.

[15] M.A. Hemer, Y. Fan, N. Mori, A. Semedo, X.L. Wang, Projected changes in wave climate from a multi-model ensemble, Nat. Clim. Change 3 (5) (2013) 471–476.

[16] J. Janjić, S. Gallagher, F. Dias, Wave energy extraction in the Northeast Atlantic: future wave climate availability, in: The 12th European Wave and Tidal Energy Conference: EWTEC 2017, 2017.

[17] D.E. Reeve, Y. Chen, S. Pan, V. Magar, D.J. Simmonds, A. Zacharioudaki, An investigation of the impacts of climate change on wave energy generation: the Wave Hub, Cornwall, UK, Renew. Energy 36 (9) (2011) 2404–2413.

[18] X.L. Wang, Y. Feng, V.R. Swail, Climate change signal and uncertainty in CMIP5-based projections of global ocean surface wave heights, J. Geophys. Res. Oceans 120 (5) (2015) 3859–3871.

[19] J.A. Church, N.J. White, Sea-level rise from the late 19th to the early 21st century, Surv. Geophys. 32 (4–5) (2011) 585–602.

[20] H.E. Pelling, J.A. Mattias Green, Sea level rise and tidal power plants in the Gulf of Maine, J. Geophys. Res. Oceans 118 (6) (2013) 2863–2873.

[21] H.E. Pelling, J.A.M. Green, Impact of flood defences and sea-level rise on the European Shelf tidal regime, Cont. Shelf Res. 85 (2014) 96–105.

[22] M.D. Pickering, K.J. Horsburgh, J.R. Blundell, J.J.-M. Hirschi, R.J. Nicholls, M. Verlaan, N.C. Wells, The impact of future sea-level rise on the global tides, Cont. Shelf Res. 142 (2017) 50–68.

[23] M. De Dominicis, R. O'Hara Murray, J. Wolf, Present and future impacts of large tidal stream turbine arrays, in: The 12th European Wave and Tidal Energy Conference: EWTEC 2017, 2017.

[24] M.R. Hashemi, M. Lewis, Wave-tide interactions in ocean renewable energy, in: Marine Renewable Energy, Springer, 2017, pp. 137–158.

[25] M.R. Hashemi, S.T. Grilli, S.P. Neill, A simplified method to estimate tidal current effects on the ocean wave power resource, Renew. Energy 96 (2016) 257–269.

[26] M.R. Hashemi, S.P. Neill, P.E. Robins, A.G. Davies, M.J. Lewis, Effect of waves on the tidal energy resource at a planned tidal stream array, Renew. Energy 75 (2015) 626–639.

[27] R.A. Dalrymple, R.G. Dean, Water wave mechanics for engineers and scientists, World Scientific, Singapore, 1991.

[28] L.C. Van Rijn, Unified view of sediment transport by currents and waves. I: initiation of motion, bed roughness, and bed-load transport, J. Hydraul. Eng. 133 (6) (2007) 649–667.

[29] R. Soulsby, Dynamics of Marine Sands: A Manual for Practical Applications, Thomas Telford, UK, 1997.

[30] N. Guillou, G. Chapalain, S.P. Neill, The influence of waves on the tidal kinetic energy resource at a tidal stream energy site, Appl. Energy 180 (2016) 402–415.

[31] International Electrotechnical Commission, IEC 62600-201 TS: marine energy—wave, tidal and other water current converters—Part 201: tidal energy resource assessment and characterization, Technical Report, 2014.

[32] M.Z. Jacobson, C.L. Archer, Saturation wind power potential and its implications for wind energy, Proc. Natl Acad. Sci. USA 109 (39) (2012) 15679–15684.

[33] S. Nash, A. Phoenix, A review of the current understanding of the hydro-environmental impacts of energy removal by tidal turbines, Renew. Sustain. Energy Rev. 80 (2017) 648–662.

[34] I.G. Bryden, S.J. Couch, ME1—marine energy extraction: tidal resource analysis, Renew. Energy 31 (2) (2006) 133–139.

[35] S.P. Neill, E.J. Litt, S.J. Couch, A.G. Davies, The impact of tidal stream turbines on large-scale sediment dynamics, Renew. Energy 34 (12) (2009) 2803–2812.

[36] C. Garrett, P. Cummins, Limits to tidal current power, Renew. Energy 33 (11) (2008) 2485–2490.

[37] P.E. Robins, S.P. Neill, M.J. Lewis, Impact of tidal-stream arrays in relation to the natural variability of sedimentary processes, Renew. Energy 72 (2014) 311–321.

[38] S.P. Neill, J.R. Jordan, S.J. Couch, Impact of tidal energy converter (TEC) arrays on the dynamics of headland sand banks, Renew. Energy 37 (1) (2012) 387–397.

[39] A.J.G. Brown, S.P. Neill, M.J. Lewis, Tidal energy extraction in three-dimensional ocean models, Renew. Energy 114 (2017) 244–257.

[40] T. Roc, D.C. Conley, D. Greaves, Methodology for tidal turbine representation in ocean circulation model, Renew. Energy 51 (2013) 448–464.

[41] International Electrotechnical Commission, IEC 62600-101 TS: marine energy—wave, tidal and other water current converters—Part 101: wave energy resource assessment and characterization, Technical Report, 2014.

[42] S.P. Neill, G. Iglesias, Impact of wave energy converter (WEC) array operation on nearshore processes, in: The 4th International Conference on Ocean Energy: ICOE 2012, 2012.

[43] The SWAN Team, SWAN User Manual. SWAN Cycle III Version 40.91, Delft University of Technology Technical Documentation, 2014, 123 pp.

[44] J. Abanades, D. Greaves, G. Iglesias, Wave farm impact on the beach profile: a case study, Coast. Eng. 86 (2014) 36–44.

[45] H.C.M. Smith, C. Pearce, D.L. Millar, Further analysis of change in nearshore wave climate due to an offshore wave farm: an enhanced case study for the Wave Hub site, Renew. Energy 40 (1) (2012) 51–64.

[46] K. Ruehl, A. Porter, C. Chartrand, H. Smith, G. Chang, J. Roberts, Development, verification and application of the SNL-SWAN open source wave farm code, in: The 11th European Wave and Tidal Energy Conference: EWTEC 2015, 2015.

[47] H.T. Özkan-Haller, M.C. Haller, J.C. McNatt, A. Porter, P. Lenee-Bluhm, Analyses of wave scattering and absorption produced by WEC arrays: physical/numerical experiments and model assessment, in: Marine Renewable Energy, Springer, 2017, pp. 71–97.

[48] V.S. Neary, M. Lawson, M. Previsic, A. Copping, K.C. Hallett, A. LaBonte, J. Rieks, D. Murray, et al., Methodology for design and economic analysis of marine energy conversion (MEC) technologies, Proceedings of the 2nd Marine Energy Technology Symposium (METS), April 15–18, 2014, Seattle, WA, 2014.

[49] R. Springer, Framework for Project Development in the Renewable Energy Sector, National Renewable Energy Laboratory (NREL), Golden, CO, 2013.

[50] S.K. Henkel, R.M. Suryan, B.A. Lagerquist, Marine renewable energy and environmental interactions: baseline assessments of seabirds, marine mammals, sea turtles and benthic communities on the Oregon shelf, in: Marine Renewable Energy Technology and Environmental Interactions, Springer, 2014, pp. 93–110.

[51] S.P. Neill, P.E. Robins, I. Fairley, The impact of marine renewable energy extraction on sediment dynamics, in: Marine Renewable Energy, Springer, 2017, pp. 279–304.

[52] I. Fairley, I. Masters, H. Karunarathna, The cumulative impact of tidal stream turbine arrays on sediment transport in the Pentland Firth, Renew. Energy 80 (2015) 755–769.

[53] S.P. Neill, A.J. Elliott, M.R. Hashemi, A model of inter-annual variability in beach levels, Cont. Shelf Res. 28 (14) (2008) 1769–1781.

[54] S.P. Neill, J.D. Scourse, The formation of headland/island sandbanks, Cont. Shelf Res. 29 (18) (2009) 2167–2177.

[55] S.P. Neill, M.R. Hashemi, A.J. Elliott, An enhanced depth-averaged tidal model for morphological studies in the presence of rotary currents, Cont. Shelf Res. 27 (1) (2007) 82–102.

[56] C.L. Provost, Generation of overtides and compound tides (review), in: Nonlinear Tidal Interactions in Shallow Water, 1991, pp. 269–295.

[57] M.N. Gallo, S.B. Vinzon, Generation of overtides and compound tides in Amazon estuary, Ocean Dyn. 55 (2005) 441–448.

[58] P.E. Speer, D.G. Aubrey, C.T. Friedrichs, Nonlinear hydrodynamics of shallow tidal inlet/bay systems, in: B.B. Parker (Ed.), Tidal Hydrodynamics, John Wiley & Sons, New York, 1991, pp. 321–339.

[59] R.D. Pingree, D.K. Griffiths, Sand transport paths around the British Isles resulting from the M_2 and M_4 tidal interactions, J. Mar. Biol. Assoc. UK 59 (1979) 497–513.

[60] C.T. Friedrichs, D.G. Aubrey, Non-linear tidal distortion in shallow well-mixed estuaries: a synthesis, Estuar. Coast. Shelf Sci. 27 (1988) 521–545.
[61] M.J. Lewis, S.P. Neill, A.J. Elliott, Interannual variability of two offshore sand banks in a region of extreme tidal range, J. Coast. Res. 31 (2) (2015) 265–275.
[62] L.S. Blunden, A.S. Bahaj, Initial evaluation of tidal stream energy resources at Portland Bill, UK, Renew. Energy 31 (2) (2006) 121–132.
[63] J. Thiébot, P.B. du Bois, S. Guillou, Numerical modeling of the effect of tidal stream turbines on the hydrodynamics and the sediment transport-Application to the Alderney Race (Raz Blanchard), France, Renew. Energy 75 (2015) 356–365.
[64] A. Cornett, J. Cousineau, I. Nistor, Assessment of hydrodynamic impacts from tidal power lagoons in the Bay of Fundy, Int. J. Mar. Energy 1 (2013) 33–54.
[65] R.A. Falconer, J. Xia, B. Lin, R. Ahmadian, The Severn barrage and other tidal energy options: hydrodynamic and power output modeling, Sci. China, Ser. E: Technol. Sci. 52 (11) (2009) 3413–3424.

FURTHER READING

[1] S.P. Neill, M.R. Hashemi, Wave power variability over the northwest European shelf seas, Appl. Energy 106 (2013) 31–46.
[2] S.P. Neill, J.D. Scourse, K. Uehara, Evolution of bed shear stress distribution over the northwest European shelf seas during the last 12,000 years, Ocean Dyn. 60 (5) (2010) 1139–1156.
[3] S.P. Neill, The role of coriolis in sandbank formation due to a headland/island system, Estuar. Coast. Shelf Sci. 79 (3) (2008) 419–428.
[4] T. Stocker, Climate Change 2013: The Physical Science Basis: Working Group I Contribution to the Fifth Assessment Report of the Intergovernmental Panel on Climate Change, Cambridge University Press, Cambridge, 2014.

Index

Note: Page numbers followed by *f* indicate figures and *t* indicate tables.

I

J

K

L

M

N

O

X

Z

Printed in Australia by Griffin Press
an Accredited ISO AS/NZS 14001:2004
Environmental Management System printer.